COMPUTER NETWORK
AND
CLOUD COMPUTING

FOR

SEMESTER – VI

**THIRD YEAR (T.Y.) B. TECH COURSE IN
ELECTRONICS AND TELECOMMUNICATION ENGINEERING**
(Also Useful for Electronics Engineering)

**Strictly According to New Revised Credit System Syllabus
of Babasaheb Ambedkar Technological University (BATU),
Lonere, (Dist. Raigad) Maharashtra,**
(w.e.f. June 2019-20)

NITIN N. SAKHARE
M. E. (Comp. Networks)
Assistant Professor, Comp. Engg. Deptt.
Global Cisco Certified CCNA Instructor
Vishwakarma Institute of Information Technology
Kondhwa (Bk.), PUNE.

V. K. KOLEKAR
M. E. (Computer)
Assistant Professor, Comp. Engg. Deptt.
Global Cisco Certified CCNA Instructor
Vishwakarma Institute of Information Technology
Kondhwa (Bk.), PUNE.

N1117

COMP. NET. & CLOUD COMPUTING (E&TC, BATU)　　　　　**ISBN : 978-93-89686-20-3**

First Edition	:	**November 2019**
©	:	**Authors**

Published By :
NIRALI PRAKASHAN
Abhyudaya Pragati, 1312 Shivaji Nagar
Off J.M. Road, PUNE 411005
Tel : (020) 25512336/37/39
Email : niralipune@pragationline.com

➤ ## DISTRIBUTION CENTRES

PUNE

Nirali Prakashan (Local) : 119 Budhwar Peth, Jogeshwari Mandir Lane, Pune 411002, Maharashtra
Tel : (020) 2445 2044, Mobile : 9657703145, Email : niralilocal@pragationline.com

Nirali Prakashan (Outstation) : S. No. 28/27 Dhayari, Near Asian College, Dhayari, Pune 411041, Maharashtra
Tel : (020) 2469 0204, Fax : (020) 2469 0316, Mobile : 9657703143
Email : bookorder@pragationline.com

MUMBAI

Nirali Prakashan : 385 S.V.P. Road, Rasadhara Co-op. Hsg. Society Ltd., Girgaum, Mumbai 400004, Maharashtra
Tel : (022) 2385 6339 / 2386 9976, Fax : (022) 2386 9976, Mobile : 9320129587
Email : niralimumbai@pragationline.com

➤ ## DISTRIBUTION BRANCHES

JALGAON

Nirali Prakashan : 34 V. V. Golani Market, Navi Peth, Jalgaon 425001, Maharashtra
Tel : (0257) 222 0395, Mob : 94234 91860, Email : niralijalgaon@pragationline.com

KOLHAPUR

Nirali Prakashan : New Mahadvar Road, Kedar Plaza 1st Floor, Opp. IDBI Bank
Kolhapur 416012, Maharashtra. Mobile : 9850046155
Email : niralikolhapur@pragationline.com

NAGPUR

Nirali Prakashan : Above Maratha Mandir, Shop No 3, Second Floor,
Rani Jhanshi Square, Sitabuldi, Nagpur 440012, Maharashtra
Tel : (0712) 254 7129, Email : niralinagpur@pragationline.com

DELHI

Nirali Prakashan : 4593/15 Basement, Agarwal Lane, Ansari Road, Daryaganj
Near Times of India Building, New DelhiV 110002 Mobile : 8505972553
Email : niralidelhi@pragationline.com

BENGALURU

Nirali Prakashan : Maitri Ground Floor, Jaya Apartments, No. 99, 6th Cross, 6th Main,
Malleswaram, Bengaluru 560003, Karnataka
Mobile : 9449043034, Email : niralibangalore@pragationline.com

niralipune@pragationline.com　|　www.pragationline.com
Also find us on 🄵 www.facebook.com/niralibooks

PREFACE

It gives us great pleasure to present the book **"Computer Network and Cloud Computing"** for the students of **Semester VI Third Year (T.Y.) B. Tech. Course Electronics and Telecommunication of Dr. Babasaheb Ambedkar Technological University (BATU), Lonere, Dist. Raigad (Maharashtra).** This book is strictly as per the new revised syllabus 2019-20 Pattern, effective from the Academic Year July 2019-20.

In New Revised Syllabus, there will Class Assessment (CA) 20 Marks, Mid Sem. Exam. (MSE) 20 Marks and End Sem. Exam. (ESE) 60 Marks. End Sem. Exam. will be based on all Six units and each unit will carry 20 Marks.

The Theory Course will have 3 Credits.

The basic objective of this book is to bridge the gap between the vast contents of the reference books, written by the renowned International Authors and the concise requirements of Undergraduate Students. This book has been written in a comprehensive manner using Simple and Lucid language, keeping in mind students' requirements. The main emphasis has been given on exploring the basic concepts rather than merely the Information. Solved Examples and Exercises have been provided throughout the book and at the end of the Unit. Also we have given **Model Question Paper** for practice at the end of book.

Our special thanks to our family members, students and all those who directly or indirectly supported us in this project.

We also take this opportunity to express our sincere thanks to Shri. Dineshbhai Furia, Shri. Jignesh Furia, Mrs. Nirali Verma, Shri. M. P. Munde and entire team of Nirali Prakashan, namely Mrs. Deepali Lachake (Co-ordinator), and her colleagues who really have taken keen interest and untiring efforts in publishing this text.

The advice and suggestions of our esteemed readers to improve the text are most welcome and will be highly appreciated.

Pune   **Authors**

SYLLABUS

Unit I : Physical Layer

Data Communications, Networks, Network types, Protocol layering, OSI model, Layers in OSI model, TCP / IP protocol suite, Addressing, Guided and Unguided Transmission media. Switching: Circuit switched networks, Packet Switching, Structure of a switch.

Unit II : Data Link Layer

Introduction to Data Link Layer, DLC Services, DLL protocols, HDLC, PPP, Media Access Control: Random Access, Controlled Access, Channelization. Wired LAN: Ethernet Protocol, Standard Ethernet, Fast Ethernet, Giagabit Ethernet, 10 Gigabit Ethernet.

Unit III : Wireless LANS & Virtual Circuit Networks

Introduction, Wireless LANS: IEEE 802.11 project, Bluetooth, Zigbee, Connecting devices and Virtual LANS: Connecting devices, Virtual LANS.

Unit IV : Network Layer

Switching, Logical addressing – IPV4, IPV6; Address mapping – ARP, RARP, BOOTP and DHCP–Delivery, Forwarding and Unicast Routing protocols.

Unit V : Transport Layer

Process to Process Communication, User Datagram Protocol (UDP), Transmission Control Protocol (TCP), SCTP Congestion Control; Quality of Service, QoS improving techniques: Leaky Bucket and Token Bucket algorithm.

Unit VI : Application Layer

Domain Name Space (DNS), DDNS, TELNET, EMAIL, File Transfer Protocol (FTP), WWW, HTTP, SNMP, Bluetooth, Firewalls, Basic concepts of Cryptography

CONTENTS

✠ ✠ ✠

1.1 ANALOG / DIGITAL, SIGNALS AND DATA

- **Analog Signal** has infinitely many levels of intensity over a period of time.
- **Digital Signal** can have only a limited number of discrete defined values (for example 0 and 1).
- **Analog Data** refers to information that is continuous in nature. Analog data takes continuous values.
- **Digital Data** refers to information that has discrete state values (for example 0 and 1).
- Prototype examples for Analog signal, Digital signal, Analog data and Digital data are as shown in Fig. 1.1.

Example of Analog Signal

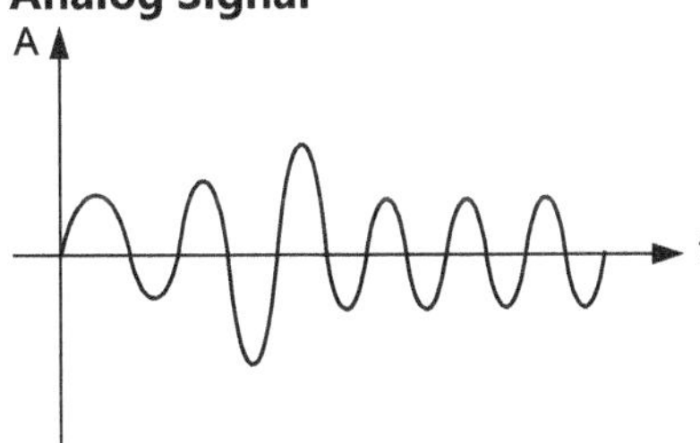

(a) Example of Analog Data

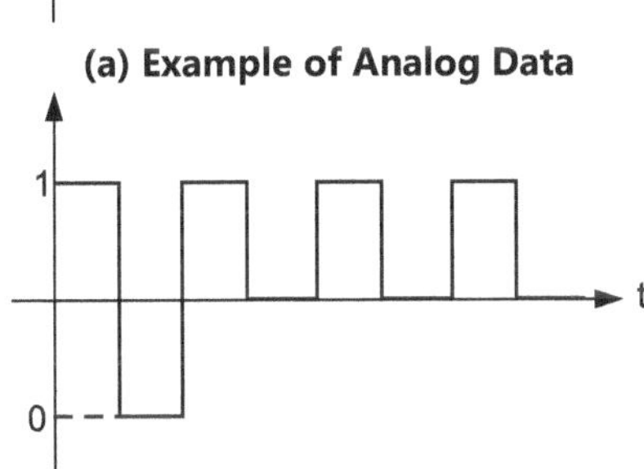

(b) Example of Digital Data

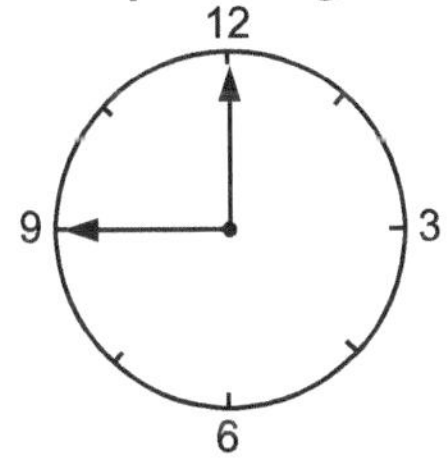

(c) (Analog Clock)

[Movement of hands are continuous]

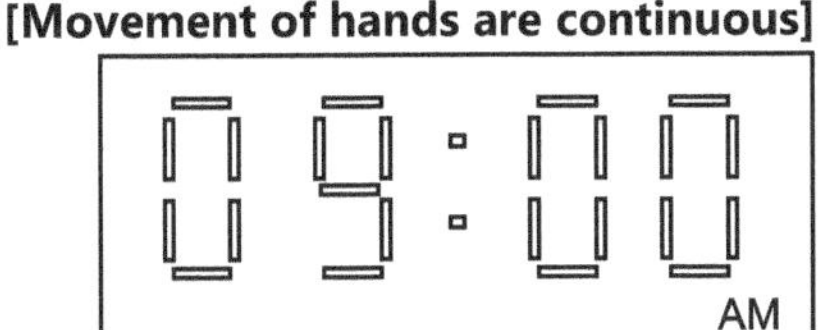

(d) (Digital Clock)

[Time change is sudden

i.e. 9:00 am to 9:01 am]

Fig. 1.1 : Analog/Digital Signals and Data Representation

1.2 PERIODIC ANALOG SIGNALS

- Periodic analog signals are classified as :

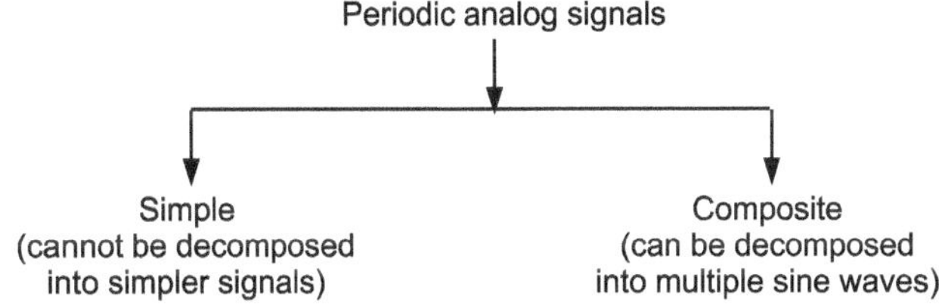

- The example of **simple** periodic analog signals which cannot be decomposed into simpler signals is sine wave.

- The examples of **composite** periodic analog signals which can be decomposed into multiple sine waves

  ➢ Square wave (duty cycle = 50%).

  ➢ Pulse wave (duty cycle ≠ 50%)

  ➢ Triangular waveform.

  ➢ RAMP waveform (positive or negative RAMP).

- Let's see the characteristics of sine wave in detail.

- Also we will discuss the remaining signal waveforms in brief.

1.2.1 Sine Wave Signal

- A sine wave has the same shape as the graph of the sine function used in trigonometry. Sine waves are produced by rotating electrical machines such as dynamos, power station turbines and electrical energy is transmitted to the consumer in this form.

- In electronics, sine waves are among the most useful of all signals in testing circuits and analyzing system performance.

- Sine wave in more detail is shown in Fig. 1.2.

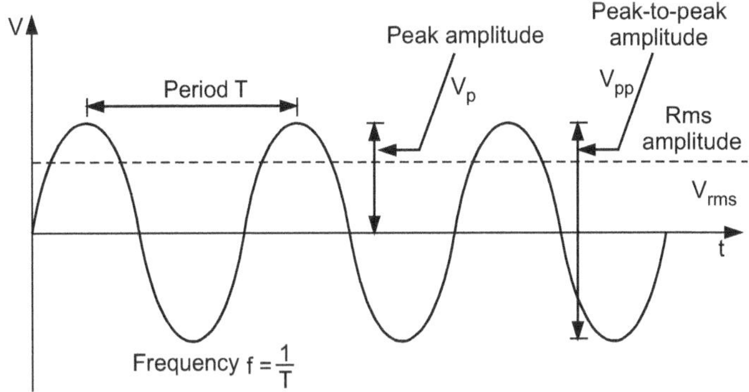

Fig. 1.2 : Periodic Sine Wave Signal

The terms defined below are needed to describe sine waves and other waveforms precisely:

1. Period (T) :

- The period is the time taken for one complete cycle of a repeating waveform.
- The period is often thought of as the time interval between peaks, but can be measured between any two corresponding points in successive cycles.

2. Frequency (f) :

- Frequency is the number of cycles completed per second.
- The measurement unit for frequency is the **hertz, Hz**. 1 Hz = 1 cycle per second.
- If you know the period, the frequency of the signal can be calculated from

$$f = \frac{1}{T}$$

Conversely, the period is given by

$$T = \frac{1}{f}$$

- Signals you are likely to use vary in frequency from about 0.1 Hz, through values in **kilohertz, kHz** (thousands of cycles per second) to values in **megahertz, MHz** (millions of cycles per second).

Table 1.1 : SI Multiples for Hertz (Hz)

Submultiples			Multiples		
Value	Symbol	Name	Value	Symbol	Name
10^{-1} Hz	dHz	decihertz	10^{1} Hz	daHz	decahertz
10^{-2} Hz	cHz	centihertz	10^{2} Hz	hHz	hectohertz
10^{-3} Hz	**mHz**	**millihertz**	10^{3} Hz	**kHz**	**kilohertz**
10^{-6} Hz	**µHz**	**microhertz**	10^{6} Hz	**MHz**	**megahertz**
10^{-9} Hz	nHz	nanohertz	10^{9} Hz	**GHz**	**gigahertz**
10^{-12} Hz	pHz	picohertz	10^{12} Hz	**THz**	**terahertz**
10^{-15} Hz	fHz	femtohertz	10^{15} Hz	PHz	petahertz
10^{-18} Hz	aHz	attohertz	10^{18} Hz	EHz	exahertz
10^{-21} Hz	zHz	zeptohertz	10^{21} Hz	ZHz	zettahertz
10^{-24} Hz	yHz	yoctohertz	10^{24} Hz	YHz	yottahertz
Common prefixed units are in bold face.					

- The **hertz** (symbol:**Hz**) is a unit of frequency.
- It is defined as the number of complete cycles per second. It is the basic unit of frequency in the International System of Units (SI). It is used worldwide in both general-purpose and scientific contexts.

- Hertz can be used to measure any periodic event; the most common uses of hertz are to describe radio and audio frequencies, more or less sinusoidal contexts in which case a frequency of 1 Hz is equal to one cycle per second.
- The unit hertz is defined by the International System of Units (SI).

3. Amplitude :

- In electronics, the amplitude, or height, of a sine wave is measured in three different ways.
- The **peak amplitude**, V_p, is measured from the X-axis, 0 V, to the top of a peak, or to the bottom of a trough. (In physics 'amplitude' usually refers to peak amplitude.)
- The **peak-to-peak amplitude**, V_{pp}, is measured between the maximum positive and negative values.
- In practical terms, this is often the easier measurement to make. Its value is exactly twice V_p.
- Although peak and peak-to-peak values are easily determined, it is often more useful to know the **root mean square**, or **rms amplitude** of the wave, where:

$$V_{rms} = \frac{V_p}{\sqrt{2}} \text{ or } V_{rms} = 0.7 \times V_p$$

and

$$V_p = \sqrt{2} \times V_{rms} \text{ or } V_p = 1.4 \times V_{rms}$$

4. Phase:

- It is sometimes useful to divide a sine wave into degrees, °, as follows:
- Remember that sine waves are generated by rotating electrical machines. A complete 360° turn of the voltage generator corresponds to one cycle of the sine wave.
- Therefore 180° corresponds to a half turn, 90° to a quarter turn and so on. Using this method, any point on the sine wave graph can be identified by a particular number of degrees through the cycle.

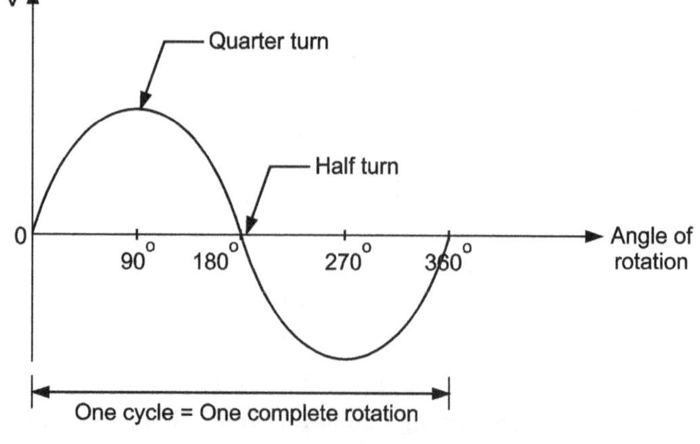

Fig. 1.3 : Phase of Sine Wave

- If two sine waves have the same frequency and occur at the same time, they are said to be **in phase**.

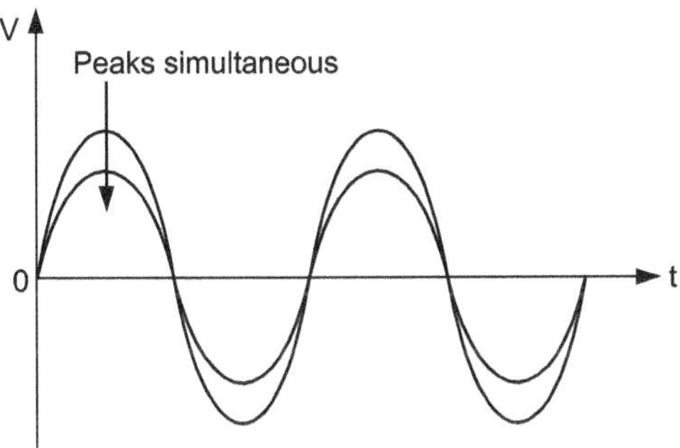

(a) Sine waves of the same frequency which are in phase

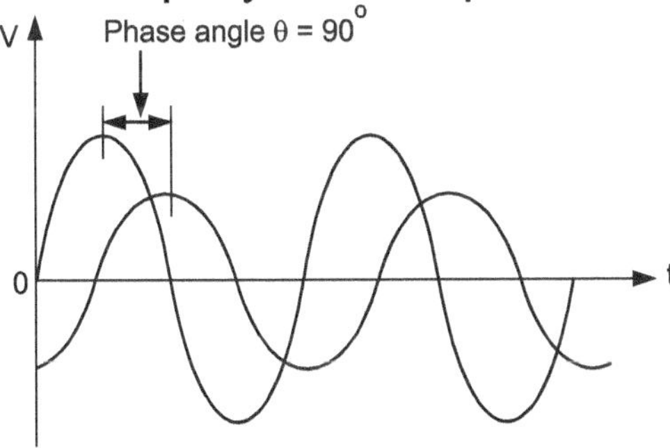

(b) Sine waves of the same frequency which are a quarter cycle (90°) out of phase

Fig. 1.4 : In Phase and Out of Phase Sine Waves

- On the other hand, if the two waves occur at different times, they are said to be **out of phase**.

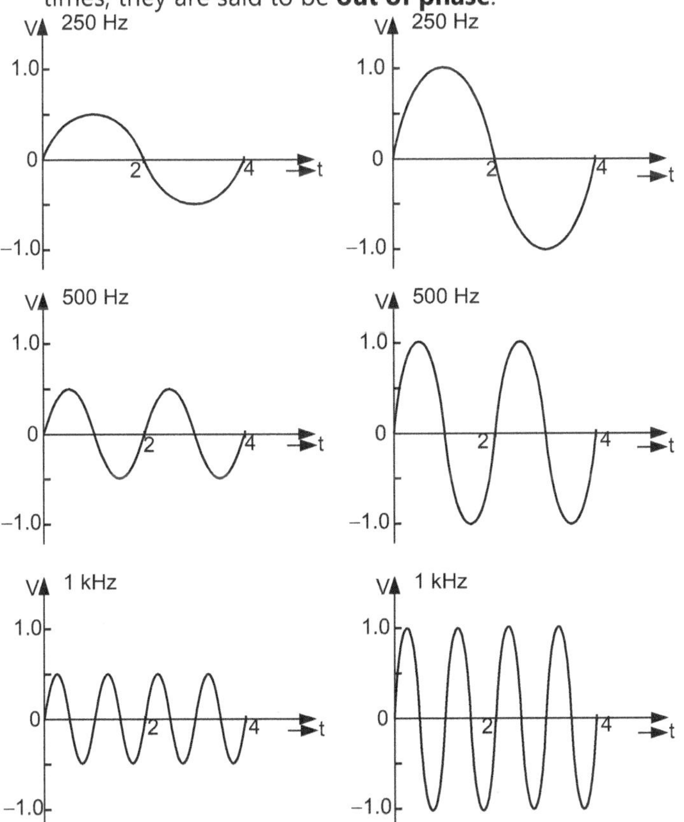

Fig. 1.5 : Sine Waves of 250 Hz, 500 Hz and 1 kHz with different amplitudes

- When this happens, the difference in phase can be measured in degrees, and is called the **phase angle**, θ. As you can see, the two waves in part (b) are a quarter cycle out of phase, so the phase angle θ = 90°.

- The graphs above show waveforms of different frequency and amplitude.

5. Wavelength :

- Wavelength is the characteristic of sine wave which binds the period or the frequency of sine wave to the propagation speed of the medium.

$$\text{Wavelength} = \text{Propagation speed} \times \text{Period} = \frac{\text{Propagation speed}}{\text{Frequency}}$$

where, propagation speed of electromagnetic signal = 3×10^8 m/s.

Making of Waves :

- Sine waves can be mixed with DC signals, or with other sine waves to produce new waveforms. Here is one example of complex waveform.

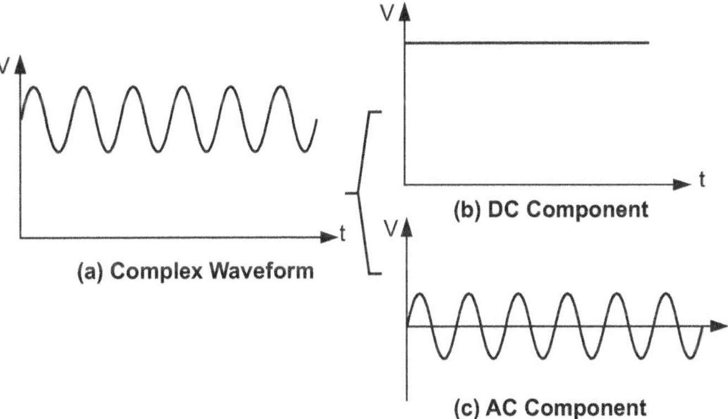

Fig. 1.6 : DC Component Superimposed with AC Component

- 'Complex' does not mean difficult to understand. A waveform like this can be thought of consisting of a DC component with a superimosed AC component. It is quite easy to separate these two components using a **capacitor.**

- More dramatic results are obtained by mixing a sine wave of a particular frequency with exact multiples of the same frequency, in other words, by adding **harmonics** to the **fundamental** frequency. The V/t graphs below show what happens when a sine wave is mixed with its 3rd harmonic (3 times the fundamental frequency) at reduced amplitude, and subsequently with its 5th, 7th and 9th harmonics.

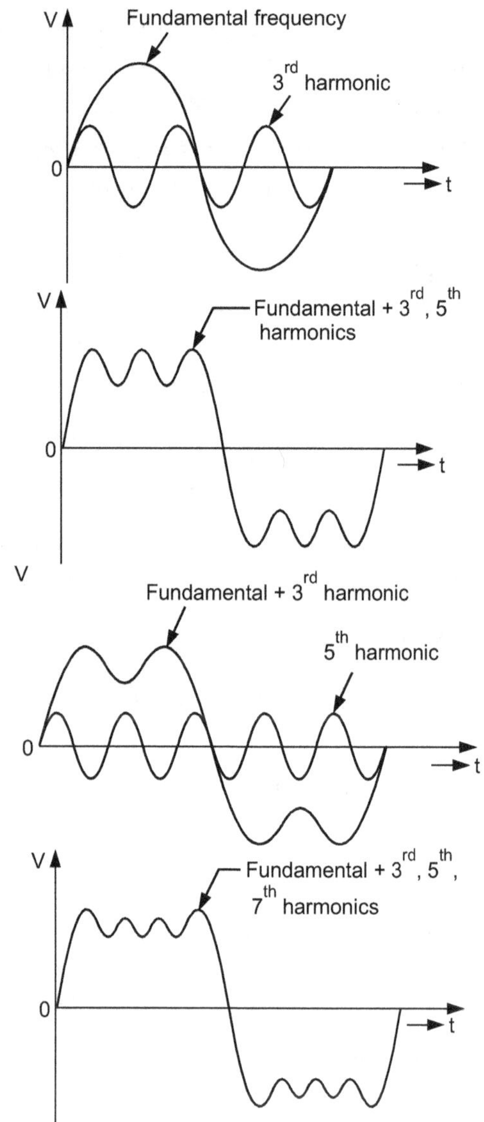

Fig. 1.7 : Fundamental and Harmonic Waveforms

- As you can see, as more odd harmonics are added, the waveform begins to look more and more like a square wave.

- This surprising result illustrates a general principle first formulated by the French mathematician Joseph Fourier, namely that *any* complex waveform can be built up from a pure sine wave plus particular harmonics of the fundamental frequency.

- Square waves, triangular waves and sawtooth waves can be produced in this way.

1.2.2 Other Signals

This part of section outlines the other types of signal you are going to meet. Circuits which generate these signals are versatile building blocks.

1. Square Waves:

- Like sine waves, square waves are described in terms of period, frequency and amplitude as shown in Fig. 1.8.

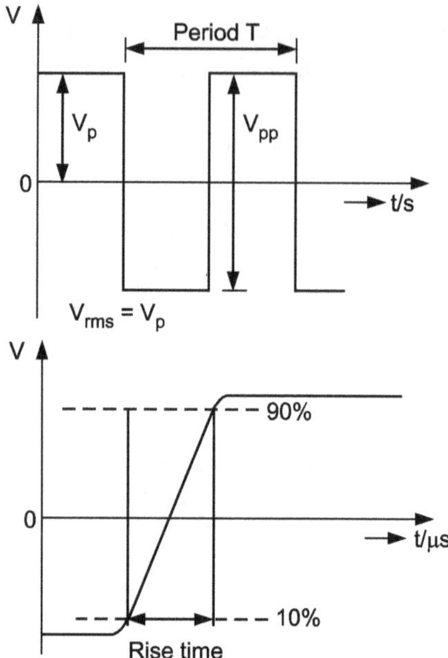

Fig. 1.8 : Square Waveform

- Peak amplitude, V_p, and peak-to-peak amplitude, V_{pp}, are measured as you might expect.

- However, the rms amplitude, V_{rms}, is greater than that of a sine wave.

- Remember that the rms amplitude is the DC voltage which will deliver the same power as the ᴵsignal. If a square wave supply is connected across a lamp, the current flows first one way and then the other.

- The current switches direction but its *magnitude* remains the same.

- In other words, the square wave delivers its maximum power throughout the cycle so that V_{rms} is equal to V_p. (If this is confusing, don't worry, the rms amplitude of a square wave is not something you need to think about very often.)

- Although a square wave may change very rapidly from its minimum to maximum voltage, this change cannot be instantaneous.

- The **rise time** of the signal is defined as the time taken for the voltage to change from 10% to 90% of its maximum value. Rise times are usually very short, with durations measured in nanoseconds (1 ns = 10^{-9} s), or microseconds (1 μs = 10^{-6} s), as indicated in the graph.

2. Pulse Waveforms:

- Pulse waveforms look similar to square waves, except that all the action takes place above the X-axis.

- At the beginning of a pulse, the voltage changes suddenly from a LOW level, close to the X-axis, to a HIGH level, usually close to the power supply voltage.

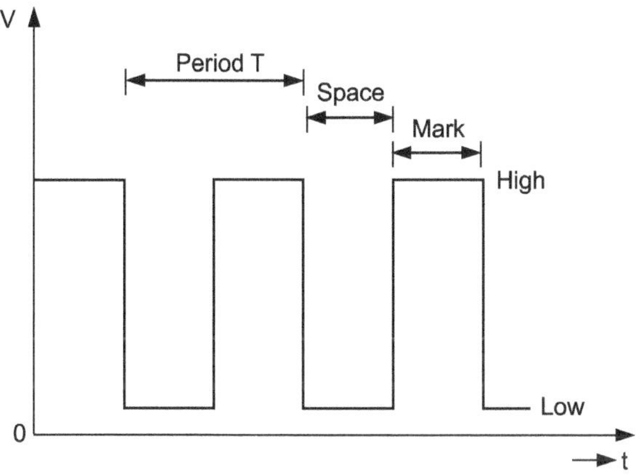

Fig. 1.9 : Pulse Waveform

- Sometimes, the 'frequency' of a pulse waveform is referred to as its **repetition rate**. This means the number of cycles per second, measured in hertz **(Hz)**.

- The HIGH time of the pulse waveform is called the **mark**, while the LOW time is called the **space**. The mark and space do not need to be of equal duration. The **mark space ratio** is given by,

$$\text{Mark space ratio} = \frac{\text{HIGH time}}{\text{LOW time}}$$

- A mark space ratio = 1.0 means that the HIGH and LOW times are equal, while a mark space ratio = 0.5 indicates that the HIGH time is half as long as the LOW time.

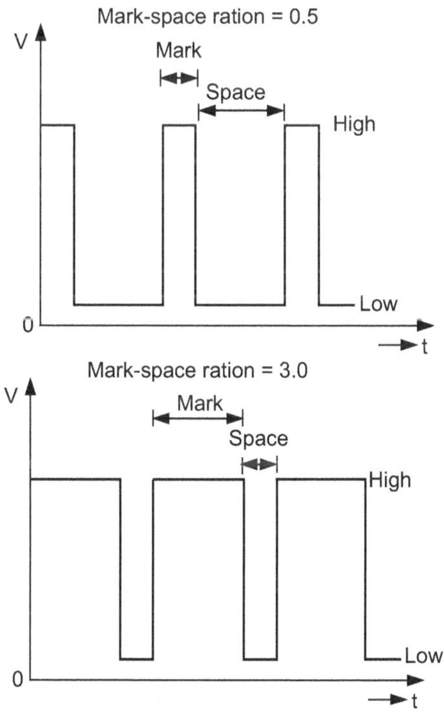

Fig. 1.10 : Pulse Waveforms with Different Duty Cycles

- A mark space ratio of 3.0 corresponds to a longer HIGH time, in this case, three times as long as the space.

- Another way of describing the same types of waveform uses the **duty cycle**, where,

$$\text{Duty cycle} = \frac{\text{HIGH time}}{\text{Period}} \times 100\%$$

- When the duty cycle is less than 50%, the HIGH time is shorter than the LOW time, and so on.

3. Ramps:

- A voltage ramp is a steadily increasing or decreasing voltage, as shown in Fig. 1.11.

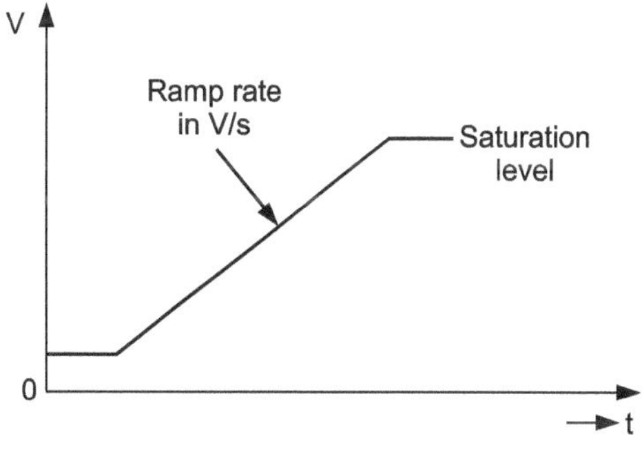

(a) Increasing Ramp

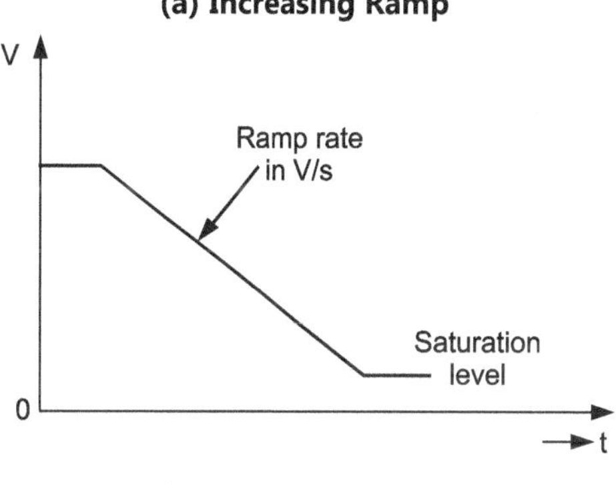

(b) Decreasing Ramp

Fig. 1.11 : Ramp Waveforms

(Increasing and Decreasing Ramp)

- The **ramp rate** is measured in units of volts per second, V/s. Such changes cannot continue indefinitely, but stop when the voltage reaches a **saturation level**, usually close to the power supply voltage.

4. Triangular and Sawtooth Waves:

- These waveforms consist of alternate positive-going and negative-going ramps.

- In a triangular wave, the rate of voltage change is equal during the two parts of each cycle, while in a sawtooth wave, the rates of change are unequal.

- Sawtooth generator circuits are an essential building block in oscilloscope and television systems.

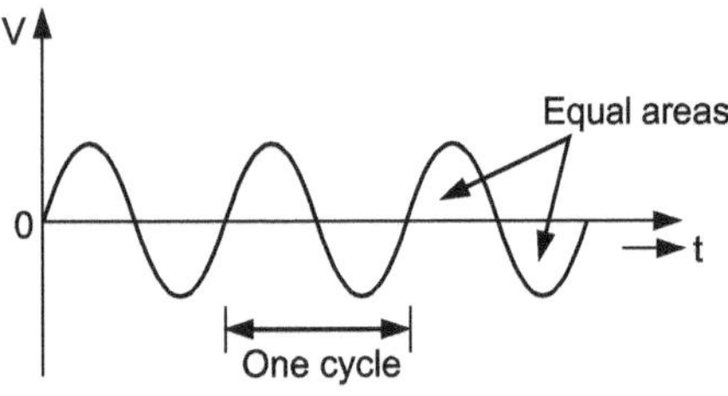

(a) Sine wave

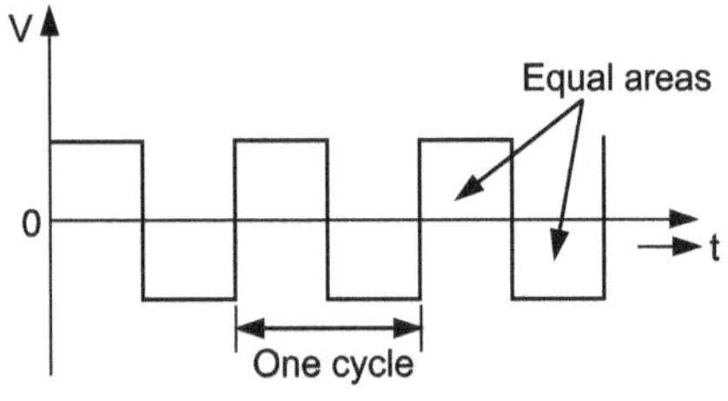

(b) Square wave

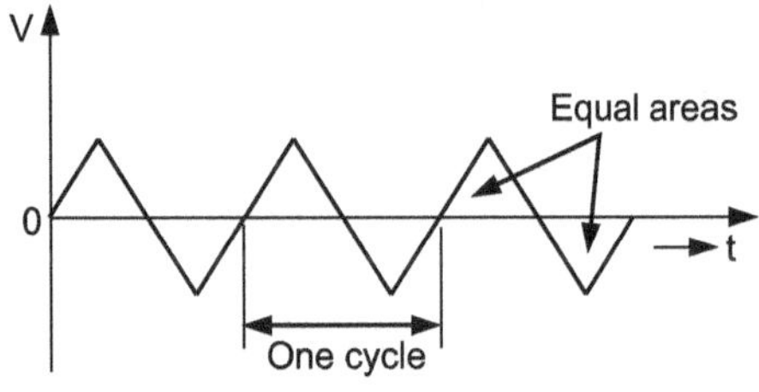

(c) Triangular wave

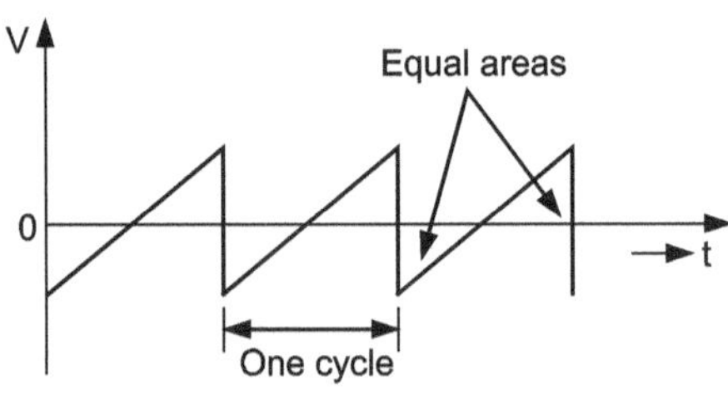

(d) Sawtooth wave

Fig. 1.12 : Other Periodic Waves

- As you can see, the voltage levels change with time and are alternate between positive values (above the X-axis) and negative values (below the X-axis).

- Signals with repeated shapes are called **waveforms** and include **sine** waves, **square** waves, **triangular** waves and **sawtooth** waves.

- A distinguishing feature of alternating waves is that equal areas are enclosed above and below the X-axis.

5. Audio Signals:

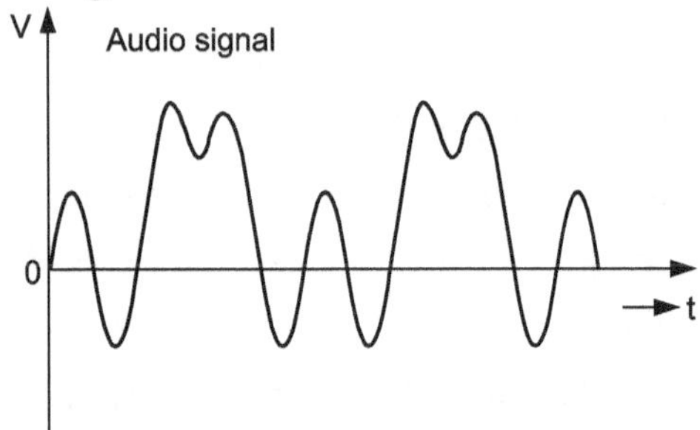

Fig. 1.13 : Audio Signal (20 Hz to 20 kHz) Waveform

- As already mentioned, sound frequencies which can be detected by the human ear vary from a lower limit of around 20 Hz to an upper limit of about 20 kHz.

- A sound wave amplified and played through a loudspeaker gives a pure audio tone.

- Audio signals like speech or music consist of many different frequencies.

- Sometimes it is possible to see a dominant frequency in the V/t graph of a musical signal, but it is clear that other frequencies are present.

6. Noise:

- A noise signal consists of a mixture of frequencies with random amplitudes as shown in Fig. 1.14.

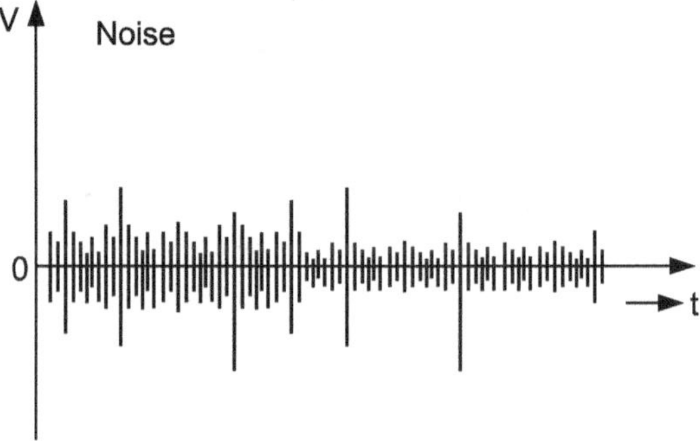

Fig. 1.14 : Noise Waveform Signal

- Noise can originate in various ways.

- For example, heat energy increases the random motion of electrons and results in the generation of **thermal noise** in all components, although some components are 'noisier' than others.

- Additional sources of noise include radio signals, which are detected and amplified by many circuits, not just by radio receivers.

- Interference is caused by the switching of mains appliances and 'spikes' and 'glitches' are caused by rapid changes in current and voltage elsewhere in an electronic system.

1.2.3 Time and Frequency Domain

- So far we have seen the amplitude, frequency and phase of the sine wave with respect to time axis, this is called as **time-domain** plot of sine wave.

- The plot between amplitude and frequency of a wave is called as **frequency-domain** plot of a wave.

- Thus, the complete sine wave of 3 Hz frequency in the time domain can be represented by one single spike in the frequency domain with (4 V peak amplitude).

- Frequency domain plot is extremely useful and compact when we are dealing with more than one sine waves.

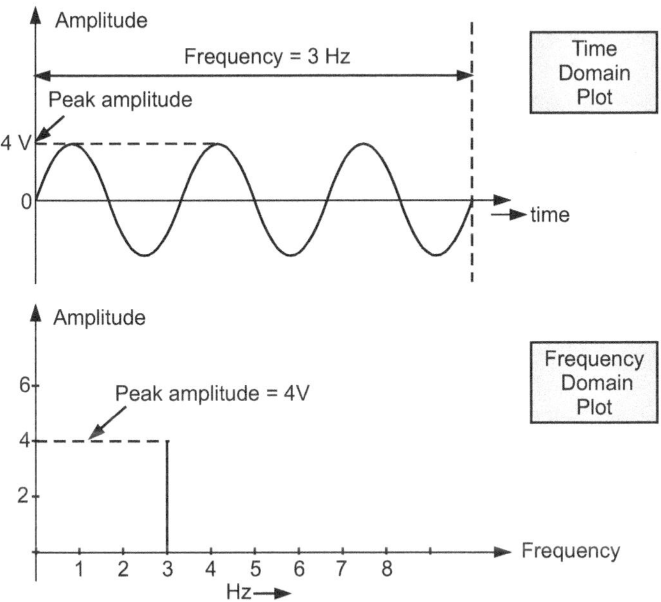

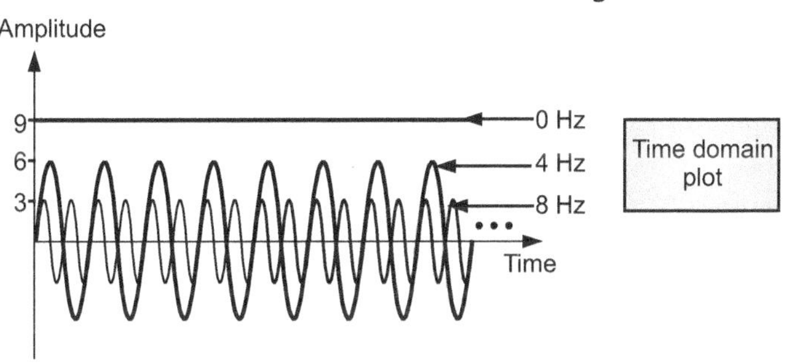

Fig. 1.15 : Time and Frequency Domain Plot of the Sine wave

- The time domain and frequency domain plot of 0 Hz, 4 Hz and 8 Hz sine waves are shown in Fig. 1.16.

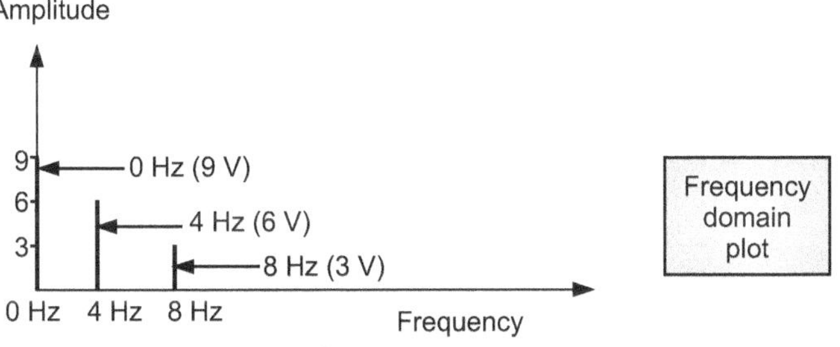

(a) Time-domain representation of three sine waves with frequencies 0, 4 and 8 Hz

(b) Frequency-domain representation of the same three signals

Fig. 1.16 : Time Domain and Frequency Domain Plot of 0 Hz, 4 Hz and 8 Hz Sine waves

- If you consider the data communication application, we are required to deal with **composite signals** like square waveform signal.

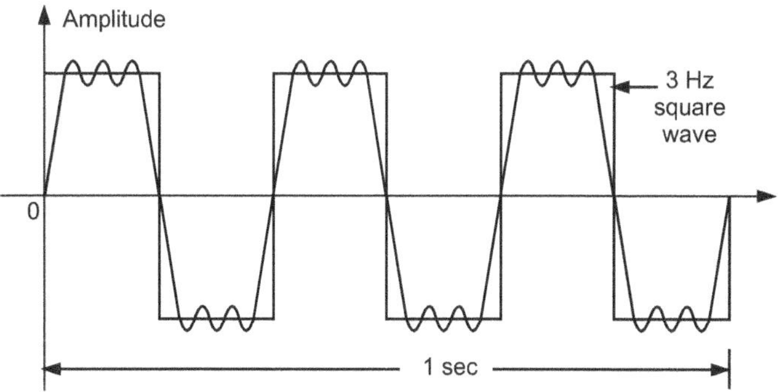

Fig. 1.17 : Square Waveform of 3 Hz (Composite and Periodic Wave is drawn)

- According to Fourier analysis, any composite signal is a combination of simple sine waves with different frequencies, amplitudes and phases.
- If the composite signal is periodic, then its decomposition gives a series of signals with discrete frequencies.
- If the composite signal is non-periodic then the decomposition gives a combination of sine waves with continuous frequencies.

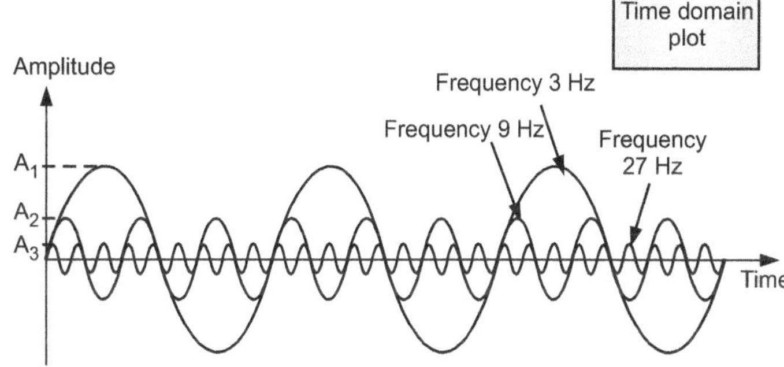

(a) Time-domain decomposition of a composite signal of 3 Hz square wave

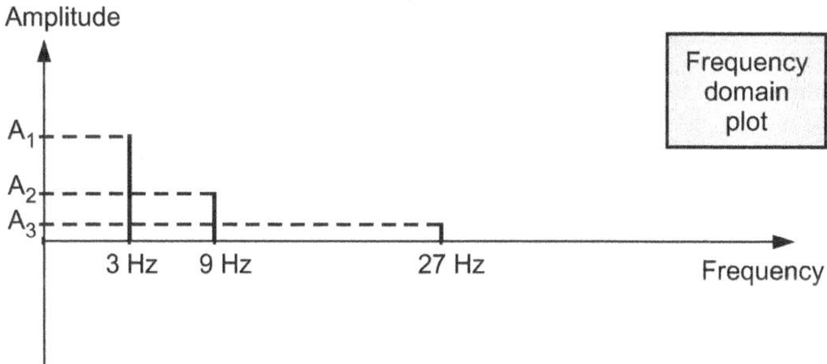

(b) Frequency-domain decomposition of a composite signal of 3 Hz square wave

Fig. 1.18 : Time Domain and Frequency Domain Plot of 3 Hz Composite Square Wave Signal

- Hence, we can say that, the square wave signal of 3 Hz is composed of the fundamental frequency = 3 Hz (or known as first harmonic), 3rd harmonic is of 9 Hz and 9th harmonic is of 27 Hz. Thus, it is integral multiple of 1, 3 and 9 but it is not float number multiple.

- Thus, for non-periodic signal, the time domain and frequency domain plots are as shown in Fig. 1.18 (c).

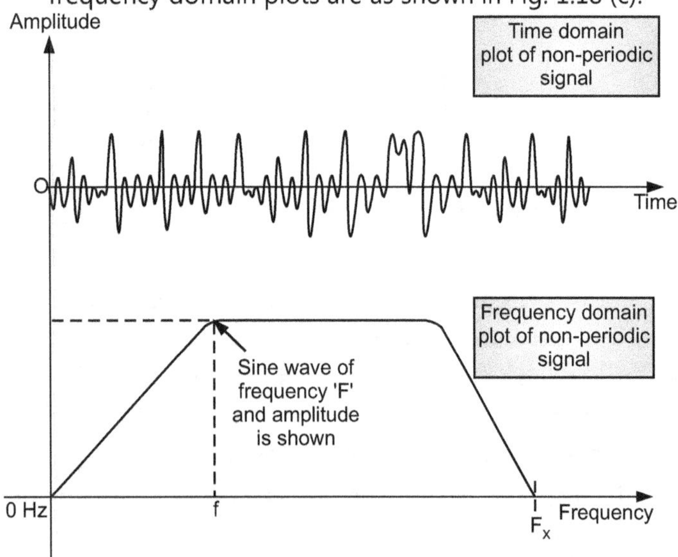

Fig. 1.18 (c) : Time Domain and Frequency Domain Plot of Non-periodic signal

- **The Bandwidth of Composite signal** can be given as the range of the frequencies contained in a composite signal.

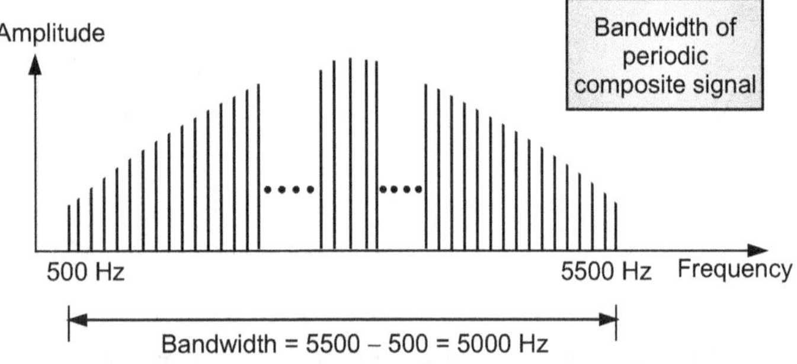

(a) Bandwidth of a periodic signal contains all integer frequencies between 500 Hz and 5500 Hz [i.e. 500, 501, 502, 503, 504 ... 5500 etc.]

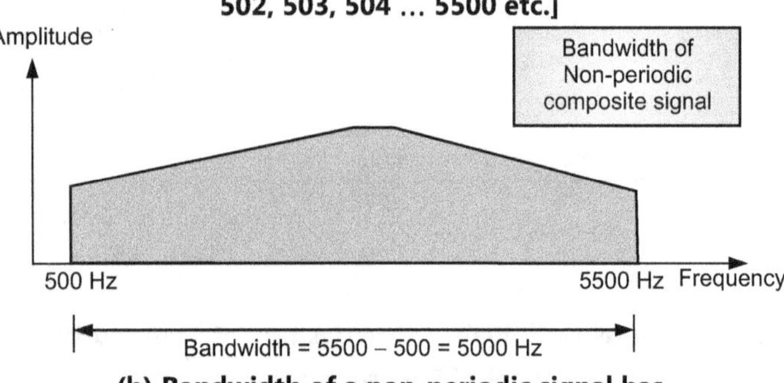

(b) Bandwidth of a non-periodic signal has same range but the frequencies are continuous

Fig. 1.19 : Bandwidth of Periodic and Non-periodic Composite Signals

- Thus, the bandwidth of a periodic composite signal contains all integer frequencies between 500 Hz and 5500 Hz (i.e. 500 Hz, 501 Hz, 502 Hz, 503 Hz, 504 Hz ... 5500 Hz, etc.).

- Also the bandwidth of a non-periodic composite signal contains same range but the frequencies are continuous as shown in Fig. 1.19.

1.3 DIGITAL SIGNALS

We know that digital signals has discrete values.

Also digital signal can have more than two levels.

Two level and four level digital signals are shown in Fig. 1.20.

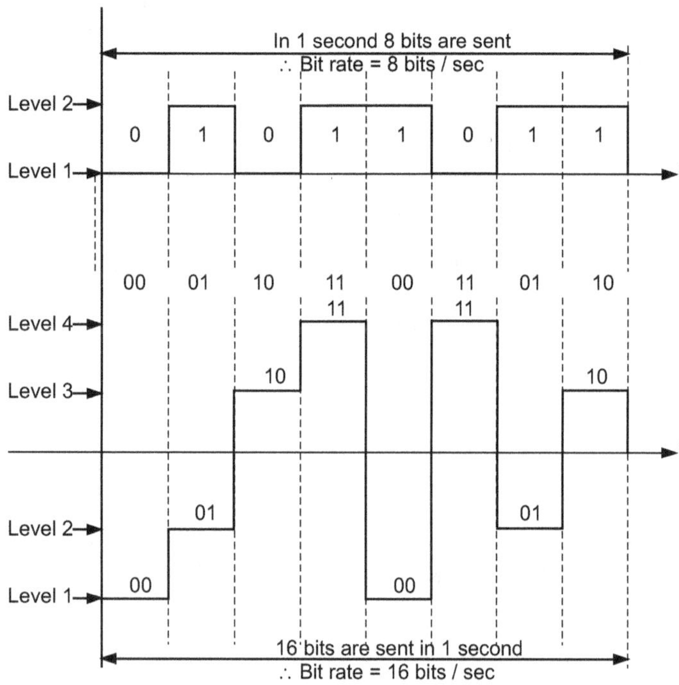

Fig. 1.20 : Two Level and Four Level Digital Signals

Bit Rate of digital signal is given as number of bits sent in one second. Hence, bit rate is given in bps.

∴ $\boxed{\text{Bit rate = Bits/sec}}$

Bit Length of digital signal is stated as the distance one bit occupies on the transmission medium and it is given as,

$\boxed{\text{Bit length = Propagation speed} \times \text{Bit duration}}$

1.3.1 Digital Signal as Composite Analog Signal

- We have already seen that fourier series analysis can be used to decompose a digital signal.

- If digital signal is periodic, decomposed signal has infinite bandwidth of discrete frequencies.

- If digital signal is non-periodic, decomposed signal has infinite bandwidth of continuous frequencies.

- The time domain and frequency domain analysis details are shown in Fig. 1.21.

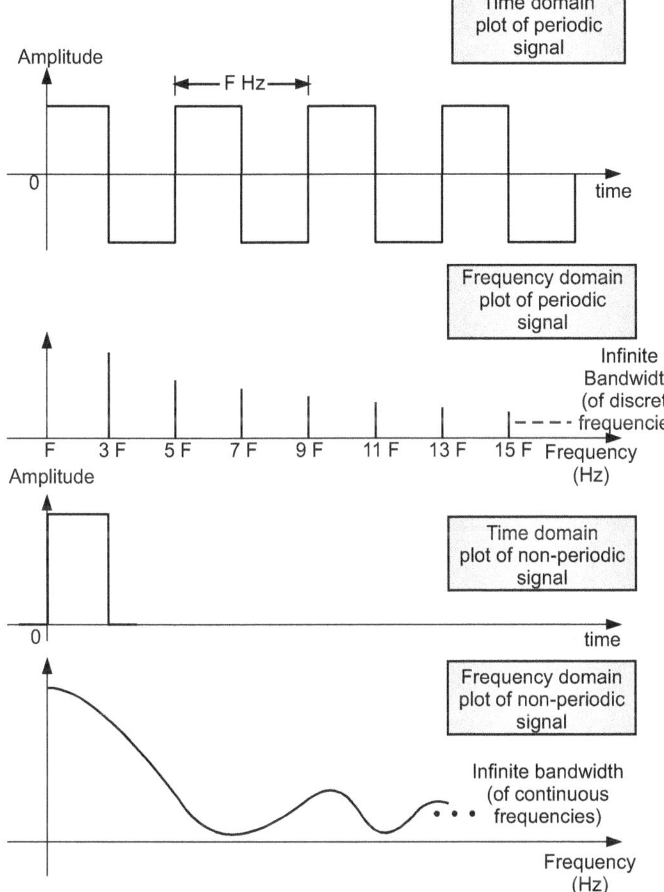

Fig. 1.21 : Time Domain/Frequency Domain Plot of Periodic Digital Signal and Non-periodic Digital Signal

1.3.2 Transmission of Digital Signals

Case 1 : Physical Medium Bandwidth is High :

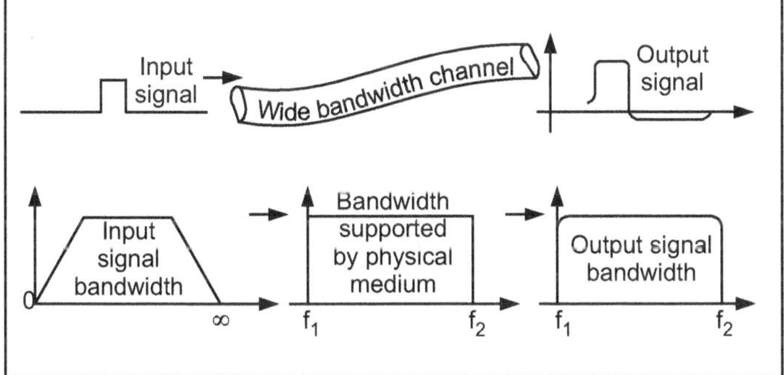

Case 2 : Physical Medium Bandwidth is Low :

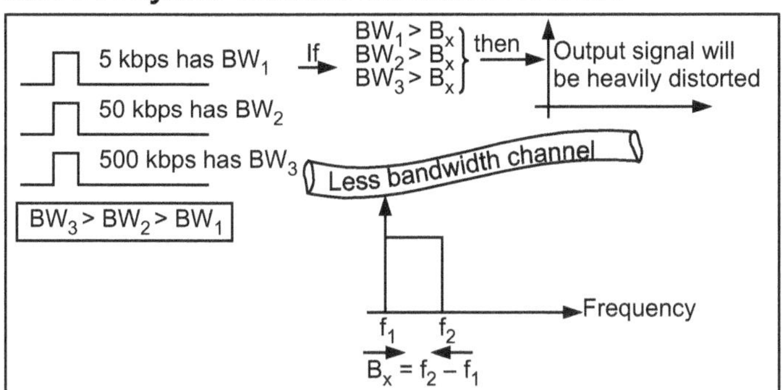

Fig. 1.22 : Digital Signal Transmission using Different Bandwidth Medium

- The transmission of digital signal is possible in two ways :

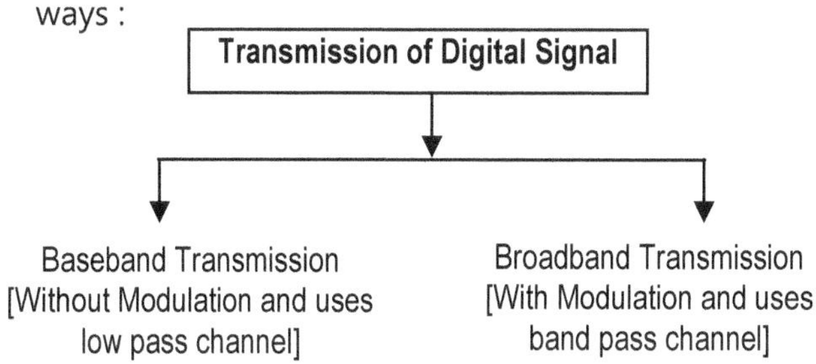

- In Fig. 1.22, we have considered the baseband transmission of the digital signal (which is always non-periodic in nature practically, in data communication).

- Here we have considered two cases :

    Case 1 : Wide bandwidth channel is used.

    Case 2 : Narrow bandwidth channel is used.

- In case 1, small amount of distortion takes place and output signal is less distorted as shown.

- Whereas in case 2, it is clearly shown that if bit rate of input baseband signal increases, then bandwidth requirement to transfer this data also increases.

- In such case 2, if the narrow bandwidth channel is used then output signal will be heavily distorted or may not be available at output.

- The bandwidth requirement for different data rates and for different harmonic values is as shown in Table 1.2.

Table 1.2 : Bandwidth requirement for different data rates and different harmonic values

Bit Rate (N)	Harmonic 1 and required BW = N/2	Harmonics 1 and 3 and required BW = 3N/2	Harmonics 1, 3 and 5 and required BW = 5N/2
N = 5 kbps	1.5 kHz	7.5 kHz	11.5 kHz
N = 50 kbps	25 kHz	75 kHz	125 kHz
N = 500 kbps	250 kHz	750 kHz	1250 kHz

- Thus, for proper digital signal transmission with more bit rate, the bandwidth requirement of medium channel increases and if not used, then signal may be heavily distorted or may even be lost.

- Now second way of digital signal transmission is with modulation. [i.e. No baseband transmission]. It is known as broadband transmission.

    Thus, Fig. 1.23 drawn is self explanatory.

- In this figure, we can see that modulation process allows us to use a bandpass channel (a channel with bandwidth which doesn't start from zero i.e. starts at f_1 and ends at f_2).

- Here, D to A conversion and A to D conversion are basically different concepts.

- D to A conversion is digital continuous wave modulation. For example, ASK (Amplitude Shift Keying), FSK (Frequency Shift Keying) or PSK (i.e. Phase Shift Keying).

- Due to this digital continuous wave modulation, which gives analog output in nature, has limited bandwidth = $f_2 - f_1$ and is less practical.

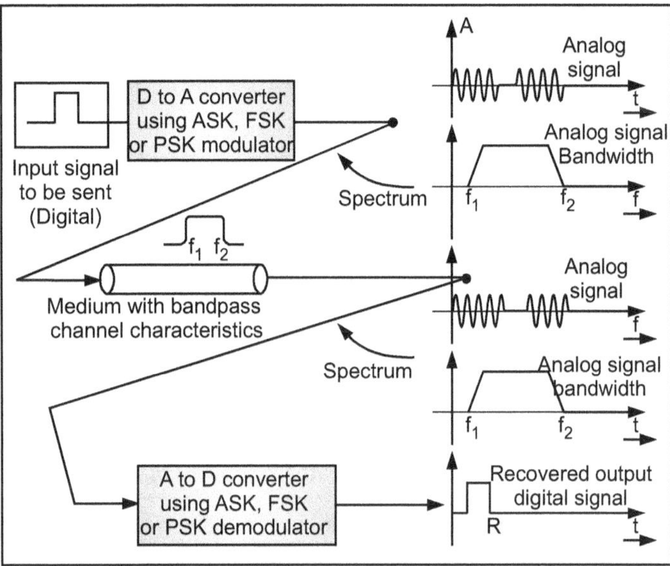

Fig. 1.23 : Transmission and Reception of

Modulated Broadband Signal over Bandpass

Channel Medium

- This bandwidth of signal is easily passed through the medium with bandpass channel characteristics and has almost negligible distortion at medium (channel) output as shown.

- Thus, now this analog signal is given to A to D converter at receiver end to recover original unmodulated digital signal as shown.

- At receiver end, A to D converter is basically a ASK, FSK or PSK demodulator.

- This ASK, FSK or PSK modulator/demodulator will be studied in Unit 2 of this book, in detail.

1.4 TRANSMISSION IMPAIRMENT

- Transmission impairment means, due to imperfections of medium, signal sent from transmitter is not equal to signal received from receiver.

- There are three causes of impairment :
  1. Attenuation   2. Distortion   3. Noise

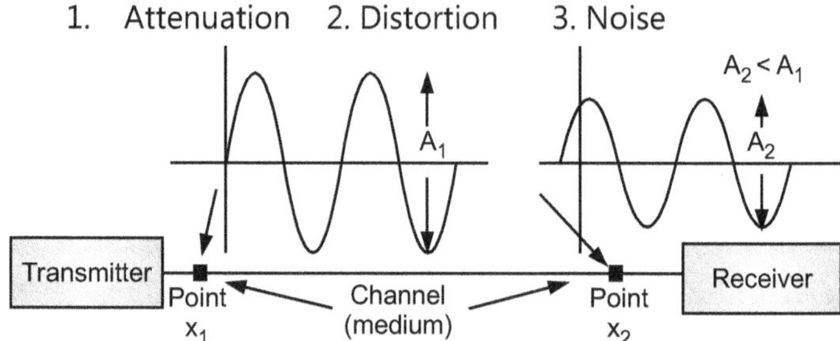

Fig. 1.24 : Attenuation Impairment

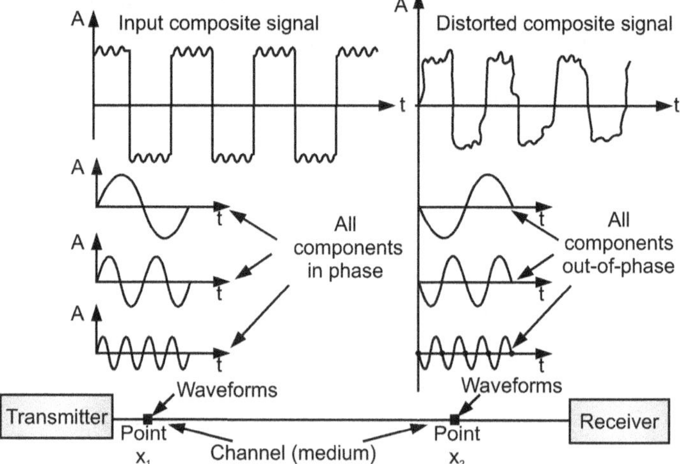

Fig. 1.25 : Distortion Impairment

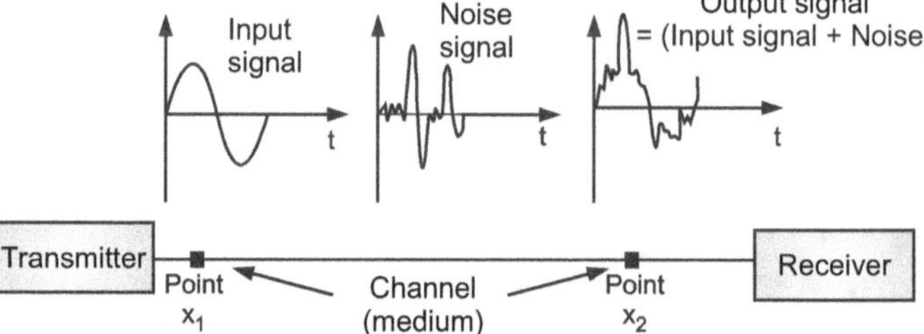

Fig. 1.26 : Noise Impairment

- Attenuation impairment is shown in Fig. 1.24. Distortion impairment is shown in Fig. 1.25 and Noise impairment is shown in Fig. 1.26.

- Attenuation means loss of electrical signal energy in the form of heat due to channel resistance. Thus, energy lost is given as,

$$E_L = I^2 (R)$$

where, I is the current flowing through medium and R = channel resistance.

Also power loss in decibels is given as,

$$P_L = \log_{10}\frac{P_2}{P_1}$$

where, P_2 is the power at point x_2 and P_1 is the power at point x_1.

- Distortion means the received signal at receiver end changes its shape or its form. This distortion is different in different frequencies due to different propagation speed of signal frequencies in medium. Thus, phases of received waveforms are changed and distortion occurs in received signal.

- Electrical disturbances interfere with the input signal and produce noise. Noise always limits the performance of communication system. In electrical terms, any unwanted introduction of energy tending to interfere with the proper reception and reproduction of transmitted signal.

- Noise is classified as external noise and internal noise. External noise is due to atmosphere, extraterrestrial noise or industrial noise.

- The different types of internal noises are thermal noise, shot noise, partition noise and low frequency or flicker noise.

- Signal to noise ratio of system is given by,

$$\text{SNR} = \frac{\text{Average signal power}}{\text{Average noise power}}$$

$$\text{SNR}_{dB} = 10 \log_{10} (\text{SNR})$$

1.5 DATA RATE LIMITS

- In data communication system, the Data Rate depends upon three factors :

  1. Available bandwidth (of medium).

  2. Signal level (amplitude of signal).

  3. Quality of medium (channel) (Noise amplitude).

- There are two formulae to calculate the data rate.

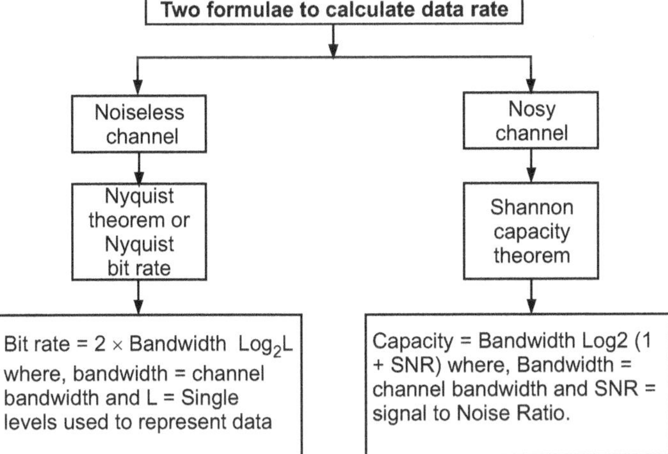

1.6 PERFORMANCE

In this section, we will discuss the following things :

   (a) QoS (Quality of Service).   (b) Bandwidth.

   (c) Throughput.   (d) Delay

   (e) Jitter.   (f) Bandwidth-delay product.

1.6.1 Problems in Data Communication

- When the Internet was first deployed many years ago, it lacked the ability to provide Quality of Service guarantees due to limits in router computing power.

- It therefore ran at default QoS level, or "best effort".

- Many things can happen to packets as they travel from origin to destination, resulting in the following problems as seen from the point of view of the sender and receiver :

(i) Dropped Packets : The routers might fail to deliver (drop) some packets if they arrive when their buffers are already full. Some, none, or all of the packets might be dropped, depending on the state of the network, and it is impossible to determine what will happen in advance. The receiving application must ask for this information to be retransmitted, possibly causing severe delays in the overall transmission.

(ii) Delay : It might take a long time for a packet to reach its destination, because it gets held up in long queues, or takes a less direct route to avoid congestion. Alternatively, it might follow a fast, direct route. Thus delay is very unpredictable.

(iii) Jitter : Packets from source will reach the destination with different delays. This variation in delay is known as jitter and can seriously affect the quality of streaming audio and/or video.

(iv) Out-of-Order Delivery : When a collection of related packets is routed through the Internet, different packets may take different routes, each resulting in a different delay. The result is that the packets arrive in a different order to the one with which they were sent. This problem necessitates special additional protocols responsible for rearranging out-of-order packets to an isochronous state once they reach their destination. This is especially important for video and VoIP streams where quality is dramatically impacted by both latency or lack of isochronicity.

(v) Error : Sometimes packets are misdirected, or combined together, or corrupted, while enroute. The receiver has to detect this and, just as if the packet was dropped, ask the sender to repeat itself.

1.7 NETWORKS

A network involves a number of devices linked together to form a communication system for information and device sharing.

- Local Area Networks (LANs) are small, limited to about 500 meters, and are commonly deployed in corporate offices to facilitate low-cost, high-bandwidth information transfer within a company.

- Cities and other metropolitan regions can be connected via Metropolitan Area Networks or (MANs), and Wide Area Networks (WANs) involve systems communicating across large geographic regions such as states or countries.

- Globally, computers in networks interlink to form what we refer to as "the Internet."

1.7.1 Advantages of Installing a Network

- **Speed :** Networks provide a very rapid method for sharing and transferring files. Without a network, files are shared by copying them to floppy disks, then carrying or sending the disks from one computer to another. This method of transferring files is very time-consuming.

- **Cost :** Networkable versions of many popular software programs are available at considerable savings when compared to buying individually licensed copies. Besides monetary savings, sharing a program on a network allows for easier upgrading of the program. The changes have to be done only once, on the file server, instead of on all the individual workstations.

- **Security :** Files and programs on a network can be designated as "copy inhibit," so that you do not have to worry about illegal copying of programs. Also, passwords can be established for specific directories to restrict access to authorized users.

- **Centralized Software Management :** One of the greatest benefits of installing a network is the fact that all of the software can be loaded on one computer (the file server). This eliminates the need to spend time and energy installing updates and tracking files on independent computers throughout the building.

- **Resource Sharing :** Sharing resources is another area in which a network exceeds stand-alone computers. Most institutes or companies cannot afford enough laser printers, fax machines, modems, scanners, and CD-ROM players for each computer. However, if these similar peripherals are added to a network, they can be shared by many users.

- **Electronic Mail :** The presence of a network provides the hardware necessary to install an e-mail system. E-mail aids in personal and professional communication for all personnel, and it facilitates the dissemination of general information to the entire users. Electronic mail on a LAN can enable users to communicate with others. If the LAN is connected to the Internet, user can communicate with others throughout the world.

- **Flexible Access :** Networks allow users to access their files from computers throughout the campus if it is a LAN. Users can begin an assignment in their LAN, save part of it on a public access area of the network, and then go to the media center after office hours to finish their work. Users can also work co-operatively through the network.

- **Workgroup Computing :** Workgroup software allows many users to work on a document or project concurrently.

1.7.2 Disadvantages of Installing a Network

- **Expensive to Install :** Although a network will generally save money over time, the initial cost of installation can be prohibitive. Cables, network cards, and software are expensive, and the installation may require the services of a technician.

- **Requires Administrative Time :** Proper maintenance of a network requires considerable time and expertise. Many companies have installed a network, only to find that they did not budget for the necessary administrative support.

- **File Server May Fail :** Although a file server is no more susceptible to failure than any other computer, when the files server "goes down," the entire network may come to a halt. When this happens, the entire company may lose access to necessary programs and files.

- **Cables May Break :** The Topology chapter presents information about the various configurations of cables. Some of the configurations are designed to minimize the inconvenience of a broken cable; with other configurations, one broken cable can stop the entire network.

1.7.3 Network Usage and Typical Computer Network

Networks are widely used in both the business and consumer landscapes.

- In the corporate environment, LANs are commonly used to share resources, including electronic files and devices such as printers.
- These LANs are generally connected to other networks via WANs and the Internet to facilitate global data access.
- In healthcare, LANs are used in the clinical environment to provide information such as patient's medical records and drug formularies for doctors and nurses.

Wireless networks provide the next step in utility and convenience for many industries, including health care.

- In general, wireless networks provide the power and freedom of mobility, with the setbacks of reduced speed and unpolished functions (as compared to wired networks).
- While wireless networks have existed for decades, only the recent boom of handheld and mobile devices has spurred the demand necessary to create robust networks.

If a home has more than one computer, then installing a computer network is a smart decision.

- Networks allow you to share an Internet connection and files among multiple PCs.

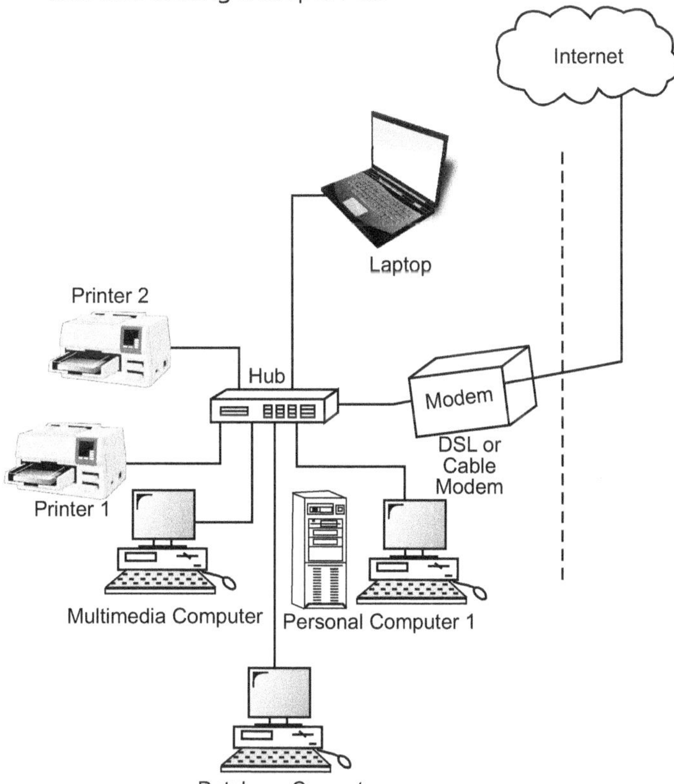

Fig. 1.27 : Typical Home Computer Network (Wired)

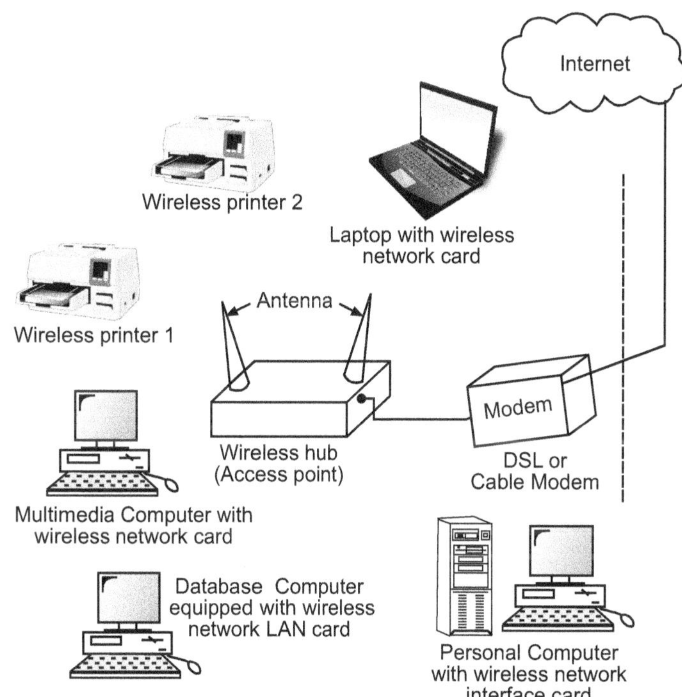

Fig. 1.28 : Typical Home Computer Network (Wireless)

- More importantly, they save time and money, and make using your computer equipment much more enjoyable for everyone in the office as well as in family.
- Lets consider the networking for the home, but all information given here applies to networking for a small business as well.
- Setting up a home network is the simplest way to get the most out of your computer equipment.
- And as your family grows or you add additional computers, expansion is no problem. Best of all, creating a network is easier than you might think.

The key benefits of networking a home include :

- Sharing a high-speed Internet connection - without anyone having to sign off, and without having another phone line installed.
- Playing games head-to-head on different computers from different rooms.
- Sharing an expensive resource like the colour photo printer in office without having to interrupt any office work.
- Everyone in the family can share files from every PC in the house - no need to put files onto floppy or zip discs and swap them.

  To go wireless or wired ? That is the question. A wireless setup uses radio waves, while wired networks communicate through data cables. Both systems have their own advantages and disadvantages.

The Following Points Decide Whether to go for Wired or Wireless Networks :

- **Range :** The range of the network is an important consideration while using a network.

- **Throughput :** The amount of data transferable using devices is important.

- **Integrity :** The network should have a stable form of communication. The robust designs of technology should provide data integrity performance equal to or better than other technologies.

- **Inter-operability :** Device should provide the ability to connect to wired or wireless LAN with ease.

- **Scalability :** Networks can be designed to be extremely simple or quite complex. Networks should support large number of nodes and/or large physical areas to boost or extend coverage.

- **Simplicity of Installation and Use :** Users should need very little new information to take advantage of LANs to be used. It should be simple and easy to install.

- **Security :** Because network technology has roots in military applications and banking applications, security has long been a design criterion for network technology.

- **Power Requirement for Networks :** End-user products should be designed to run with less power and accordingly the networking technique will be decided.

- **Safety :** The used technology should be safe for human and nature. Network must meet stringent government and industry regulations for safety.

Thus Network can be Briefly Explained as Follows :

- A network is a group of two or more computers that are able to communicate with one another and share data (text, sound, images), files, programs, and operations.

- The computers are able to communicate and exchange information because they use software that observes the same set of parameters, or protocol.

- There are several different types of networks.

- All networks operate using the same basic principle : Whenever a computer on network sends information to another computer or peripheral, the information is in the form of a "packet." When the packet reaches the designated station, the information is transferred to the computer.

- The basic equipment you need to set up a network includes network cards, cables, and networking software. You also need a "hub" into which all the cables are connected.

1.8.1 Local Area Network (LAN)

- Networks used to interconnect computers in a single room, rooms within a building or buildings on one site are called Local Area Network (LAN).

- LAN transmits data with a speed of several megabits per second (106 bits per second). The transmission medium used is normally coaxial cables.

- LAN links computers, i.e., software and hardware, in the same area for the purpose of sharing information.

- Usually LAN links computers within a limited geographical area because they must be connected by a cable, which is quite expensive.

- People working in LAN get more capabilities in data processing, work processing and other information exchange compared to stand-alone computers.

- Because of this information exchange, most of the business and government organizations are using LAN.

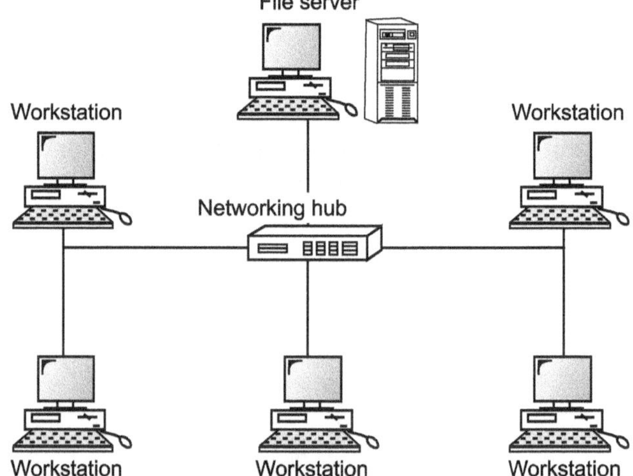

Fig. 1.29 : Typical LAN

Major Characteristics of LAN

- Every computer has the potential to communicate with any other computers of the network.

- High degree of interconnection between computers.

- Easy physical connection of computers in a network.

- Inexpensive medium of data transmission.

- High data transmission rate.

Components of LAN

1. Workstations

- In LAN, a workstation refers to a machine that will allow users to access a LAN and its resources while providing intelligence on board allowing local execution of applications.

- It may allow data to be stored locally or remotely on a file server.
- Diskless workstations require all data to be stored remotely, including that data necessary for the diskless machine to boot up.
- Executable files may reside locally or remotely as well, meaning a workstation can run its own programs or those copied off the LAN.

2. Servers

- A server is a computer that provides the data, software and hardware resources that are shared on the LAN.
- A LAN can have more than one server; each has its unique name on the network and all LAN users identify the server by its name.
- **Dedicated Server:** A server that functions only as a storage area for data and software and allows access to hardware resources is called a dedicated server. Dedicated servers need to be powerful computers.
- **Non-Dedicated Server:** In many LANs, the server is just another workstation. Thus, there is a user networking on the computer and using it as a workstation, but part of the computer also doubles up as a server. Such a server is called a non-dedicated server since it is not completely dedicated to serving. LANs do not require a dedicated server since resource sharing amongst a few workstations is proportionately on a smaller scale.
- **Other Types of Servers:** In large installations which have hundreds of workstations sharing resource, a single computer is often not sufficient to function as a server.

Some of the other servers have been discussed here :

- **File Server :** A file server stores files that workstations can access and it also decides on the rights and restrictions that the users need to have while accessing files on LAN.
- **Printer Server :** A Printer server takes care of the printing requirement of number of workstations.
- **Modem Server :** It allows LAN users to use the modem to transmit long distance messages. Server attached to one or two modems would serve the purpose.

3. Clients

- A client is any machine that requires something from a server.
- In the more common definition of a client, the server supplies files and sometimes processing power to the smaller machines connected to it.

- Each machine is a client.
- Thus, a typical ten PC local area network may have one large server with all the major files and databases on it and all the other machines connected as clients.
- This type of terminology is common with TCP/IP networks, where no single machine is necessarily the central repository.

4. Nodes

- Small networks that comprise of a server and number of PCs.
- Each PC on the network is called a node.
- A node essentially means any device that is attached to the network. Because each machine has a unique name or number (so the rest of the network can identify it), you will hear the term node name or node number quite often.

5. Network Interface Cards

- The Network Interface card, or LAN adapter, functions as an interface between the computer and the network cabling, so it must serve two masters.
- Inside the computer, it controls the flow of data to and from the Random-Access Memory (RAM).
- Outside the computer, it controls the flow of data in and out of the network cable system.
- An interface card has a specialized port that matches the electrical signaling standards used on the cable and the specific type of cable connector.
- One must select a network interface card that matches your computer's data bus and the network cable.
- Token ring LANs require token ring NICs, Ethernet LANs require Ethernet NICs, etc.
- The peripheral component interface bus (PCI) has emerged as a new standard for adapter card interfaces.
- It is advisable to use bus PCI-equipped computers and PCI LAN adapters wherever possible.
- Software is required to interface between a particular NIC and an operating system called as Network Interface Card Driver.

6. Connectors

- Connectors used with TP included RJ-11 and RJ-45 modular connectors in current used by phone companies.
- Occasionally other special connectors, such as IBM's Data Connector, are used.
- RJ-11 connectors accommodate 4 wires or 2 twisted pairs, while RJ-45 houses 8 wires or 4 twisted pairs.

7.   The Network Operating System (NOS):

- The Network Operating System software acts as the command center, enabling all of the network hardware and all other network software to function together as one cohesive, organized system. In other words, the network operating system is the heart of the network.

- The special functions of NOS are connecting computers and devices into a LAN or Inter-network. Some examples of NOS are Novell Netware, Windows NT/2000, Linux, Sun Solaris, UNIX, and IBM OS/2 etc.

- NOS provide connectivity among a number of autonomous computers.

Features of NOS:

- Provide fundamental operating system features like support for processors, routine hardware detection, protocols and multi-processing of applications.

- Provide security by authentication, access control and authorization.

- Availability of name and directory services also file, print, web and back-up services.

- Support to internetworking such as routing and WAN ports.

- Provides system management, administration, User management, remote access and support for logon and logoff.

- Primarily there are two types of NOS named as peer-to-peer and client / server.

- Peer-to-peer NOS allow users to share and access resources, files located on their and other computers. In this network, all computers are considered equal with same privileges to use the resources available on the network. Windows for Workgroups is an example of the program that can function as peer-to-peer NOS.

- Client/server NOS allow the network to centralize functions and applications in one or more dedicated file servers. The file servers turn into coral part of the system which allows access to resources with security. The clients have access to the resources available on the file servers.

- The NOS allows multiple users to simultaneously share the same resources irrespective of physical location. Novell Netware and Windows 2000Server are examples of client-server NOS.

Advantages of LAN

- The reliability of network is high because the failure of one computer in the network does not affect the functioning for other computers.

- Addition of new computer to network is easy.

- High rate of data transmission is possible.

- Peripheral devices like magnetic disk and printer can be shared by other computers.

Uses of LAN

Followings are the major areas where LAN is normally used:

- File transfer and Access
- Word and text processing
- Electronic message handling
- Remote database access
- Personal computing
- Digital voice transmission and storage
- Office automation
- Factory automation
- Distributed Computing
- Fire and Security Systems
- Process Control
- Document Distribution.

1.8.2 Wide Area Network (WAN)

- The term Wide Area Network (WAN) is used to describe a computer network spanning a regional, national or global area.

- For example, for a large company the headquarters might be at Delhi and regional branches at Mumbai, Chennai, Bengaluru and Kolkata.

- Here regional centers are connected to headquarters through WAN.

- The distance between computers connected to WAN is larger. Therefore, the transmission mediums used are normally telephone lines, microwaves and satellite links.

Characteristics of WAN

Following are the major characteristics of WAN.

1.   Communication Facility

- For a big company spanning over different parts of the country, the employees can save long distance phone calls and it overcomes the time lag in overseas communications.

- Computer conferencing is another use of WAN where users communicate with each other through their computer system.

2. Remote Data Entry

- Remote data entry is possible in WAN. It means sitting at any location you can enter data, update data and query other information of any computer attached to the WAN but located in other cities.

- For example, suppose you are sitting at Chennai and want to see some data of a computer located at Delhi, you can do it through WAN.

3. Centralized Information

- In modern computerized environment, you will find that big organizations go for centralized data storage.

- This means if the organization is spread over many cities, they keep their important business data in a single place.

- As the data are generated at different sites, WAN permits collection of this data from different sites and save at a single site.

WAN diagram

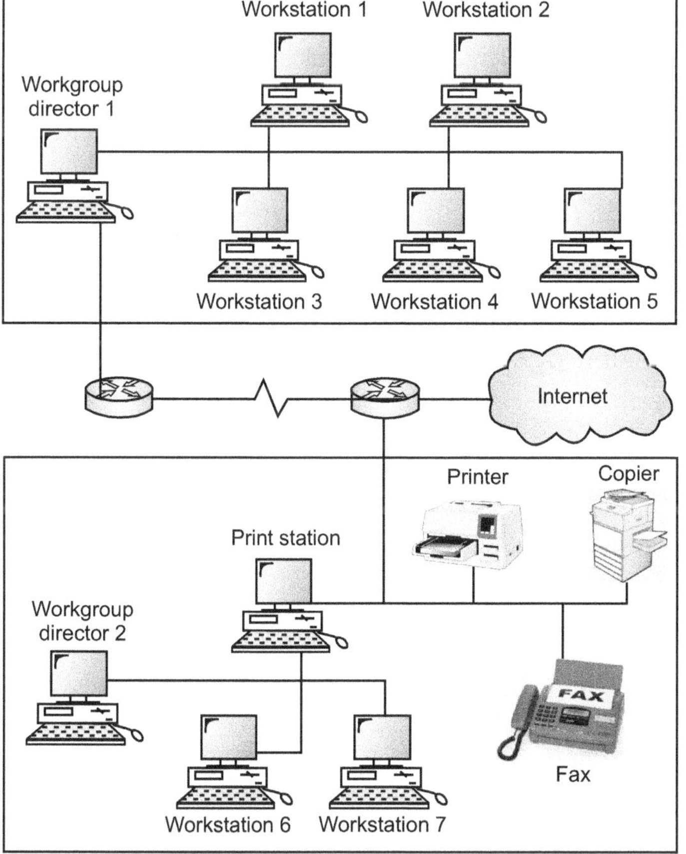

Fig. 1.30 : Typical WAN

Table 1.3 : Difference between LAN and WAN

Sr. No.	LAN	WAN
1.	LAN is restricted to limited geographical area of few kilometers	WAN covers great distance and operates nationwide or even worldwide.
2.	In LAN, the computer terminals and peripheral devices are connected with wires and coaxial cables.	In WAN, there is no physical connection. Communication is done through telephone lines and satellite links.
3.	Cost of data transmission in LAN is less because the transmission medium is owned by a single organization.	In case of WAN the cost of data transmission is very high because the transmission mediums used are wired, either telephone lines or satellite links.
4.	The speed of data transmission is much higher in LAN than in WAN. The transmission speed in LAN varies from 0.1 to 100 megabits per second.	In case of WAN the speed ranges from 1800 to 9600 bits per second (bps).
5.	Few data transmission errors occur in LAN compared to WAN. It is because in LAN the distance covered is negligible.	Relatively high chance of transmission errors
6.	LANs are limited to a small area networking such as in a home, business, school, etc.	WANs cover larger areas, such as cities, and even allow computers in different nations to connect.

1.8.3 Metropolitan Area Network (MAN)

- A Metropolitan Area Network (MAN) is a bigger version of a Local Area Network (LAN) and usually uses similar technology.

- A MAN can cover a group of corporate offices or a town or city, and can be either privately or publicly owned. A MAN can support both data and voice, and may be related to the local cable television network (CATV).

- A MAN employs one or two cables, and does not contain switching elements, which simplifies the design.

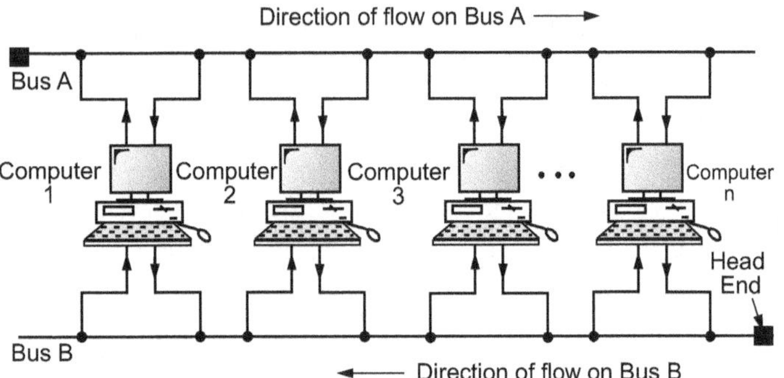

Fig. 1.31 : Typical MAN network
(also known as 802.6 DQDB network)

- A standard has been adopted for MANs called Distributed Queue Dual Bus (DQDB) and is defined by IEEE 802.6.

- DQDB consists of two unidirectional buses (cables) to which all of the computers on the network are connected.

- Each bus has a head-end that initiates transmission activity.

- In the following diagram, traffic that is intended for a computer to the right of the source computer uses the upper bus, while traffic intended for a computer to the left uses the lower bus.

The network is based on fiber-optic cable in a dual-bus topology, and traffic on each bus is unidirectional, providing a fault-tolerant configuration.

Bandwidth is allocated using time slots, and both synchronous and asynchronous modes are supported.

1.8.4 Comparison between LAN, MAN, WAN

Table 1.4 : Comparison of LAN, MAN, WAN

Basis of Comparison	LAN	MAN	WAN
Expands to	Local Area Network	Metropolitan Area Network	Wide Area Network
Meaning	A network that connects a group of computers in a small geographical area.	It covers relatively large region such as cities, towns.	It spans large locality and connects countries together. Example Internet.
Ownership of Network	Private	Private or Public	Private or Public
Design and maintenance	Easy	Difficult	Difficult
Propagation Delay	Short	Moderate	Long
Speed	High	Moderate	Low
Fault Tolerance	More Tolerant	Less Tolerant	Less Tolerant
Congestion	Less	More	More
Used for	College, School, Hospital.	Small towns, City.	Country / Continent.

1.8.5 Personal Area Network (PAN)

- A Personal Area Network (PAN) - is a computer network organized around an individual person. Personal area networks typically involve a mobile computer, a cell phone and/or a handheld computing device such as a PDA. You can use these networks to transfer files including email and calendar appointments, digital photos and music.

- Personal area networks can be constructed with cables or be wireless.

- Wireless PANs typically use Bluetooth or sometimes infrared connections. Bluetooth PANs are also sometimes called piconets.

- Personal area networks generally cover a range of less than 10 meters (about 30 feet). PANs can be viewed as a special type (or subset) of local area network (LAN) that supports one person instead of a group.

- A Personal Area Network – (PAN) is a personal devices network equipped at a limited area. PAN has mobile devices like cell phone, tablet, laptop. That type of network could also be wirelessly connected to Internet.

- A personal area network handles the interconnection of IT devices at the surrounding of a single user. Generally, PAN contains such appliances: cordless mice and keyboards, cordless phone, Bluetooth.

Advantages of PAN

- PAN is convenient, beneficial and handy.

Disadvantages of PAN

- Sometimes PAN has bad connection to other networks at the same radio bands.

- Bluetooth networks have slow data transfer speed, but comparatively safe.

- Bluetooth has distance limits.

1.8.6 AD-HOC Network

- Ad-hoc networking describes a mode of connecting electronic devices to one another without the use of a central device like a router that conducts the flow of communications. Devices connected to an ad-hoc network called nodes, forward data to other nodes.

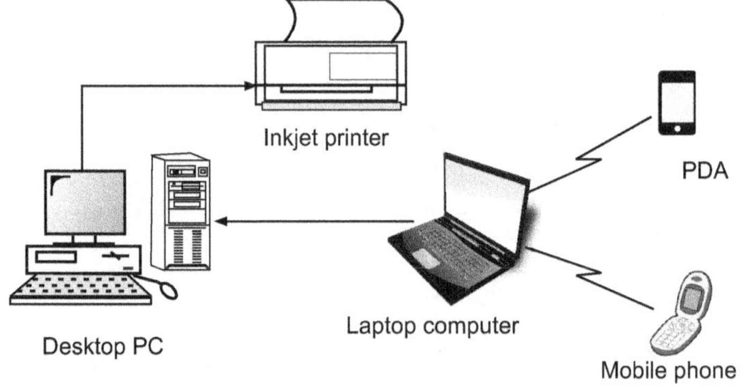

Fig. 1.32 : Ad-hoc network

- An ad-hoc network is a Local Area Network (LAN) that requires minimal configuration and can be deployed quickly, usually for specific or temporary needs.
- Ad-hoc networking takes its name from the Latin *ad hoc,* meaning "for this."
- An ad-hoc network tends to feature a small group of devices all in very close proximity to each other. Performance suffers as the number of devices grows, and a large ad-hoc network quickly becomes difficult to manage.

Creating Ad-Hoc Wireless Networks

- To set up an ad-hoc wireless network, each wireless adapter must be configured for ad-hoc mode instead of infrastructure mode, which is the mode used in networks where there is a central device like a router or server that manages network traffic. In addition, all wireless adapters on the ad-hoc network must use the same Service Set Identifier, or SSID, and the same wireless channel number.
- Wireless ad-hoc networks cannot bridge to wired LANs or to the internet without installing a special-purpose network gateway.
- Ad hoc networks make sense when needing to build a small, all-wireless LAN quickly for a minimum amount of money spent on equipment.
- Ad-hoc networks also work well as a temporary fallback mechanism if equipment for an infrastructure mode network, such as a router or access point, fails.

Ad-Hoc Network Security

- Ad-hoc networks are often secured given their usually temporary or impromptu nature. Without network access control, for example, ad-hoc networks can be open to attacks.
- Devices in an ad-hoc network cannot disable SSID broadcasting in the way that devices in infrastructure mode can. Attackers generally will have little difficulty finding and connecting to an ad-hoc device if they get within signal range.
- Authentication can help with this, but ad-hoc networks are inherently more vulnerable to security threats.

1.9 ISO OSI MODEL

1.9.1 Layered Task

- The basic need/use/application of a networking is to transfer the data from system to another system.
- It is not always data, but it can be voice, video or data.
- This is simply indicated in Fig. 1.33, that every system which is involved in data communication process is represented by a stack of layers like:
  (i)   Higher layers
  (ii) Middle layers
  (iii) Lower layers

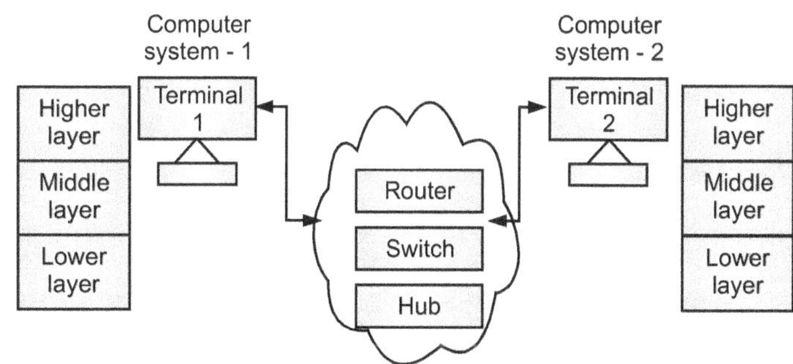

Fig. 1.33 : Network communication system represented by stack of layers

- The significance of each layer and related tasks can be explained with the simple example of sending a letter by one person, which is being received by another person as shown in Fig. 1.34.

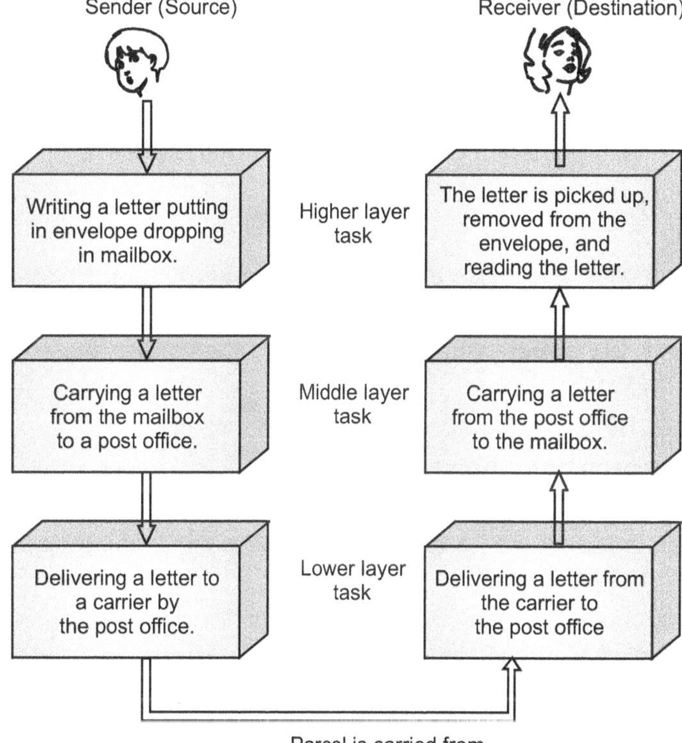

Fig. 1.34 : Tasks involved [source to destination] in journey of letter

- Thus, transporting of letter from source to destination or between sender and receiver is done by carrier.
- Also, each layer at the source (sender side) uses the services of the layer below it.
- The higher layer uses services of middle layer, middle layer uses the services of the lower layer and lower layer uses services of the carrier.
- Thus, in data communication system, each system is represented by a stack of different layers. This reduces the design complexity in the networks.

This issue is discussed in detail, in following sections.

1.9.2 Protocol Fundamentals

- In computing, a protocol is a set of rules which is used by computers to communicate with each other across a network.

- A protocol is a convention or standard that controls or enables the connection, communication, and data transfer between computing endpoints. In its simplest form, a protocol can be defined as the rules governing the syntax, semantics, and synchronization of communication.

- Protocols may be implemented by hardware, software, or a combination of the two. At the lowest level, a protocol defines the behavior of a hardware connection.

Typical Properties

Detection of the underlying physical connection (wired or wireless), or the existence of the other endpoint or node:

- Handshaking.
- Negotiation of various connection characteristics.
- How to start and end a message.
- Procedures on formatting a message.
- What to do with corrupted or improperly formatted messages (error correction).
- How to detect unexpected loss of the connection, and what to do next.
- Termination of the session and/or connection.

Importance of Protocols

- The protocols in human communication are separate rules about appearance, speaking, listening and understanding.

- All these rules, also called protocols of conversation, represent different layers of communication.

- They work together to help people successfully communicate. The need for protocols also applies to network devices. Computers have no way of learning.

Terms and Definitions

- **Protocol :** Protocol is agreement between the communication - communicating parties on how communication is to proceed.

OR

- **Protocol :** Protocol is strict procedure and sequence of actions to be followed in order to achieve orderly exchange of information among peer entities.

OR

- **Protocol :** Protocol is a set of rules governing the format and meaning of the frames, packets or messages that are exchanged by the peer entities within a layer.

- **Protocol Stack :** A list of protocols used by a certain system, one protocol per layer is called a protocol stack.

- **Interface :** Between each pair of adjacent layers, there is an interface. The interface defines which primitive operations and services the lower layers offers to the upper one.

- **Network Architecture :** A set of layers and protocols is called as network architecture.

- **Service :** Services and protocols are distinct concepts although they are frequently confused.

Service is a set of primitives (operations) that a layer provides to the layer above it.

The service defines what operations the layer is prepared to perform on behalf of its users, but it says nothing at all about how these operations are implemented.

A service relates to an interface between two layers, with the lower layer being the service provider and the upper layer being the service user.

1.9.3 ISO OSI Reference Model

- This model is based on a proposal developed by the International Standards Organization (ISO) as a first step towards International Standardization of Protocols used in various layers.

- This model is called as ISO-OSI (Open Systems Interconnection) reference model because it deals with connecting open systems, that is, systems that are open for communication with other systems.

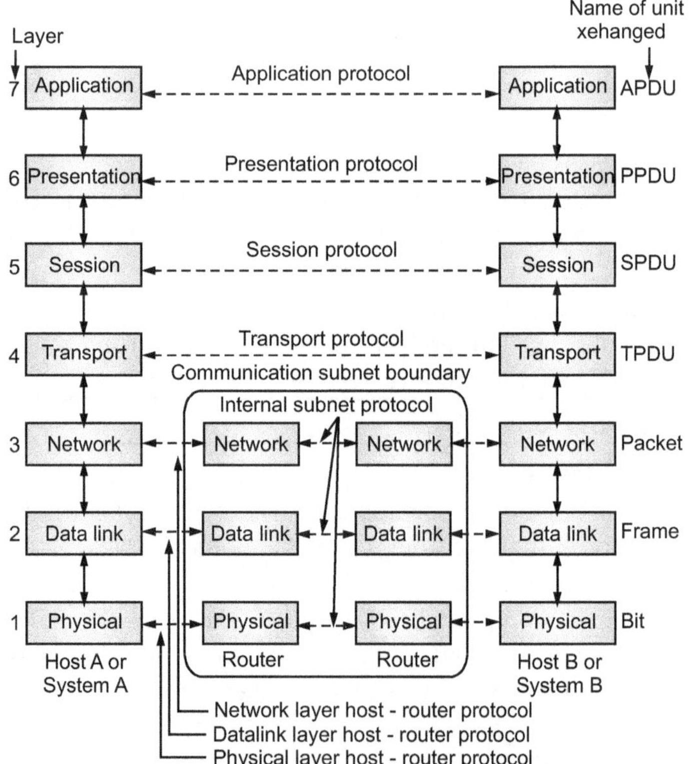

Fig. 1.35: ISO OSI reference model

- OSI model has seven layers. The OSI model defines a layered architecture as pictured. The protocols defined in each layer are responsible for following:

- Communicating with the same peer protocol layer running in the opposite computer.

- Providing services to the layer above it (except for the top-level application layer).

- Peer layer communication provides a way for each layer to exchange messages or other data.

Characteristics of the OSI Layers

- The seven layers of the OSI reference model can be divided into two categories: upper layers and lower layers.

- The upper layers of the OSI model deal with application issues and generally are implemented only in software.

- The highest layer, the application layer, is closest to the end user. Both users and application layer processes interact with software applications that contain a communication component.

- The term upper layer is sometimes used to refer to any layer above another layer in the OSI model.

- The lower layers of the OSI model handle data transport issues. The physical layer and the data link layer are implemented in hardware and software.

- The lowest layer, the physical layer, is closest to the physical network medium (the network cabling, for example) and is responsible for actually placing information on the medium.

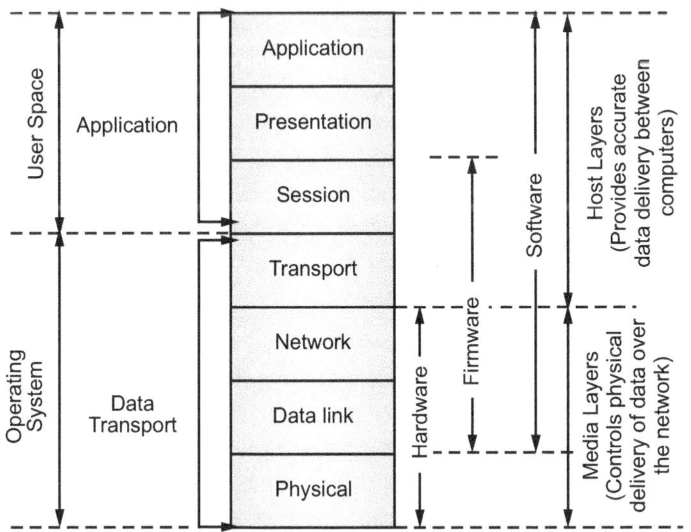

Fig. 1.36 : Illustrates the division between the upper and lower OSI layers

1.9.4 OSI Model Layers and Information Exchange

- The seven OSI layers use various forms of control information to communicate with their peer layers in other computer systems. This control information (headers) consists of specific requests and instructions that are exchanged between peer OSI layers.

- Control information typically takes one of two forms : headers and trailers.

- Headers are prepended to data that has been passed down from upper layers.

- Trailers are appended to data that has been passed down from upper layers.

- An OSI layer is not required to attach a header or a trailer to data from upper layers.

- Headers, trailers and data are relative concepts, depending on the layer that analyzes the information unit.

- At the network layer, for example, an information unit consists of a layer 3 header and data.

- At the data link layer, however, all the information is passed down by the network layer (the layer 3 header and the data) is treated as data.

- In other words, the data portion of an information unit at a given OSI layer potentially can contain headers, trailers and data from all the higher layers. This is known as encapsulation.

Information Exchange Process

- The information exchange process occurs between peer OSI layers. Each layer in the source system adds control information to data, and each layer in the destination system analyzes and removes the control information from that data.

- If System A has data from software application to send to System B, the data is passed to the application layer.

- The application layer in System A then communicates any control information required by the application layer in System B by prepending a header to the data.

- The resulting information unit (a header and the data) is passed to the presentation layer, which prepends its own header containing control information intended for the presentation layer in System B.

- The information unit grows in size as each layer prepends its own header (and, in some cases, a trailer) that contains control information to be used by its peer layer in System B.

- At the physical layer, the entire information unit is placed onto the network medium.

- The physical layer in System B receives the information unit and passes it to the data link layer.

- The data link layer in System B then reads the control information contained in the header prepended by the data link layer in System A.

- The header is then removed, and the remainder of the information unit is passed to the network layer.

- Each layer performs the same actions: The layer reads the header from its peer layer, strips it off, and passes the remaining information unit to the next highest layer.

- After the application layer performs these actions, the data is passed to the recipient software application in System B, in exactly the form in which it was transmitted by the application in System A.

1.9.5 OSI Layers in Detail

The following is a description of just what each layer does :

- **Physical Layer :** The Physical layer provides the electrical and mechanical interface to the network medium (the cable). This layer gives the data-link layer (layer 2) its ability to transport a stream of serial data bits between two communicating systems. It conveys the bits that move along the cable. It is responsible for making sure that the raw bits get from one place to another, no matter what shape they are in, and deals with the mechanical and electrical characteristics of the cable.

- **Data-Link Layer :** The Data-Link layer handles the physical transfer, framing (the assembly of data into a single unit or block), flow control and error-control functions (and retransmission in the event of an error) over a single transmission link; it is responsible for getting the data packaged and onto the network cable. The data link layer provides the network layer (layer 3) reliable information-transfer capabilities. The data-link layer is often subdivided into two parts – Logical Link Control (LLC) and Medium Access Control (MAC) depending on the implementation.

- **Network Layer :** The Network layer establishes, maintains, and terminates logical and/or physical connections. The network layer is responsible for translating logical addresses or names into physical addresses. It provides network routing and flow-control functions across the computer-network interface.

- **Transport Layer :** The Transport layer ensures that data is successfully sent and received between the two computers. If data is sent incorrectly, this layer has the responsibility to ask for retransmission of the data. Specifically, it provides a network-independent, reliable message-independent, reliable message-interchange service to the top three application-oriented layers. This layer acts as an interface between the bottom and top three layers. By providing the session layer (layer 5) with a reliable message-transfer service, it hides the detailed operation of the underlying network from the session layer.

- **Session Layer :** The Session layer decides when to turn communication on and off between two computers - it provides the mechanism that controls the data-exchange process and co-ordinates the interaction between them. It sets up and clears communication channels between two communicating components. Unlike the network layer (layer 3), it deals with the programs running in each machine to establish conversations between them.

- **Presentation Layer :** The Presentation layer performs code conversion and data reformatting (syntax translation). It is the translator of the network, making sure that the data is in the correct form for the receiving application. Of course, both the sending and receiving applications must be able to use data subscribing to one of the available abstract data syntax forms.

- **Application Layer :** The Application layer provides the user interface between the software running in the computer and the network. It provides functions to the user's software, including file transfer access, management and electronic mail.

Thus, the OSI or Open Systems Interconnection model defines a networking frame-work for implementing protocols in seven layers. This can be summarized in Table 1.5.

Table 1.5 : Functions of Layers

Application (Layer 7)	This layer supports application and end-user processes. Communication partners are identified, quality of service is identified, user authentication and privacy are considered, and any constraints on data syntax are identified. Everything at this layer is application-specific. This layer provides application services for file transfers, e-mail, and other network software services.
Presentation (Layer 6)	This layer provides independence from differences in data representation (e.g., encryption) by translating from application to network format, and vice versa. The presentation layer works to transform data into the form that the application layer can accept. This layer formats and encrypts data to be sent across a network, providing freedom from compatibility problems. It is sometimes called the syntax layer.

Session (Layer 5)	This layer establishes, manages and terminates connections between applications. The session layer sets up, co-ordinates, and terminates conversations, exchanges and dialogues between the applications at each end. It deals with session and connection co-ordination.
Transport (Layer 4)	This layer provides transparent transfer of data between end systems, or hosts, and is responsible for end-to-end error recovery and flow control. It ensures complete data transfer.
Network (Layer 3)	This layer provides switching and routing technologies, creating logical paths, known as virtual circuits, for transmitting data from node to node. Routing and forwarding are functions of this layer, as well as addressing, internetworking, error handling, congestion control and packet sequencing.
Data Link (Layer 2)	At this layer, data packets are encoded and decoded into bits. It furnishes transmission protocol knowledge and management and handles errors in the physical layer, flow control and frame synchronization. The data link layer is divided into two sub layers : The Media Access Control (MAC) layer and the Logical Link Control (LLC) layer. The MAC sub layer controls how a computer on the network gains access to the data and permission to transmit it. The LLC layer controls frame synchronization, flow control and error checking.
Physical (Layer 1)	This layer conveys the bit stream electrical impulse, light or radio signal through the network at the electrical and mechanical level. It provides the hardware means of sending and receiving data on a carrier, including defining cables, cards and physical aspects. Fast Ethernet, RS-232 and ATM are protocols with physical layer components.

Table 1.6 : Application Oriented Explanation

ISO-OSI Reference Model Layers	Functions
Application Layer (Layer 7)	Computer Applications like Word processor, Presentation Graphics, Spreadsheet, Database. Network Applications like Email, FTP, Remote Access, Client-Server, Peer-to-Peer, Network management. Internetwork Applications like WWW, Data Exchange, Email Gateways, Finance transactions, Conferencing.
Presentation Layer (Layer 6)	Provides data formats, translations and code conversion. Data compression and Data encryption. Text/data (ASCII or EBCDIC). Sound/Video (MP3, Wave, Mpeg, Quick time). Graphics/Images (JPEG, Gif, BMP).
Sessional Layer (Layer 5)	Network file System (NFS), X-Windows System. Re-establishment of connection in case of failure. Connection Permission Half-Duplex, Full Duplex. Dialog control. Synchronization. Process to process delivery.
Transport Layer (Layer 4)	Establishes reliable End-to-End transport connection. Flow control, error control, connection control. Data error detection, recovery for end-to-end connection.
Network Layer (Layer 3)	Routing Algorithm (Routing). Logical addressing. Congestion Control Algorithm. Internetworking.
Data Link Layer (Layer 2)	NIC (Network Interface Card) Driver has LLC (Logical Link Control)-Framing, Flow Control, Error Control, etc. MAC (Media Access Control)-802.3, 802.4, 802.5, etc.
Physical Layer (Layer 1)	Handles Voltages and Electrical Pulses. Specifies Cables, Connectors and Media Interface Component.

1.10 TCP/IP PROTOCOL SUITE   (Feb. 16, 3M)

1.10.1 Introduction to TCP/IP

- TCP/IP is a suite of protocols, also known as the Internet Protocol Suite. It should not be confused with the OSI reference model, although elements of TCP/IP exist in OSI.

- The Transmission Control Protocol and the Internet Protocol are fundamental to the suite, hence the TCP/IP title.

- TCP/IP is a set of protocols developed to allow co-operating computers to share resources across a network.

- A community of researchers centered around the ARPANET developed this TCP/IP.

- The Internet protocol suite is the set of communication protocols that implement the protocol stack on which the internet and most commercial networks run.

- The internet protocol suite like many protocol suites can be viewed as a set of layers, each layer solves a set of problems involving the transmission of data, and provides a well-defined service to the upper layer protocols based on using services from some lower layers.

- Upper layers are logically closer to the user and deal with more abstract data, relying on lower layer protocols to translate data into forms that can eventually be physically transmitted.

- The Transmission Control Protocol/Internet Protocol (TCP/IP) protocol suite is the engine for the Internet and networks worldwide.

- Its simplicity and power has led to its becoming the single network protocol of choice in the world today. In this chapter, we give an overview of the TCP/IP protocol suite.

Application Layer
Transport Layer
Network/Internet Layer
Network Access Layer

Fig. 1.37 : Typical four-layer TCP/IP model

OSI-ISO Reference Model	TCP/IP Model
Application Layer (7)	Application Layer
Presentation Layer (6)	
Session Layer (5)	
Transport Layer (4)	Transport Layer
Network Layer (3)	Network/Internet Layer
Data Link Layer (2)	Network Access Layer
Physical Layer (1)	

Fig. 1.38 : 7-Layer OSI Model and 4-Layer TCP/IP model

Socket Application	← Application Layer
TCP and UDP	← Transport Layer
IP, ICMP, IGMP, ARP, RARP	← Network Layer
LAN Technologies [802.3, 802.4, 802.5] and WAN Technologies [PPP, Frame relay, ATM]	← Lower layers

Fig. 1.39 : TCP/IP 4 layers and main protocols

- The main design goal of TCP/IP was to build an interconnection of networks, referred to as an Internetwork, or Internet, that provided universal communication services over heterogeneous physical networks.

- The clear benefit of such an internetwork is the enabling of communication between hosts on different networks, perhaps separated by a large geographical area.

1.10.2 Layers in the Internet Protocol Suite Stack

The IP suite uses encapsulation to provide abstraction of protocols and services. Generally a protocol at a higher level uses a protocol at a lower level to help accomplish its aims. The internet protocol stack can be roughly fitted into the four fixed layers and are shown before.

Application Layer

- This layer is broadly equivalent to the application, presentation and session layers of the OSI model.

- It gives an application access to the communication environment.

- Examples of protocols found at this layer are Telnet, FTP (File Transfer Protocol), SNMP (Simple Network Management Protocol), HTTP (Hyper Text Transfer Protocol) and SMTP (Simple Mail Transfer Protocol).

- An application is a user process co-operating with another process usually on a different host (there is also a benefit to application communication within a single host).

- The interface between the application and transport layers is defined by port numbers and sockets.

Transport Layer

- The transport layer is similar to the OSI transport model, but with elements of the OSI session layer functionality.

- This layer provides an application layer delivery service.

- The two protocols found at the transport layer are TCP (Transmission Control Protocol) and UDP (User Datagram Protocol).

- Either of these two protocols are used by the application layer process, the choice depends on the application's transmission reliability requirements.

- Transport layer provides the end-to-end data transfer by delivering data from an application to its remote peer.

- Multiple applications can be supported simultaneously.

- The most-used transport layer protocol is the Transmission Control Protocol (TCP), which provides

- connection-oriented reliable data delivery, duplicate data suppression, congestion control, and flow control.
- TCP is a reliable, connection-oriented protocol that provides error checking and flow control through a virtual link that it establishes and finally terminates.
- This gives a reliable service, therefore TCP would be utilized by FTP and SNMP File transfer and email delivery have to be accurate and error free.
- UDP is an unreliable, connectionless protocol that provides data transport with lower network traffic overheads than TCP. UDP does not error check or offer any flow control, this is left to the application process.
- SNMP uses UDP. SNMP is used to monitor network performance, so its operation must not contribute to congestion.

Network Layer or Internet Layer

- This layer is responsible for the routing and delivery of data across networks.
- It allows communication across networks of the same and different types and carries out translations to deal with dissimilar data addressing schemes.
- Internetwork layer, also called the internet layer or the network layer, provides the "virtual network" image of an internet (this layer shields the higher levels from the physical network architecture below it).
- Internet Protocol (IP) is the most important protocol in this layer.
- It is a connectionless protocol that doesn't assume reliability from lower layers. IP does not provide reliability, flow control, or error recovery. These functions must be provided at a higher level.
- A message unit in an IP network is called an IP datagram.
- This is the basic unit of information transmitted across TCP/IP networks.
- Other internetwork layer protocols are IP, ICMP, IGMP, ARP and RARP.
- With the advent of the concept of Internetworking, additional functionality was added to this   layer, namely getting data from the source network to the destination network.
- This generally involves routing the packet across a network of networks, known as an internet.
- In the internet protocol suite, IP performs the basic task of getting packets of data from source to destination.

- IP can carry data for a number of different upper layer protocols; these protocols are each identified by a unique protocol number.
- ICMP and IGMP are protocols 1 and 2, respectively.
- Some of the protocols carried by IP, such as ICMP (used to transmit diagnostic information about IP transmission) and IGMP (used to manage multicast data) are layered on top of IP but perform internetwork layer functions, illustrating an incompatibility between the internet and the IP stack and OSI model.
- All routing protocols, such as BGP, OSPF, and RIP are also really part of the network layer, although they might seem to belong higher in the stack.

Layers 2 and 1 (Network Access Layers)

- The combination of data link and physical layers deals with pure hardware (wires, satellite links, network interface cards, etc.) and access methods such as CSMA/CD (carrier sensed multiple access with collision detection).
- Ethernet exists at the network access layer it's hardware operates at the physical layer and its medium access control method (CSMA/CD) operates at the data link layer.
- Network interface layer, also called the link layer or the data-link layer, is the interface to the actual network hardware.
- This interface may or may not provide reliable delivery, and may be packet or stream oriented.
- In fact, TCP/IP does not specify any protocol here, but can use almost any network interface available, which illustrates the flexibility of the IP layer.
- The link layer is not really part of the internet protocol suite, but is the method used to pass packets from the network layer on two different hosts.
- This process can be controlled both in the software device driver for the network card, as well as on firmware or specialist chipsets.
- These will perform data link functions such as adding a packet header to prepare it for transmission, and then actually transmit the frame over a physical medium.
- The link layer can also be the layer where packets are intercepted to be sent over a virtual private network.
- When this is done, the link layer data is considered the application data and proceeds back down the IP stack for actual transmission.
- On the receiving end, the data goes up the IP stack twice (once for the VPN and the second time for routing).

- The physical layer is made up of the actual physical network components (hubs, repeaters, network cable, fiber optic cable, coaxial cable, network cards, Host Bus Adapter cards and the associated network connectors : RJ-45, BNC, etc).

1.11 ADDRESSING

- In the data communication, like Internet communication following types of addresses are used.

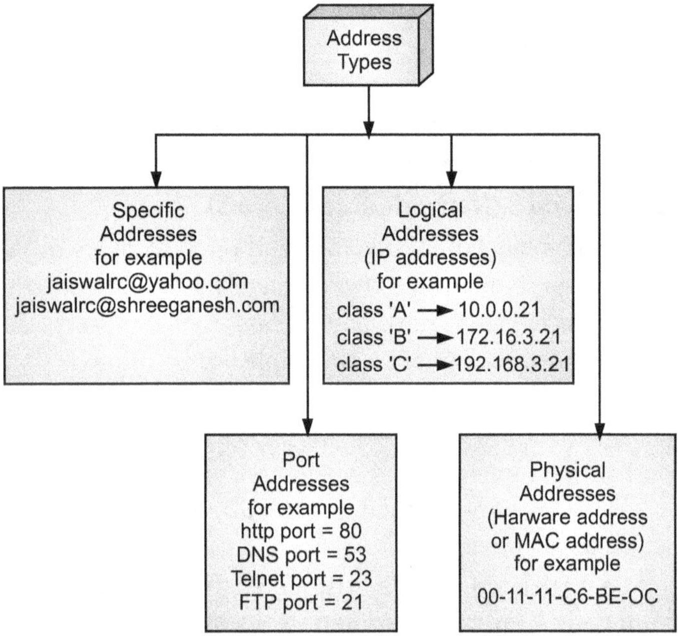

Fig. 1.40 (a) : Types of Addresses in Networking

- These addresses are related to specific layer of the TCP/IP layered architecture.

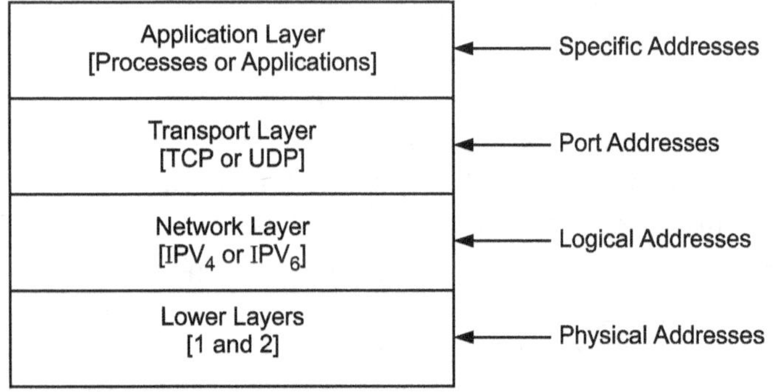

Fig. 1.40 (b) : Layer Specific Addresses are indicated

1.11.1 MAC Address

- The MAC address is a unique value associated with a network adapter. MAC addresses are also known as **hardware** addresses or **physical** addresses. They uniquely identify an adapter on a LAN.

- MAC addresses are 12-digit hexadecimal numbers (48 bits in length). By convention, MAC addresses are usually written in one of the following two formats:

      MM:MM:MM:SS:SS:SS

      MM-MM-MM-SS-SS-SS

- The first half of a MAC address contains the ID number of the adapter manufacturer. These IDs are regulated by an Internet standards body (see sidebar). The second half of a MAC address represents the serial number assigned to the adapter by the manufacturer.

- In the example,

      00:A0:C9:14:C8:29

      The prefix

            00A0C9

      indicates the manufacturer in Intel Corporation.

      00000C-For CISCO

      000011- for Tektronics

      00001B- For Novell

      000048- For Epson

        0000C6- For HP

      08003E- For Motorola

- MAC addresses allow computers to uniquely identify themselves on a network at relatively low level.

- Whereas MAC addressing works at the data link layer, IP addressing functions at the network layer (layer 3).

- It is a slight oversimplification, but one can think of IP addressing as supporting the software implementation and MAC addresses as supporting the hardware implementation of the network stack.

- The MAC address generally remains fixed and follows the network device, but the IP address changes as the network device moves from one network to another.

- IP networks maintain a mapping between the IP address of a device and its MAC address. This mapping is known as the **ARP cache** or **ARP table**.

- ARP, the Address Resolution Protocol, supports the logic for obtaining this mapping and keeping the cache up to date.

- DHCP also usually relies on MAC addresses to manage the unique assignment of IP addresses to devices.

- In Windows OS, At the command prompt, type 'ipconfig /all' without quotes and you can get MAC address of the LAN card or if using Windows XP, you can use the command 'getmac'.

1.11.2 IP Address

- Every machine on the Internet has a unique number assigned to it, called an IP address. Without an unique IP address on your machine, you will not be able to communicate with other devices, users, and computers on the Internet. You can look at your IP address as if it were a telephone number, each one being unique and used to identify a way to reach you and only you.

- An IP address always consists of 4 numbers separated by periods, with the numbers having a possible range of 0 through 255. An example of how an IP address appears is : **192.168.1.10**.

- This representation of an IP address is called decimal notation and is what is generally used by humans to refer to an IP address for readability purposes. With the ranges for each number being between 0 and 255 there are a total 4,294,967,296 possible IP addresses (4 Billions).

- Out of these addresses there are 3 special ranges that are reserved for special purposes. The first is the 0.0.0.0 address and refers to the default network and the 255.255.255.255 address which is called the broadcast address. These addresses are used for routing. The third address, 127.0.0.1, is the loopback address, and refers to your machine. Whenever you see, 127.0.0.1, you are actually referring to your own machine.

- There are some guidelines to how IP address can appear, though. The four numbers must be between 0 and 255, and the IP address of 0.0.0.0 and 255.255.255.255 are reserved, and are not considered usable IP addresses.

- IP addresses must be unique for each computer connected to a network. That means that if you have two computers on your network, each must have a different IP address to be able to communicate with each other. If by accident the same IP address is assigned to two computers, then those computers would have what is called an "IP Conflict" and not be able to communicate with each other.

- **IP Address Classes :** These IP addresses can further be broken down into classes. These classes are A, B, C, D, E and their possible ranges can be seen in Table 1.7.

Table 1.7

Class	Start address	Finish address
A	0.0.0.0	126.255.255.255
B	128.0.0.0	191.255.255.255
C	192.0.0.0	223.255.255.255
D	224.0.0.0	239.255.255.255
E	240.0.0.0	255.255.255.255

- If you look at the table you may notice something strange. The range of IP address from Class A to Class B skips the 127.0.0.0-127.255.255.255 range. That is because this range is reserved for the special addresses called Loopback addresses that have already been discussed above.

- The rest of classes are allocated to companies and organizations based upon the amount of IP addresses that they may need. Listed below are descriptions of the IP classes and the organizations that will typically receive that type of allocation.

- **Default Network :** The special network 0.0.0.0 is generally used for routing.

- **Class A :** From the table above you see that there are 126 class A networks. These networks consist of 16,777,214 possible IP addresses that can be assigned to devices and computers. This type of allocation is generally given to very large networks such as multi-national companies.

- **Loopback :** This is the special 127.0.0.0 network that is reserved as a loopback to your own computer. These addresses are used for testing and debugging of your programs or hardware.

- **Class B :** This class consists of 16,384 individual networks, each allocation consisting of 65,534 possible IP addresses. These blocks are generally allocated to Internet Service Providers and large networks, like a college or major hospital.

- **Class C :** There is a total of 2,097,152 Class C networks available, with each network consisting of 255 individual IP addresses. This type of class is generally given to small to mid-sized companies.

- **Class D :** The IP addresses in this class are reserved for a service called Multicast.

- **Class E :** The IP addresses in this class are reserved for experimental use.

- **Broadcast :** This is the special network of 255.255.255.255, and is used for broadcasting messages to the entire network that your computer resides on.

Private IP Addresses

- There are also blocks of IP addresses that are set aside for internal private use for computers not directly connected to the Internet.

- These IP addresses are not supposed to be routed through the Internet, and most service providers will block the attempt to do so.

- These IP addresses are used for internal use by company or home networks that need to use TCP/IP but do not want to be directly visible on the Internet. These IP ranges are:

Table 1.8

Class	Private Start Address	Private End Address
A	10.0.0.0	10.255.255.255
B	172.16.0.0	172.31.255.255
C	192.168.0.0	192.168.255.255

- If you are on a home/office private network and want to use TCP/IP, you should assign your computers/devices IP addresses from one of these three ranges. That way your router/firewall would be the only device with a true IP address which makes your network more secure.

1.11.3 Port Address

- In internet communication, the actual data communication is done between two processes of the system 1 and system 2.

- For example, web browser is communicating with webserver on the internet. Hence, on one computer system web-browsing process is running and on other computer system webserver process is running.

- Hence, at both ends the logical port numbers are assigned by operating system and TCP/IP protocol stack.

- IANA (Internet Assigned Number Authority) has divided port numbers into three ranges as shown in Fig. 1.41.

Well - known ports	Registered ports	Dynamic ports
0000 to 1023	1024 to 49, 151	49, 151 to 65, 535

Fig. 1.41 : IANA Ports Address Range

- Thus, for web-browsing process port no. 1023 above numbers are used whereas for webserver process well known port – 80 is used.

- Well known port for webserver process is 80, for FTP is 21, Telnet – 23, DNS – 53 and for SMTP is 25.

- Port addresses used by computer systems can be checked using command **netstat – n – a** on command prompt.

1.12 PHYSICAL MEDIA

In computers, media refers to whatever medium is used to communicate data.

Media is usually the copper or fiber optic glass cables but data can also be sent through the air via electromagnetic frequencies such as infrared, micro waves or radio waves.

Media is important because it is often half the cost of the network.

Important Factors in Determining Media Include :

- Required speed.

- Distance.

- Ease of installation and maintenance access.

- Technical expertise required to install and utilize.

- Resistance to internal EMI (Electromagnetic Interference) inside the cable, especially the cross talk of parallel wires.

- Resistance to external EMI outside the cable.

- Resistance to other environmental hazards such as workers carelessly drilling into walls, fire and the weather.

- *Bandwidth* : The range of frequencies that the cable can accommodate. LANs generally carry data rates of 1 to 100 megabits per second and require moderately high bandwidth.

- *Attenuation* characteristics : Attenuation describes how cables reduce the strength of a signal with distance. Resistance is one factor that contributes to signal attenuation.

- Cost.

When data is sent across the network it is converted into electrical signals.

These signals are generated as electromagnetic waves (analog signaling) or as a sequence of voltage pulses (digital signaling).

To be sent from one location to another, a signal must travel along a physical path.

The physical path that is used to carry a signal between a signal transmitter and a signal receiver is called the transmission medium.

There are Two Types of Transmission Media :

1. Guided   2. Unguided.

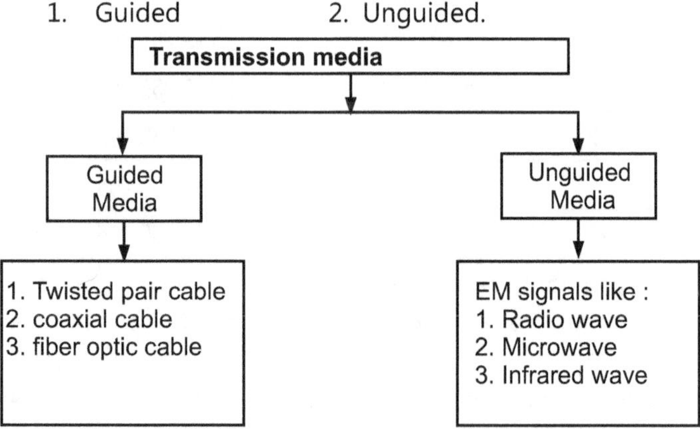

Fig. 1.42 : Classification of transmission media

1.13 GUIDED MEDIA

Guided media are manufactured so that signals will be confined to a narrow path and will behave predictably.

The three most commonly used types of guided media are twisted-pair wiring, coaxial cable, and optical fiber cable. Each type is suited to specific applications and network topologies.

1.13.1 Unshielded Twisted Pair (UTP) Cable and Shielded Twisted Pair (STP) Cable

1. Unshielded Twisted Pair (UTP) Cable :

- Twisted pair has become the most popular network cabling media today.
- Twisted pair cabling is used in a star or star tree topology for Ethernet networks.
- Maximum number of network devices is 1,024, with a maximum cable length of 100 meters for individual devices and a total distance of 500 meters of cabling between the farthest two devices, including links between data closets.
- The signal from a network hub can be repeated three times, giving you a maximum of four data closets.
- The distance between closets can be extended by switching to a star bus topology and using fiber optic cable for links between closets.
- **Twisted Pair Copper Cable** is perhaps the oldest and certainly still the most commonly used transmission medium.
- A twisted pair consists of two insulated copper cables, typically about 1 mm in diameter, twisted together to reduce electrical interference between adjacent pairs of wires (two pairs of parallel wires can act as a crude antenna).

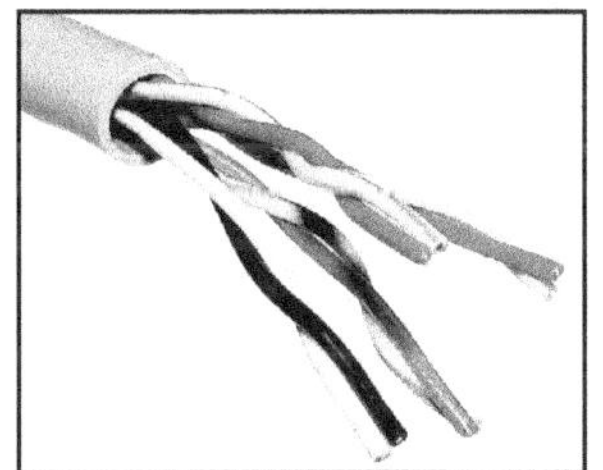

Fig. 1.43 : Unshielded Twisted Pair cable

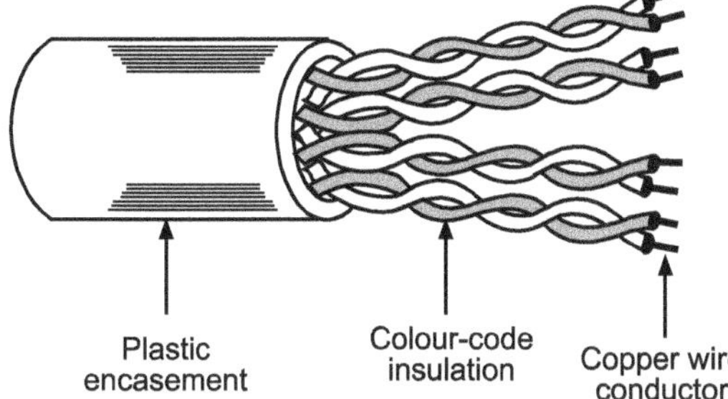

Fig. 1.44 : Four pair UTP cable used in LAN (Local Area Network)

- Twisted pair cable is still used in the public telephone system, specifically in the subscriber loop, the link from a domestic or business telephone subscriber to the local telephone exchange.
- These links are good for several kilometers without amplification, but longer runs need repeaters.
- The local subscriber loop is essentially an analogue transmission medium, but twisted pair cables can also be used for digital transmission.
- The data rate (or *bandwidth*) for twisted pair depends on factors such as the diameter of wire used and the length of the transmission line, but several megabits per second (Mbps) can be achieved over a few kilometers.
- Low cost and ease of installation have kept twisted pair in widespread use both in the telephone system and in Local Area Networks (LANs).
- The type of twisted pair cables used in LANs fall into two main categories.

Category 1 twisted pair cable is the type you will normally find connected to your domestic telephone outlet and consists of two insulated wires twisted gently twisted together. Typically, four pairs are grouped together within a plastic sheath which serves both as protection and to keep the eight wires together.

Category 2 twisted pairs, introduced in the late eighties, are similar to category 1 twisted pair but with more twists per centimeter and Teflon-based insulation.

This results in a further reduction in crosstalk and a better quality signal over long distances, which makes them more suitable for high-speed data communication.

Both types are referred to as *Unshielded Twisted Pair* (UTP).

A Summary of the Twisted Pair Categories is Given Below :

Category 1 : Used for traditional telephone voice communication (but not data) - most telephone cable used before 1983 were category 1.

Category 2 : Four twisted pairs - suitable for data rates of up to 4 Mbps.

Category 3 :

- Four twisted pairs with three twists per foot - suitable for data rates of up to 10 Mbps.
- In the beginning of twisted pair technology, some networks were set up utilizing spare pairs on existing phone systems or cabled with Category 3.
- These networks are only capable of 10Base-T (10 megabits per second) data transfer, and most of them

- set up on spare pairs of existing phone wiring run for slower than that.
- These networks have been obsolete for some time and cannot match the network speeds of today. New Category 3 should only be installed for phone systems.

Category 4 : Supports data transmission of up to 20 Mbps.

Category 5 :

- Four twisted pairs with a higher number of twists per foot than previous categories and Teflon based outer coating.
- Category 5 is generally accepted as the cable to install because of its higher transmission rate and better noise immunity.
- Testing of these cables assumes that only two pairs will be used - one to transmit (T_X) and one to receive (R_X).
- Falling costs in recent years have made category 5 twisted pair a more cost-effective option.
- The predominant type of twisted pair installed in the majority of commercial buildings is unshielded Category 5.
- It is most commonly used for 100 Base T Ethernet networks, giving data transfer rates of 100 megabits per second.
- In addition, the IEEE has approved a network standard for 1000 BASE T Ethernet networks (data transfer of 1,000 megabits per second) which can utilize most existing Category 5 cabling when it has been properly installed and certified.
- In addition to unshielded Category 5, there is also a shielded version, which provides some protection against electromagnetic interference.
- A typical application might be for a heavy manufacturing plant where interference from large electric motors could present a problem.
- For the vast majority of existing offices and smaller industrial plants, unshielded Category 5 is the most commonly found cable.

Category 5e :

- **Enhanced Category 5 :** More comprehensive testing is carried out on all four pairs to measure the effect of transmitting data, particularly with regard to crosstalk.
- This category is primarily intended for use in Gigabit Ethernet networks.
- Over the last several years, Category 5e has become the replacement for Category 5.

- There are two main types, known as "Little e" and "Big E", capable of 155 and 350-megabit transmission respectively.
- Although there are no network standards to support these speeds, the increased bandwidth does enhance this cable's ability to run gigabit Ethernet.
- With the price drop of Category 5e over the last few years, it has become the most common choice for new network installation.

Category 6 :

- A proposed standard for cable having a transmission frequency of 200 MHz, with all components coming from one manufacturer (i.e. no "mixing").
- With the ever increasing speed of networks today, Category 6 is becoming more common in new office installations that demand reliable gigabit network speed.
- It is a viable choice today for a new network installation in a commercial space where the tenants plan to stay for an extended length of time.
- In addition, gigabit switches and network cards are also beginning to drop in price, so the cost of the hardware necessary to set up a true gigabit network is becoming less expensive as well.

Category 7 :

- A proposed standard for cable having a transmission frequency of 600 MHz using fully shielded cables, i.e. shielding is to be provided for both individual pairs and for the grouped pairs. A new connection type is also proposed.
- The maximum recommended cable run for unshielded twisted pair is 100 meters.
- UTP cables are terminated with RJ45 connectors, similar in design to the connectors used to connect telephones into a wall socket outlet (RJ11).
- Twisted pair cables are most commonly used to connect workstations to hubs or MAUs.
- The standard connector for unshielded twisted pair cabling is RJ-45 connector.
- This is a plastic connector that looks like a large telephone-style connector.
- A slot allows the RJ-45 to be inserted only one way.
- RJ stands for Registered Jack, implying that the connector follows a standard borrowed from the telephone industry.
- This standard designates which wire goes with each pin inside the connector.

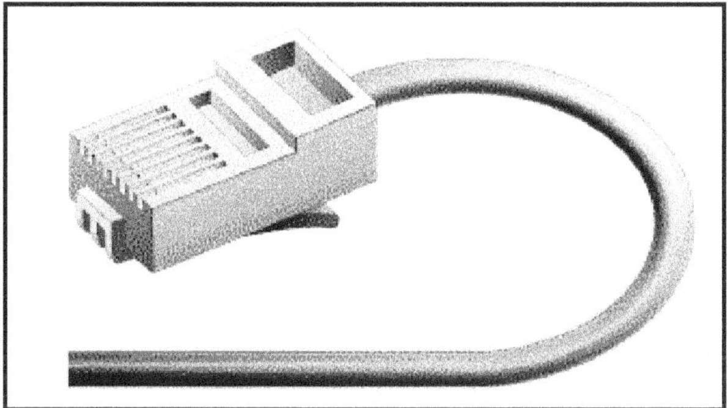

Fig. 1.45 : RJ-45 Connector and UTP Cable

2. Shielded Twisted Pair Cable :

- *Shielded Twisted Pair* (STP) cable was introduced in the 1980s by IBM as the recommended medium for their Token Ring network technology, and has a characteristic impedance of 150 ohms.

- Each cable consists of two pairs, with each pair individually foil shielded, and an overall braided shield.

- Because STP was specified by IBM, many users thought that it was required for reliable data transfer.

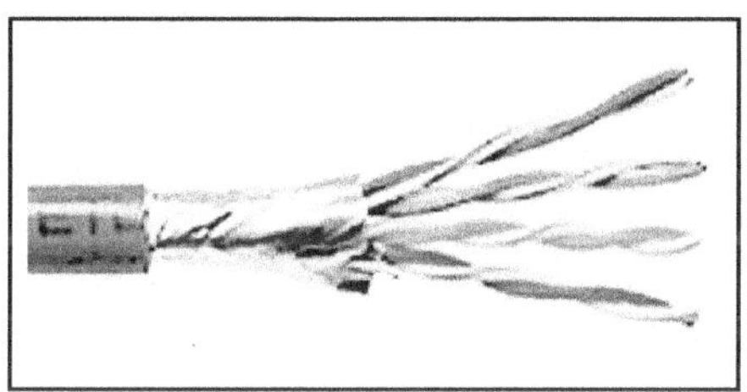

Fig. 1.46 : Shielded Twisted Pair Cable

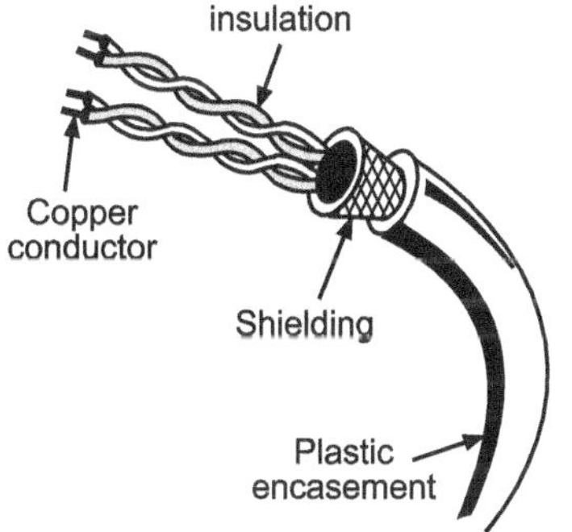

Fig. 1.47 : Typical two pair Shielded Twisted Pair Cable (STP Cable)

- Since this is not in fact the case its popularity has declined due to :

- The high cost of the cable and connectors (much more complex than UTP).

- The increased bulk of cable and connectors compared to UTP.

- The increased time required for installation compared to UTP.

- **Ground Loops :** These arise when the ground voltage at each end of a cable run is different, causing a current to flow in the cable's shield and creating a magnetic field, which induces, current (noise) in the same cable the shielding is designed to protect.

- The same cable length restrictions apply (100 meters maximum) as for UTP.

- STP is limited for data communication to IBM machines and Token Ring networks - there is no standard for STP for Ethernet, ISDN or analog telephones.

- Shielded twisted pair is now manufactured to the same standard as Unshielded Twisted Pair.

Thus Following Points Summarize the Features of STP Cable :

- Speed and throughput—10 to 100 Mbps
- Average cost per node—Moderately expensive
- Media and connector size—Medium to large
- Maximum cable length—100 m (short)

When Comparing UTP and STP, Keep the Following Points in Mind :

- The speed of both types of cable is usually satisfactory for local-area distances.

- These are the least-expensive media for data communication. UTP is less expensive than STP.

- Because most buildings are already wired with UTP, many transmission standards are adapted to use it, to avoid costly rewiring with an alternative cable type.

Table 1.9 : Categories of Unshielded Twisted Pair

Type	Bandwidth	Use
Category 1	< 1 MHz	Voice Only (Telephone Wire).
Category 2	1 MHz	Data to 4 Mbps (Local Talk) and Telephone, T_1 lines etc.
Category 3	16 MHz	Data to 10 Mbps (Ethernet), Telephone, 10Base-T, Token Ring, LAN applications.
Category 4	20 MHz	Data to 20 Mbps (16 Mbps Token Ring), 10Base-T, LAN application.
Category 5	100 MHz	Data to 100 Mbps (Fast Ethernet), 10Base-T, 100Base-T, LAN applications.
Category 5e	350 MHz	125 Mbps, Data Networks.
Category 6	550 MHz	Proposed standard for cable having a data rate of 200 Mbps, LAN applications.
Category 7	600 MHz	Proposed standard for cable having a data rate of 600 Mbps using fully shielded cables, LAN applications.

Advantages of Twisted Pair :

- Reasonable cost.
- High speed.
- Easy to add additional network devices.
- Supports large number of network devices.
- Telephone cable standards are mature and well established. Materials are plentiful, and a wide variety of cable installers are familiar with the installation requirements.
- It may be possible to use in-place telephone wiring if it is of sufficiently high quality.
- UTP represents the lowest cost cabling. The cost for STP is higher and is comparable to the cost of coaxial cable.

Disadvantages of Twisted Pair :

- High attenuation (signal loss) limits individual runs to 100 meters.
- Susceptible to EMI/RFI (except shielded type).
- STP can be expensive and difficult to work with.
- Compared to fiber optic cable, all Twisted Pair cable is more sensitive to EMI. UTP especially may be unsuitable for use in high-EMI environments.
- Twisted Pair cables are regarded as being less suitable for high-speed transmissions than coaxial cable or fiber optic. Technology advances, however, are pushing upward the data rates possible with Twisted Pair. Cable segment lengths are also more limited with Twisted Pair.

1.13.2 Coaxial Cable

- A coaxial cable consists of a central copper wire core, which is surrounded by an insulating material.
- The insulator is surrounded by braided metal shielding which helps to absorb external electronic signals (noise) and prevents it from interfering with the data signal.
- A plastic sheath protects the outer conductor. A durable plastic or Teflon jacket coats the cable to prevent damage. Fig. 1.48 diagram illustrates the basic construction of a coaxial cable.

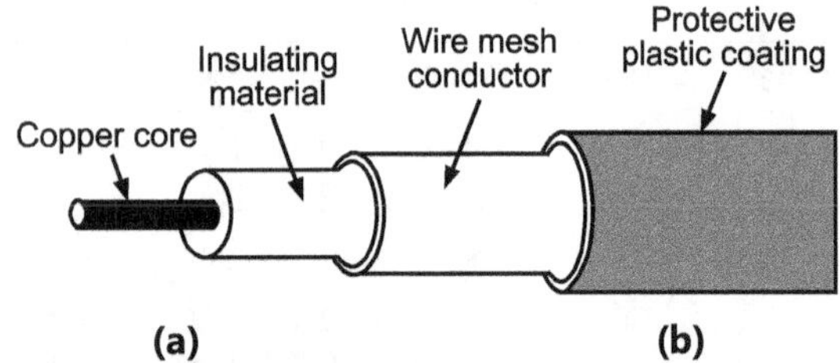

Fig. 1.48 : Coaxial cable construction

- The construction and shielding of coaxial cable provides a high degree of immunity to noise, and coaxial cable can be used over longer distances (upto 500 meters) than twisted pair cable.
- Coaxial cable runs are used to provide the network backbone cable segments in networks having a **bus topology**, and require a terminating resistor at each end of the cable in order to prevent interference due to signal reflection.
- Coaxial cable has many desirable characteristics. It is highly resistant to EMI and can support high *bandwidths*.
- Some types of coaxial cable have heavy shields and center conductors to enhance these characteristics and to extend the distances, so that signals can be transmitted reliably.
- A wide variety of coaxial cable cable is available. You must use cable that exactly matches the requirements of a particular type of network.
- Coaxial cable cables vary in a measurement known as the *impedance* (measured in a unit called the ohm), which is an indication of the cable's resistance to current flow.
- The specifications of a given cabling standard indicate the required impedance of the cable.
- Two types of coaxial cable can be used in computer networks :
  1. Thinnet or Cheapernet (also known as Baseband Coax - RG-58)
  2. Thicknet

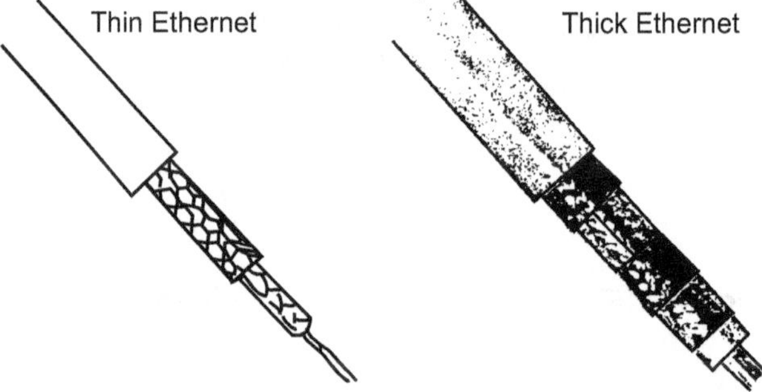

Fig. 1.49 : Thin Ethernet and Thick Ethernet

- Thinnet (10Base2) is so called because of the thin, inexpensive coaxial cabling it uses, and is fairly flexible, being 0.25 inches in diameter.
- The IEEE specification refers to this type of cable as 10Base2, referring to its main specifications of 10 Mbps data rate, baseband transmission type, and 185 (nearly 200) meter maximum segment length.

- The cable between computers must be at least 0.5 meters (20 inches) long.
- An IEEE standard for Thinnet doesn't allow a drop cable to be used from the bus T-connector to a workstation.
- Instead, the T-connector fits directly onto the network adapter card using a BNC connector.
- **A Thinnet network** can support a maximum of 30 nodes per cable segment, and up to five segments can be connected using repeaters, of which three segments may be populated, allowing up to 90 nodes to be supported (based on the IEEE 802.3 specification).
- **Thicknet cable** (also known as **Standard Ethernet**) is relatively rigid, being 0.5 inches in diameter.
- The IEEE specification refers to this type of cable as 10Base5, referring to its main specifications of 10 Mbps data rate, baseband transmission type, and 500-metre maximum segment length.
- Thicknet is generally used to provide the network backbone and can support up to 100 nodes per backbone segment.
- The minimum cable length between connections (or taps) on a Thicknet cable segment is 2.5 meters (about 8 feet).
- Thicknet cable has a data rate of 10 Mbps and can carry a signal for 500 meters before a repeater is required.
- The grade of coaxial cable used will depend on where it is used.
- Normal PVC coaxial cable is flexible, easy to work with, and may be used in exposed areas of offices, but because it gives of poisonous fumes when it burns, it is against the fire regulations in many countries for it to be installed in floor and ceiling voids which are also used to allow air to circulate around the building.
- Thick coax was the transmission medium originally used by Xerox for their Ethernet network, although it was later superceded by thin coax.
- Although still used in many networks, coaxial cable is gradually being replaced by fiber optic and UTP cable - fiber optic is normally used for the network backbone, with UTP being used to connect workstations to hubs or MAUs.

- Here are some common examples of coaxial cables used in LANs, along with their impedances, and the LAN standards with which they are associated :
  - ➢ RG-8 and RG-11 are 50-ohm cables required for thick wire Ethernet. (10Base5 - ThikNet).
  - ➢ RG-58 is a smaller 50-ohm cable required for use with thin wire Ethernet. (10Base2 – ThinNet).
  - ➢ RG-59 is a 75-ohm cable most familiar when used to wire cable TV. RG-59 is also used to cable broadband 802.3 Ethernet.
  - ➢ RG-62 is a 93-ohm cable used for ARCnet. It is also commonly employed to wire terminals in an IBM SNA network.

Advantages of Coaxial Cable :

- Low cost due to less total footage of cable, hubs not needed.
- Lower attenuation than twisted pair.
- Good immunity to EMI/RFI / Highly insensitive to EMI.
- Supports high bandwidths.
- Heavier types of coax are sturdy and can withstand harsh environments.
- Represents a mature technology that is well understood and consistently applied among vendors.

Disadvantages of Coaxial Cable :

- Limited in network speed.
- Limited in size of network.
- One bad connector can take down entire network.
- Although fairly insensitive to EMI, coax remains vulnerable to EMI in harsh conditions such as factories.
- Coax can be bulky.
- Coax is among the most expensive types of wire cables.

1.13.3 Fiber Optic Cable

- Data transmission over optical fiber has greatly increased over the last few years, although fiber to the desktop has not really caught on as expected.
- However, fiber optic plays an important role in many networks.
- In addition, it has some outstanding advantages over copper cabling for certain applications.
- There are a number of network topologies and standards based on fiber optic, such as 10 BASE FL and FDDI, which apply mainly to the backbone cabling of very large facilities and campus environments.
- The discussion will be limited here to the uses of fiber optic in star bus Ethernet network topologies.

- When used as a link in a star bus topology, multi-mode fiber optic cable can transmit a maximum distance of 2,000 meters between all data closets, using a less expensive LED light source.

- While single mode fiber can transmit up to 3,000 meters, it requires a more expensive laser light source.

- By using fiber optic to link closets, it is possible to greatly extend the distance limitations in Ethernet networks using twisted pair only.

- Fiber optic is an outstanding choice for linking buildings together.

- In addition to the much greater distances possible, it is completely immune to over currents from lightning strikes and ground potential problems.

- There is literally nothing metallic in a fiber optic cable to conduct current. It is an excellent choice for heavy manufacturing environments, such as a foundry, due to its immunity to EMI/RFI.

- Finally, it is the best choice where data of a highly sensitive nature is being transmitted.

- Fiber optic cable radiates no electrical signal at all, and the cable would be down for quite some time if someone tried to splice into it.

Construction of Optical Fiber :

- Optical fiber cable carries light signals instead of electric signal.

- Each fiber has inner core of either plastic or glass that carries light.

- The inner core is surrounded by cladding, a layer of plastic or glass that reflects the light back into core.

- Fiber optic cable can have single fiber or bundle of fibers at the centre of the cable. The refractive index of the core is relatively high.

- Refractive index is low. Cladding material is lossy.

- This entire optical core-cladding assembly is then coated with protective inner jacket and outer plastic jacket as shown in Fig. 1.50.

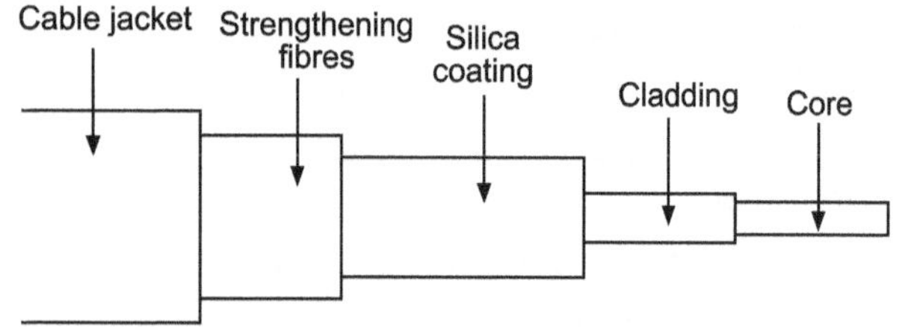

Fig. 1.50 : Construction of Fiber Optic Cable

- The cross-section of a fiber illustrating the different layers is as shown in Fig. 1.51.

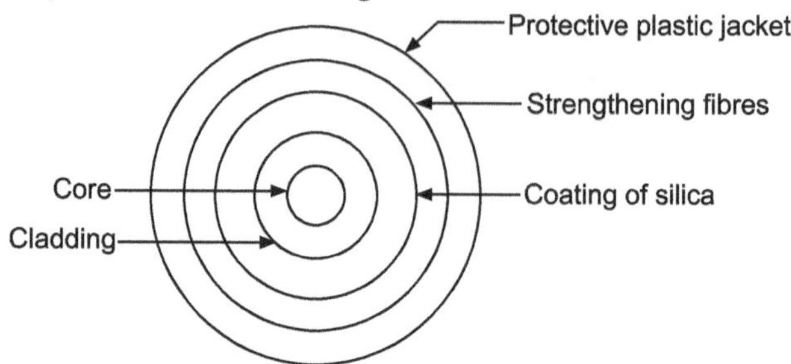

Fig. 1.51 : A Cross-Section of a Fiber illustrating the Different Layers

- The core is surrounded by cladding surface and entire thing is coated with silicon oil and organic material with silica. External plastic jacket is to provide mechanical strength and optical fiber is protected from mechanical wear and tear.

Typical optical fiber communication is shown in Fig. 1.52.

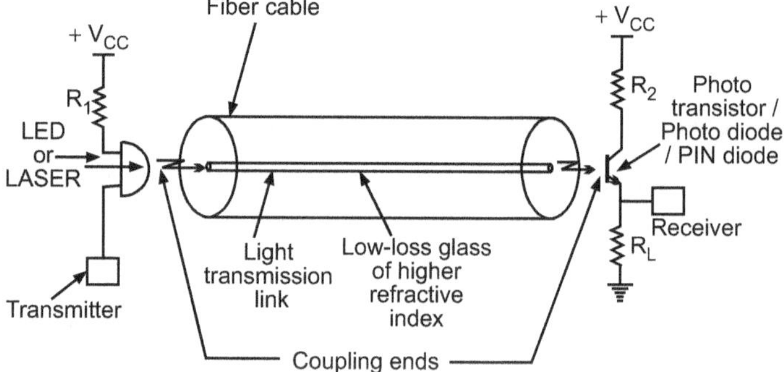

Fig. 1.52: Typical Fiber Optic Communication System

The basic point-to-point fiber optic communication system consists of three basic units.

1. Optical transmitter unit

2. Fiber optic cable unit

3. Optical receiver unit.

1. Optical Transmitter Unit :

The transmitter converts applied electrical, analog or digital signal into a corresponding light signal. The source of the light signal can be either a light emitting diode, or a solid-state laser diode.

2. Fiber Optic Cable Unit :

Converted signal travels in the form of light from one end of fiber to other end of fiber. This is also called as light transmission link. If the distance between transmitter and receiver is in kilometers, then two or more fiber optic cables can be joined together.

3. Optical Receiver :

The receiver converts the optical signal into the original electrical signal. This optical receiver uses the photodetectors like avalanche type photodiode or PIN type of photodiode (P type-Intrinsic-N type material is used).

Optical fibers can be classified in two ways as shown.

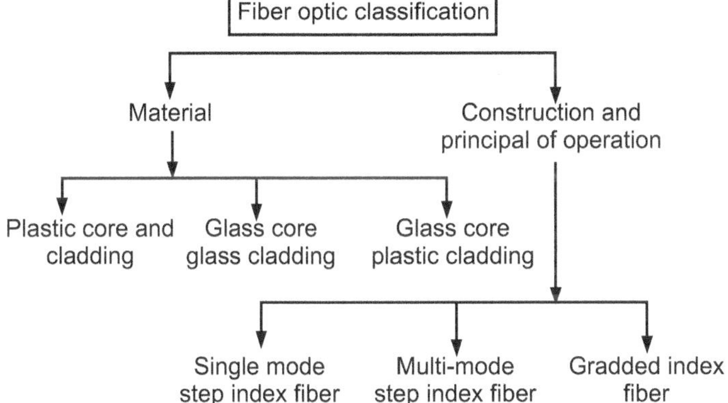

1. Material :

(a) Plastic Core and Cladding :

These type of fibers are more rugged than glass. They are less expensive. They provide high attenuation characteristics and can be used within a single building or a building complex. Less attenuation when exposed to external radiation.

(b) Glass Core with Plastic Cladding :

These type of fibers are more effectively used in military applications.

(c) Glass Core and Glass Cladding :

These type of fibers are least rugged and are more susceptible to increase in attenuation when exposed to external radiation compared to above both optical fibers.

2. Construction and Principle of Operation :

(a) Single Mode Step Index Fiber has Following Properties :

- Support only one mode of operation.
- Use of LASER is must, makes power launching difficult.
- Fiber coupling is difficult due to less diameter size of core which is approximately $\approx$ 8 to 12 μm.

(b) Multi-Mode Step Index Fiber has following Properties :

- Supports hundreds of modes of propagation.
- LED can be used as optical transmitter, so power launching becomes easy.
- Fiber coupling is easy because moderate core diameter which is approximately $\approx$ 50 to 200 μm.

- In this fiber the pulse which is sent at transmitter end spreads in time, when it is received at receiver end. This pulse distortion is called as intermodal dispersion.

(c) Graded Index Fiber :

- The intermodal dispersion of pulse is reduced in the graded index fiber due to its construction itself.
- Graded index fibers have larger bandwidth than step index fibers.
- Construction of the monomode or single mode step index fiber is given in Fig. 1.53 (a).

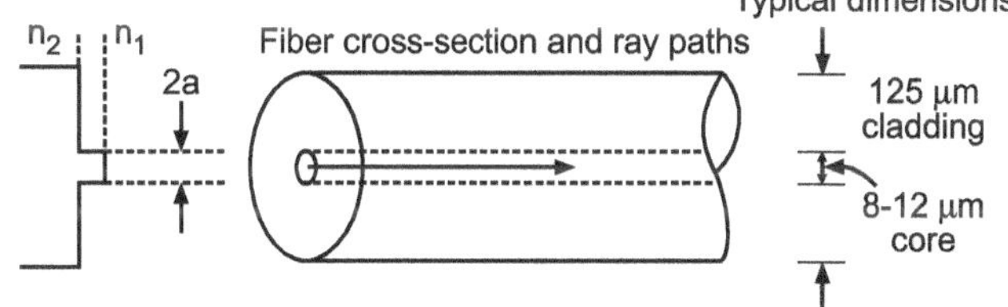

Fig. 1.53 (a) :Monomode or Single Mode Step Index Fiber

- Construction of the multi-mode step index fiber is given in Fig. 1.53 (b).

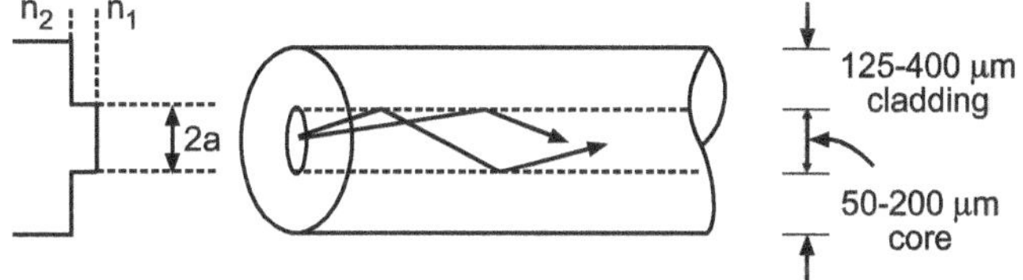

Fig. 1.53 (b) : Multi-mode Step Index Fiber

- Construction of the graded index fiber is given in Fig. 1.53 (c).

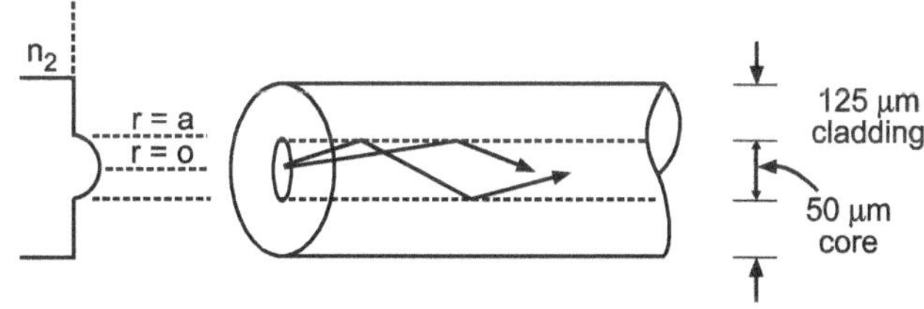

Fig. 1.53 (c) : Graded Index Fiber

- Thus fiber optic transmission line confines light energy within its surface and guides the light in a direction parallel to its axis.

In previous figures,

$$n_1 = \text{refractive index of the core}$$
$$n_2 = \text{refractive index of cladding}$$

Advantages of Fiber Optic Cable :

- High data rate and wide bandwidth.
- Immunity to EMI/RFI and lightning damage.
- No ground loops.
- Low attenuation (Low data loss).
- Longer distance - 2 and 5 km with multi-mode fiber and over 25 km with single mode fiber.

- Small cable diameter fits anywhere.
- Light weight.
- No sparks if cut.
- No shock hazard.
- Secure communication.
- Safe and easy installation.
- Low system cost.
- Longer life expectancy than copper or coaxial cable.
- Cabling of the future.

Disadvantages of Fiber Optic Cable :

Cost :

- Despite the fact that the raw material for making optical fibers is abundant and cheap, optial fibers are still more expensive per meter than copper.

Special Skills :

- Optical fibers cannot be joined together as easily as copper. It requires additional training for person. Expensive precision splicing and measurement equipment are also required.

Installation and Maintenance Cost :

- Initial installation of the fiber optic system is more and maintenance is expensive.
- Thus choosing the correct type of cabling depends on what type of network you have or intend to have, the number of network devices used, expected future growth, the speed requirements of your applications and the physical layout of your facility.
- Make this decision with the assistance of a professional, licensed and insured network cabling company and a good information technology consultant.

Table 1.10 : Summary of Cable Characteristics

Cable Type	Cable Cost	Installation Cost	EMI Sensitivity	Data Bandwidth
UTP	Lowest	Lowest	Highest	Lowest
STP	Medium	Moderate	Low	Moderate
Coax	Medium	Moderate	Low	High
Fiber Optic	Highest	Highest	None	Very high

Table 1.11 : Characteristic Comparison of Guided Media

Twisted Pair	Coaxial Cable	Fiber Optic Cable
It uses electrical signal for transmission.	It uses electrical signal for transmission.	It uses optical signal for transmission.
Affected by EMI and noise.	Less affected by EMI and noise.	Not affected by EMI and noise.
Bandwidth is low which is 3 to 4 MHz.	Bandwidth is high which is 300 to 400 MHz.	Bandwidth is very high which is 2 to 3 GHz.
Used for analog and digital transmission.	Used for analog and digital transmission.	Used for analog and digital transmission.
Supports low data rates upto 4 Mbps.	Supports high data rates upto 400 to 500 Mbps.	Supports very high data rates upto 3 Gbps.
Cost is very less.	Cost is moderate.	More costly.
For long distance communication, repeaters are required after every 2 km distance.	For long distance communication, repeaters are required after every 1 km distance.	For long distance communication, repeaters are required after every 10 km distance.
Signal attenuation is more.	Signal attenuation is moderate.	Signal attenuation is least.
Installation is easiest.	Installation is easy.	Installation is difficult.
Signal to noise ratio is less.	Signal to noise ratio is moderate.	Signal to noise ratio is very high.
Crosstalk is more.	Crosstalk is moderate.	No crosstalk is present.
Losses like copper losses and radiation losses are present.	Losses like copper losses and radiation losses are present.	Losses like microbending and macrobending losses are present.

1.14　UNGUIDED MEDIA

1.14.1 Introduction

- Unguided media are natural parts of the earth's environment that can be used as physical paths to carry electrical signals.
- The atmosphere and outer space are examples of unguided media that are commonly used to carry signals.
- These media can carry such electromagnetic signals as microwaves, infrared light waves, and radio waves.

- Network signals are transmitted through all transmission media as a type of waveform.
- When transmitted through wire and cable, the signal is an electrical waveform.
- When transmitted through fiber-optic cable, the signal is a light wave : either visible or infrared light.
- When transmitted through earth's atmosphere or outer space, the signal can take the form of waves in the radio spectrum, including VHF and microwaves, or it can be light waves, including infrared or visible light (for example, lasers).
- Recent advances in radio hardware technology have produced significant advancements in wireless networking devices : the cellular telephone, wireless modems, and wireless LANs.
- These devices use technology that in some cases has been around for decades but until recently was too impractical or expensive for widespread consumer use.
- The next few sections explain technologies unique to unguided media that are especially of concern to networking.
- There are a variety of wireless network media, each of which uses a different transmission protocol. Typically, a wireless network uses infrared light or radio transmissions to distribute data.
- **Infrared Networks** communicate by using beams of infrared light. They have a maximum range of 100 meters. Theoretically, they can transmit at 10 Mbps, but 1-3 Mbps is more typical.

- **Narrow Band Radio Networks** can cover an area up to 5,000 square meters at up to 4.8 Mbps. Their disadvantage is that they offer little security.
- **Spread-Spectrum Radio Networks** use multiple frequencies. These multiple channels provide network security. They can transmit data at up to 1 Mbps at a range of 800 feet indoors, though 300 kbps is more typical.
- Some common applications of wireless data communication include the following :
  - ➢ Accessing the Internet using a cellular phone.
  - ➢ Establishing a home or business Internet connection over satellite.
  - ➢ Beaming data between two hand-held computing devices.
  - ➢ Using a wireless keyboard and mouse for the PC.

1.14.2 The Electromagnetic Spectrum

- All electromagnetic waves travel at the speed of light (300,000,000 metres per second) in a vacuum, whatever their frequency (in copper or fibre), the speed drops to approximately two thirds of this value, and is slightly frequency dependent.
- The relationship between frequency, wavelength and the speed of light (C) in a vacuum is given by :
$$F \lambda = C$$
- Since C is a constant, if wavelength is known, then frequency can be calculated and vice versa.
- Thus, a frequency of 1 MHz would give a wavelength of approximately 300 meters, and a 1 cm wavelength would give a frequency of approximately 30 GHz. The Electromagnetic Spectrum is shown in Fig. 1.54.

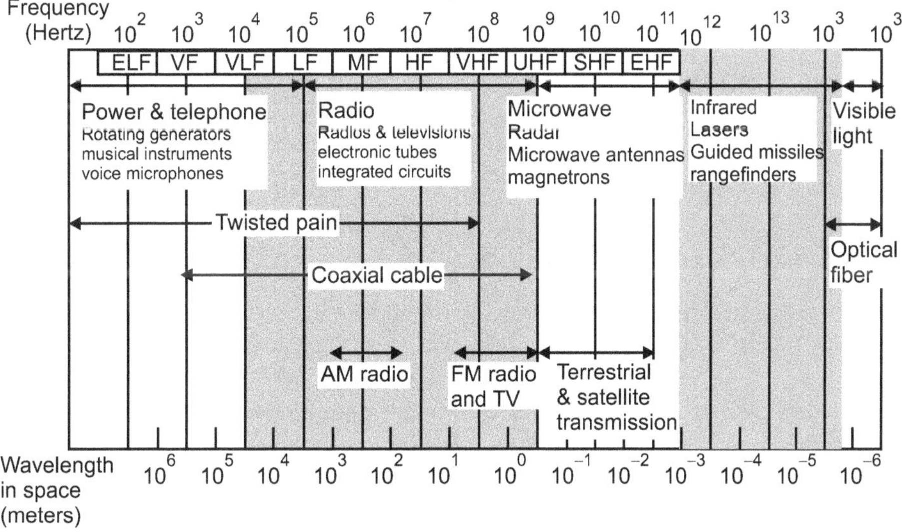

Fig. 1.54 : The Electromagnetic Spectrum

- The parts of the electromagnetic spectrum which can be used for transmitting information using amplitude, frequency or phase modulation are shown using a darker shading and include radio, microwave, infrared and visible light.

1.15 SWITCHING INTRODUCTION

- When multiple devices want to communicate with each other, then simple solutions are :
  - ➤ Mesh network formation.
  - ➤ Bus network formation.

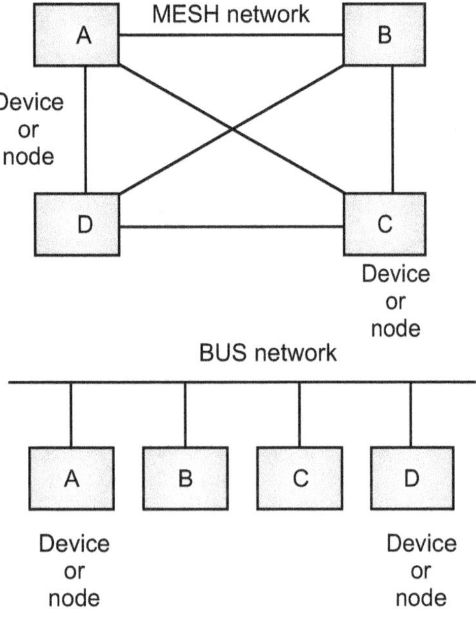

Fig. 1.55 : Interdevice Communication

- When above techniques are employed, it becomes impractical and wasteful when network size increases.
- Also network cost and maintenance becomes difficult for large network of devices.
- Then the solution to above problems is to use the switching system.
- Typical use of switch and switching system is as shown in Fig. 1.56.

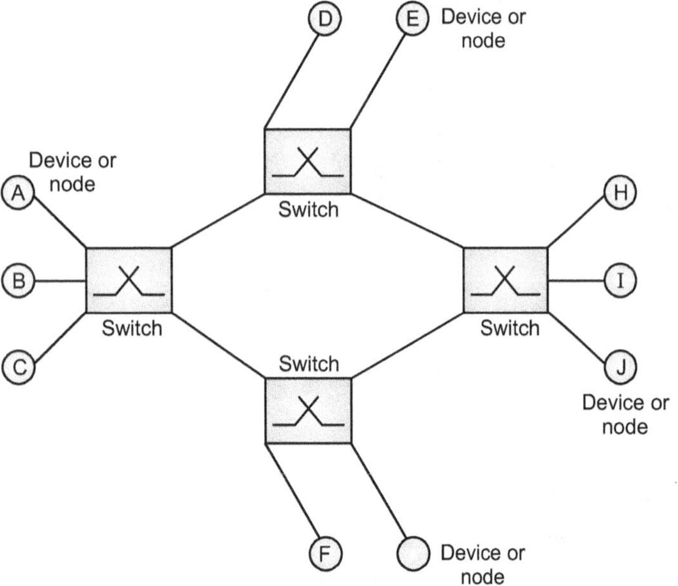

Fig. 1.56 : Typical Switch Based Network

- Thus, different devices/nodes can communicate with each other with the help of switches connected in the typical switch based network.
- The switches work on the principle of switching system.

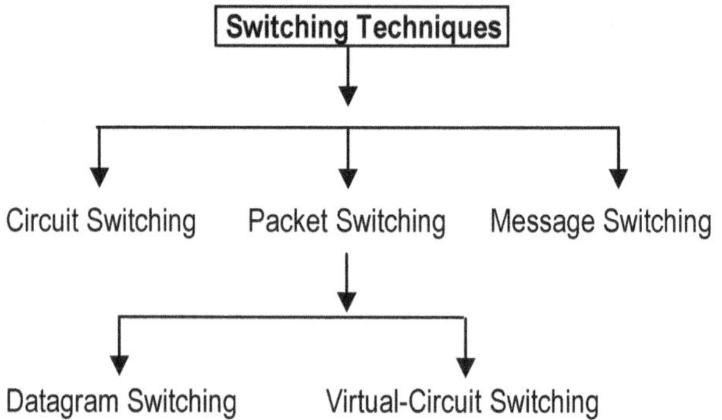

1.15.1 Circuit Switching Networks

- In circuit switching networks, nodes or devices are connected to each other by physical links via switches.
- Typical circuit switched network is as shown in Fig. 1.57.

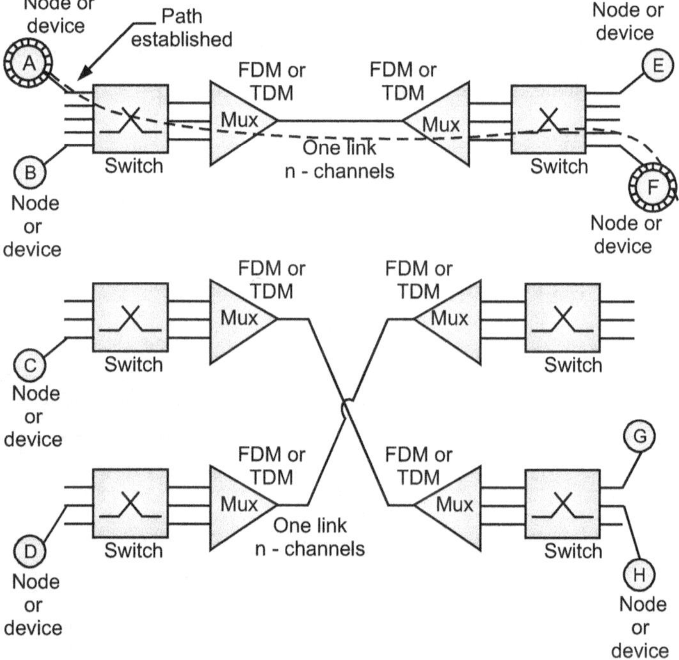

Fig. 1.57 : Typical Circuit Switched Network and Communication between Two End Systems A and F

- In this diagram, multiplexer (MUX) symbol is explicitly shown and it is implicitly included in switch fabric itself.
- The end system can be computer node or device like telephone set.
- In Fig. 1.57, the communication between two end systems A and F is highlighted.

- In circuit switched network, the communication between node 'A' and node 'F' is done through 3 phases as follows :
  - ➤ Set-up phase (also known as connection establishment).
  - ➤ Data transfer phase.
  - ➤ Teardown phase (also known as connection release).

- **In Set-Up Phase a Dedicated Circuit** (i.e. combination of channels in links) needs to be established. For this node 'A' sends set-up request through switch fabric including multiplexer, to node 'F'. Then node 'F' receives this request and sends acknowledgement to node 'A' through same dedicated path. Thus, only after receiving this acknowledgement from node 'F', we can say that connection is established or set-up phase is completed. In this circuit switched network, end systems use addresses in TDM network whereas use telephone numbers in FDM network.

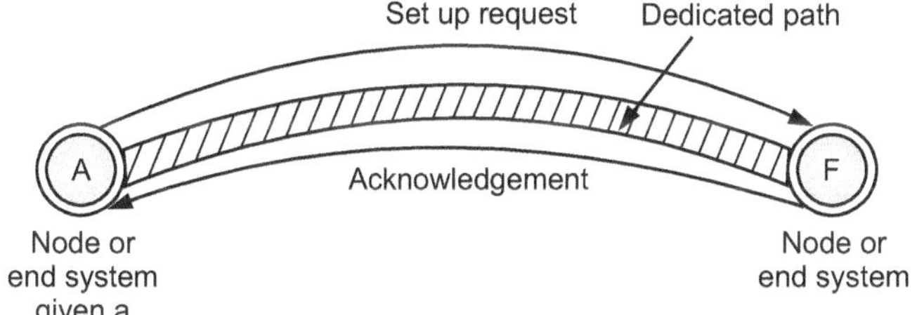

Fig. 1.58 : Typical Set-up Phase between End Systems A and F

- **In data transfer phase**, end systems 'A' and 'F' can transfer data (or communication between two end systems is done).

- **In teardown phase (or connection release process)**, either 'A' system or 'F' system can stop the communication and release the common resources like dedicated link and switch etc.

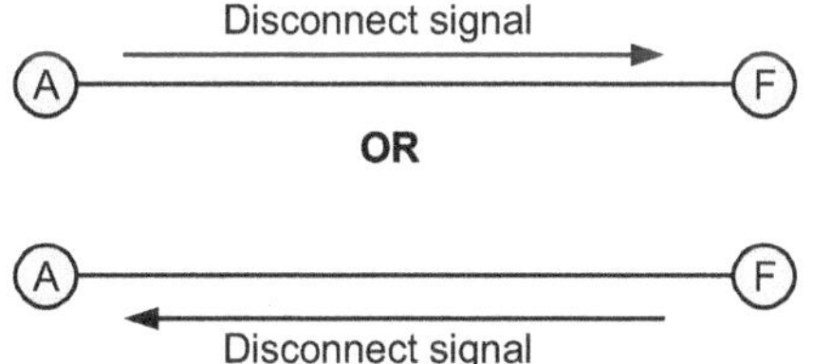

Fig. 1.59 : Teardown Phase (or Connection Release Process)

- **The circuit switched network is less efficient as compared to others** because network resources (switch and link) are allocated to 'A' and 'F' node and cannot be used by others or other connections are deprived.

- The total delay in communication between 'A' and 'F' end systems is given as :

$$\text{Total delay} = \text{Connection establishment delay} + \text{Data transfer delay} + \text{Connection release delay}$$

- Data transfer delay is also given as :

$$\text{Data transfer delay} = \text{Propagation time delay} + \text{Data transfer time delay}$$

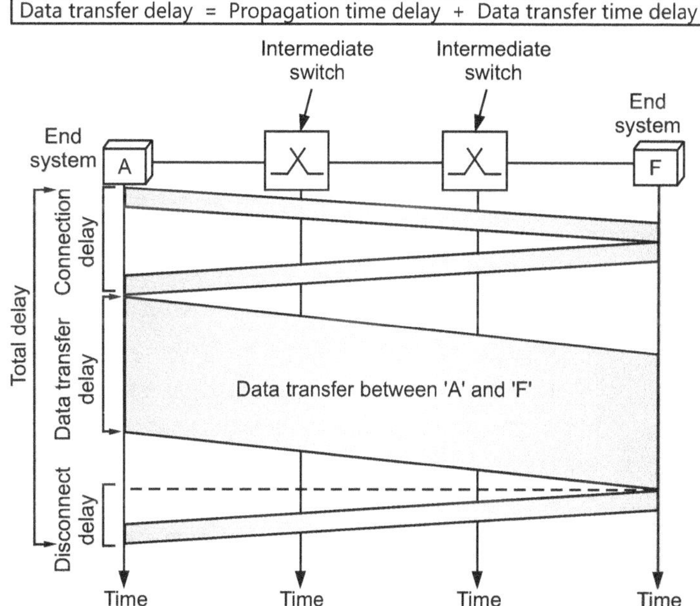

Fig. 1.60 : Typical Delays in Circuit Switched Network

(Here between two end systems like 'A' and 'F')

- Thus, typical example or application of circuit switched network in telephone communication is as shown in Fig. 1.61.

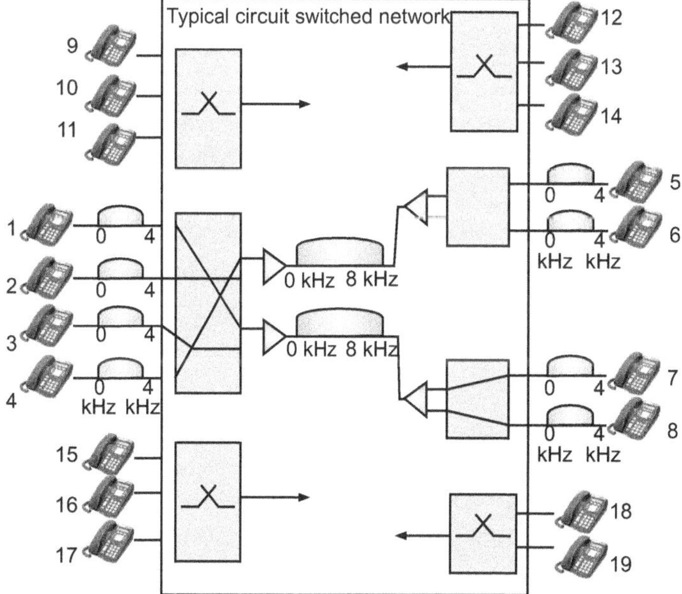

Fig. 1.61 : Typical Circuit Switched Network Example

- There is a common misunderstanding that circuit switching is used only for connecting voice circuits (analog or digital). The concept of a dedicated path persisting between two communicating parties or nodes can be extended to signal content other than voice. Its advantage is that it provides for non-stop transfer without requiring packets and without most of the overhead traffic usually needed, making maximal and optimal use of available bandwidth for that communication. The disadvantage of inflexibility tends to reserve it for specialized applications, particularly with the overwhelming proliferation of Internet-related technology.

- For call setup and control (and other administrative purposes), it is possible to use a separate dedicated signalling channel from the end node to the network. ISDN is one such service that uses a separate signalling channel while Plain Old Telephone Service (POTS) does not.

- The method of establishing the connection and monitoring its progress and termination through the network may also utilize a separate control channel as in the case of links between telephone exchanges which use SS7 packet-switched signalling protocol to communicate the call setup and control information and use TDM to transport the actual circuit data. **Signalling System Number** 7 (SS7) is a set of telephony signalling protocols which are used to set up most of the world's public switched telephone network telephone calls. The main purpose is to set up and tear down telephone calls

- Early telephone exchange is a suitable example of circuit switching. The subscriber would ask the operator to connect to another subscriber, whether on the same exchange or via an inter-exchange link and another operator. In any case, the end result was a physical electrical connection between the two subscribers's telephones for the duration of the call. The copper wire used for the connection could not be used to carry other calls at the same time, even if the subscribers were in fact not talking and the line was silent.

- Thus, generally, resources are frequency intervals in a Frequency Division Multiplexing (FDM) scheme or more recently time slots in a Time Division Multiplexing (TDM) scheme. The set of resources allocated for a connection is called a circuit. A path is a sequence of links located between nodes called *switches*. The path taken by data between its source and destination is determined by the circuit on which it is flowing, and does not change during the lifetime of the connection. The circuit is *terminated* when the connection is closed.

- In circuit switching, resources remain allocated during the full length of a communication, after a circuit is established and until the circuit is terminated and then the allocated resources are freed. Resources remain allocated even if no data is flowing on a circuit, hereby wasting link capacity when a circuit does not carry as much traffic as the allocation permits. This is a major issue since frequencies (in FDM) or time slots (in TDM) are available in finite quantity on each link, and establishing a circuit consumes one of these frequencies or slots on each link of the circuit. As a result, establishing circuits for communication that carry less traffic than allocation permits can lead to resource exhaustion and network saturation, preventing further connections from being established. If no circuit can be established between a sender and a receiver because of a lack of resources, the connection is *blocked*.

- A second characteristic of circuit switching is the time cost involved when establishing a connection. In a communication network, circuit-switched or not, nodes need to lookup in a *forwarding table* to determine on which link to send incoming data, and to actually send data from the input link to the output link. Performing a lookup in a forwarding table and sending the data on an incoming link is called *forwarding*. Building the forwarding tables is called *routing*. In circuit switching, routing must be performed for each communication, at circuit establishment time. During circuit establishment, the set of switches and links on the path between the sender and the receiver is determined and messages are exchanged on all the links between the two end hosts of the communication in order to make the resource allocation and build the routing tables. In circuit switching, forwarding tables are hardwired or implemented using fast hardware, making data forwarding at each switch almost instantaneous. Therefore, circuit switching is well suited for long-lasting connections where the initial circuit establishment time cost is balanced by the low forwarding time cost.

- The circuit identifier (a range of frequencies in FDM or a time slot position in a TDM frame) is changed by each switch at forwarding time so that switches do not need to have a complete knowledge of all circuits established in the network but rather only local knowledge of available identifiers at a link. Using local identifiers instead of global identifiers for circuits also enables networks to handle a larger number of circuits.

- *Traffic Engineering* (TE) consists in optimizing resource utilization in a network by choosing appropriate paths followed by flow of data, according to static or dynamic constraints. A main goal of traffic engineering is to balance the load in the network, i.e., to avoid congestion on links on a network while other links are under-utilized. To achieve such goals, traffic engineering methods can vary from offline capacity planning algorithms to automatic, dynamic changes. Since circuit switching allocates a fixed path for each flow, circuits can be established according to traffic engineering algorithms.

- On the other hand, circuit switching networks are not reactive when a network topology change occurs. For instance, on a link failure, all circuits on a failed link are cut and communication is interrupted. Special mechanisms that handle such topological changes have to be devised. Traffic engineering can alleviate the consequences of a link failure by pre-planning failure recovery. A backup circuit can be established at the same time or after the primary circuit used for a communication is set up, and traffic can be rerouted from the failed circuit to the backup circuit if a link of the primary circuit fails. Circuit switching networks are intrinsically sensitive to link failures and rerouting must be performed by additional traffic engineering mechanisms.

Examples of Circuit Switched Networks

- Public Switched Telephone Network (PSTN).
- ISDN B-channel.
- Circuit Switched Data (CSD) and High-Speed Circuit-Switched Data (HSCSD) service in cellular systems such as GSM.
- Datakit [It supports file transfers, remote login, remote printing, and remote command execution. At the physical layer, it can operate over multiple media, from slow speed EIA-232 to 500Mbit fiber optic links (called FIBERKIT)].
- X.21 (Used in the German DATEX-L and Scandinavian DATEX circuit switched data network).

1.15.2 Datagram Switching Networks

- In packet switching networks, voice, video or data is converted into packet. Packet can be of fixed size or variable size, decided by network used and protocol used at both ends.
- In datagram switching, there is no resource allocation for a packet travelling from sender to receiver.
- This means there is no bandwidth reservation on links and no scheduled processing time for each packet.
- Thus, resources are allocated on demand and this allocation is done on a first come first serve basis.

- As a simple analogy consider two hotels (or restaurants). One which requires reservation and another that neither require reservation and nor accept them.
- For the hotel (or restaurant) which requires reservation, we have to go through the hassle of calling person of restaurant before we leave home and reach to restaurant.
- But when we arrive at restaurant we get table, can communicate with waiter and order for food.
- Thus, for the other restaurant which does not require reservation, we don't need to bother to reserve anything.
- In this restaurant when we arrive, we may have to wait for table, we may have to wait for communicating with waiter to order the food.
- Thus, restaurant with reservation and without reservation, this analogy is applicable for circuit switched network and datagram switched network respectively.
- The typical packet flow in datagram packet switched network is as shown in Fig. 1.62.

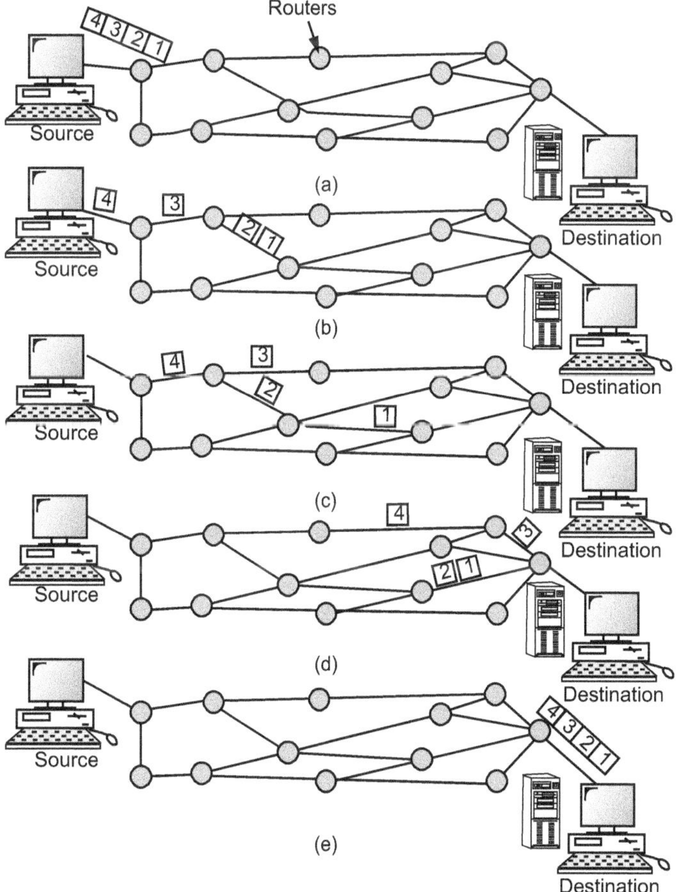

Fig. 1.62 : Datagram Network Packet Transfer from Source to Destination

- Thus, in datagram packet networks, following characteristics are important :
  - ➢ Each packet is treated independently.
  - ➢ Packet can take any practical route.
  - ➢ Packets may arrive out of order at destination router.
  - ➢ Packets may go missing in the datagram network journey.
  - ➢ Receiver at another end is responsible to reorder the packets and recover the missing packets.
  - ➢ Transport layer at both (sender and receiver) ends is responsible to reorder the packet sequence and recover the missing packets.
- We have already discussed the connection oriented and connectionless services in first chapter. Datagram based networks are also known as connectionless networks.
- There is no set-up phase or teardown phase present in datagram switching network. When data is ready, it is transferred with full source and destination address from sender to receiver. For this each intermediate router maintains the routing table as shown in Fig. 1.63.

Routing Table maintained by Router Device

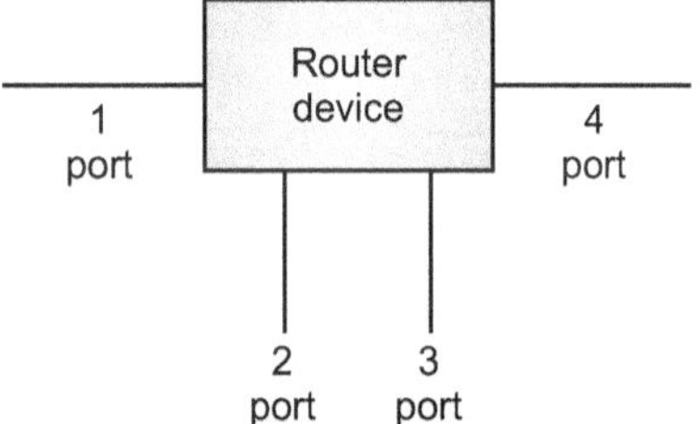

Fig. 1.63 : Router Device uses Routing Table based on Destination Address

Table 1.12

Destination Address	Output Port
12347	1
34569	2
22130	3
⋮	⋮
⋮	⋮
⋮	⋮
75759	4

- The routing tables are dynamic in nature and are updated periodically. (Specific time is set by router administrator).

- ●. The destination address is carried by the header among other information (or control information) of packet. This address remains same during the entire journey of the packet from source to destination or sender to receiver.
- Efficiency of datagram switch network is better than circuit switched network, because network resources are allocated only when there are packets to be transferred from source to destination.
- If source to destination packet transfer is finished or delayed then these resources can be used by other nodes or systems connected to this network.
- The delay between sender and receiver is given by,

$$\text{Total delay } (T_D) = \text{Transmission delay } (T_t) + \text{Propagation delay } (T_p) + \text{Waiting delay } (T_w)$$

- Typical datagram based packet switching network uses two routers in between sender and receiver as shown in Fig. 1.64.

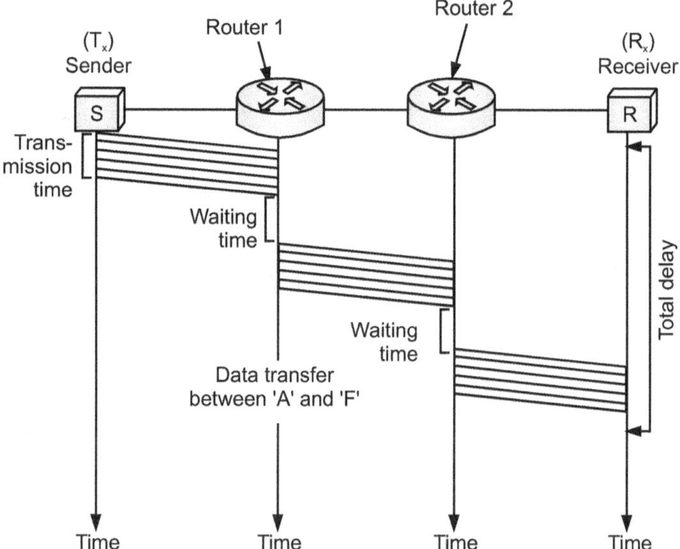

Fig. 1.64 : Delay present in Datagram Network

- Thus, total delay in above typical datagram network is given as,

$$T_D = 3T_t + 3T_p + T_{w1} + T_{w2}$$

where,

$3T_t$ = Three transmission times available (hence $3T_t$)

$3T_p$ = Three propagation delays available (hence $3T_p$)

$\left.\begin{array}{c} T_{w1} \\ T_{w2} \end{array}\right\}$ = Two waiting time delays available (hence T_{w1} and T_{w2})

- The best example of datagram network is Internet or TCP/IP protocol based LAN or Internet communication.
- Thus, a routing table contains a mapping between the possible final destination of packets and the outgoing link on their path to the destination. Routing tables can be very large because they are indexed by possible

destinations, making lookups and routing decisions computationally expensive, and the full forwarding process relatively slow compared to circuit switching. In datagram packet switching networks, each packet must carry the address of the destination host and use the destination address to make a forwarding decision. Consequently, routers do not need to modify the destination addresses of packets when forwarding packets.

- Since each packet is processed individually by a router, all packets sent by a host to another host are not guaranteed to use the same physical links. If the routing algorithm decides to change the routing tables of the network between the instants two packets are sent, then these packets will take different paths and can even arrive out of order.

- Second, on a network topology change such as a link failure, the routing protocol will automatically recompute routing tables so as to take the new topology into account and avoid the failed link. As opposed to circuit switching, no additional traffic engineering algorithm is required to reroute traffic. Since routers make routing decisions locally for each packet, independent of the flow to which a packet belongs. Therefore, traffic engineering techniques, which heavily rely on controlling the route of traffic, are more difficult to implement with datagram packet switching than with circuit switching.

- There are three primary types of datagram packet switches:

  1. **Store and Forward :** Buffers data until the entire packet is received and checked for errors. This prevents corrupted packets from propagating throughout the network but increases switching delay.

  2. **Fragment Free :** Filters out most error packets but doesn't necessarily prevent the propagation of errors throughout the network. It offers faster switching speeds and lower delay than store-and-forward mode.

  3. **Cut Through :** Does not filter errors; it switches packets at the highest throughput, offering the least forwarding delay.

- A datagram network is a best effort network. Delivery is not guaranteed. Reliable delivery must be provided by the end systems (i.e. user's computers) using additional protocols.

- The most common datagram network is the Internet, which uses the IP network protocol. Applications which do not require more than a best effort service can be supported by direct use of packets in a datagram network, using the User Datagram Protocol (UDP) transport protocol.

- Applications like voice and video communications and notifying messages to alert a user that she/he has received new email are using UDP. Applications like e-mail, web browsing and file upload and download need reliable communications, such as guaranteed delivery, error control and sequence control. This reliability ensures that all the data is received in the correct order without errors. It is provided by a protocol such as the Transmission Control Protocol (TCP) or the File Transfer Protocol (FTP).

1.15.3 Virtual Circuit Networks (VC Networks)

- Virtual circuit network is another type of packet switched network.

- Virtual circuit network is a cross between datagram switching network and circuit switching network. It has characteristics of both the networks.

Table 1.13

Characteristics of Circuit Switched Network	Characteristics of Datagram Network
1. It has three phases like : • Set-up phase (connection establishment). • Data transfer. • Teardown phase (connection release).	1. Resource allocation can be on demand in VC networks.
2. Resources can be allocated during the set-up phase or connection establishment phase. 3. All packets follow the same path established during set-up phase or connection establishment phase.	2. In datagram packet header, destination IP addresses are mentioned, whereas in VC packet header next switch VCI (Virtual Circuit Identifier) number is mentioned.
In todays technology • Circuit switched network is implemented in physical layer. • Datagram switched network is implemented in network layer. • VC switched network is implemented in data link layer.	

- In VC switched networks, two types of addressing used are as follows :

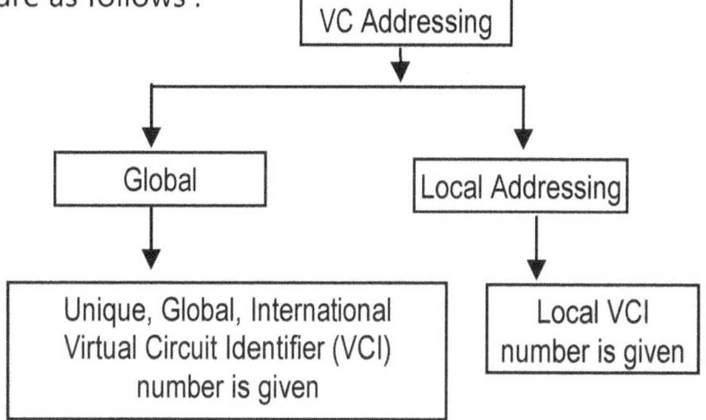

Virtual Circuit Packet Network

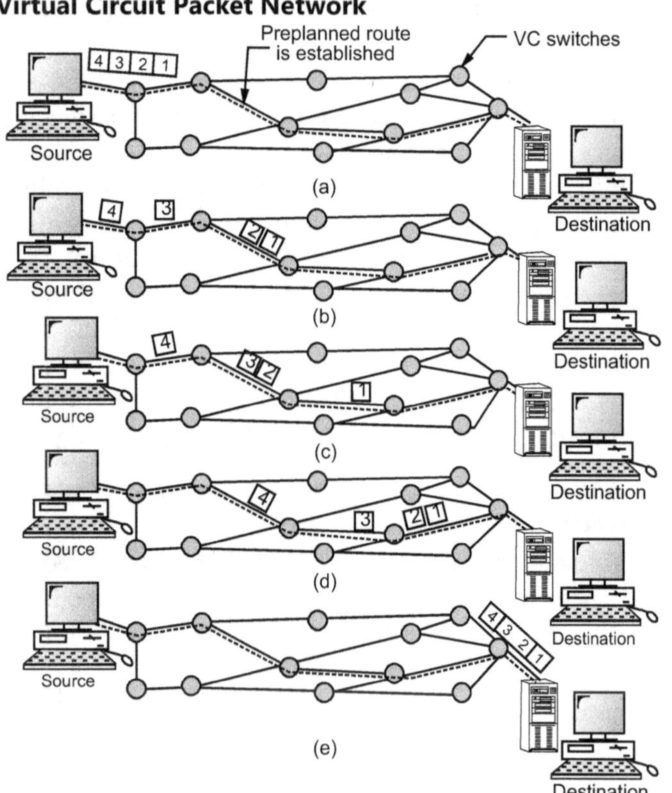

Fig. 1.65 (a) : Packet flow in Typical Virtual Circuit Packet Switched Network

- Typical VC switched network is as shown in Fig. 1.65 (b).
- **Virtual Circuit Identifier (VCI)** is used for data transfer from one end to another.
- VCI is used by frames between two VC switches. When frame arrives at one VC switch, it has a VCI, when it leaves it has a different VCI.

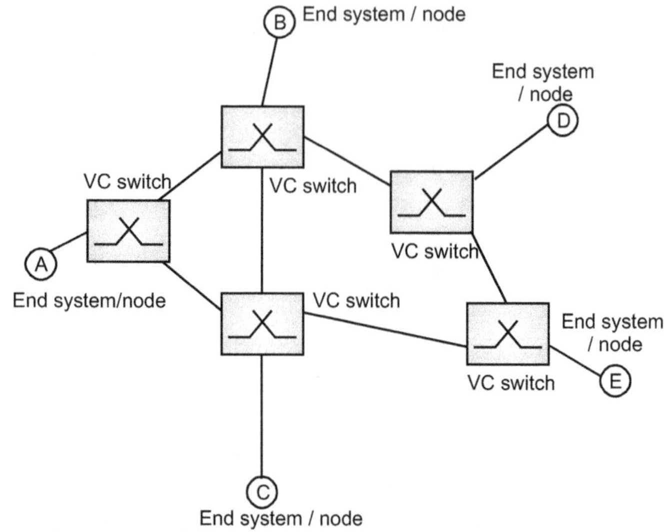

Fig. 1.65 (b) : Typical VC Switched Network

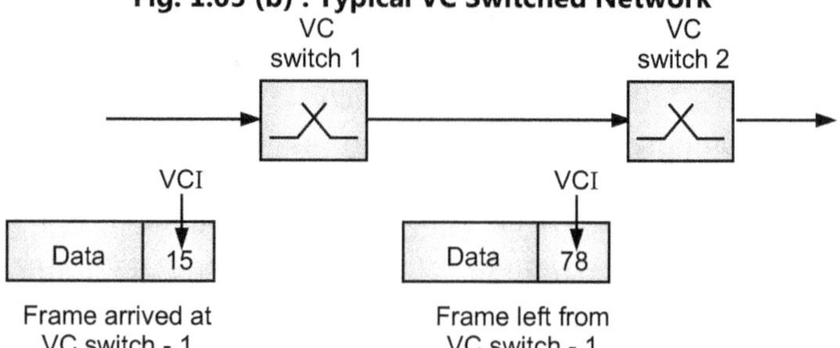

Fig. 1.66 : VC Switch decides VCI Number of Frames

- The VC switched network uses three steps for communication :
  - ➤ Set-up phase (connection establishment phase).
  - ➤ Data transfer phase.
  - ➤ Teardown phase (connection release phase).

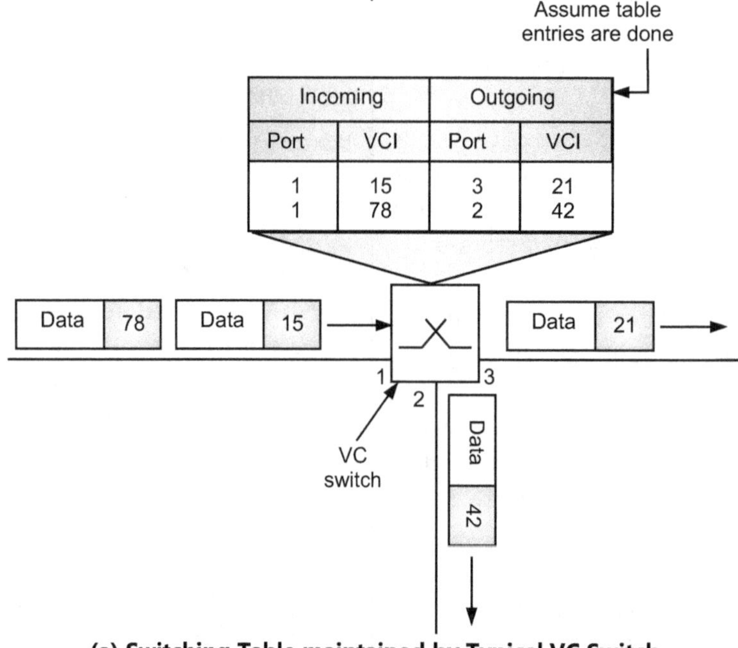

(a) Switching Table maintained by Typical VC Switch (Assume that entries are done initially)

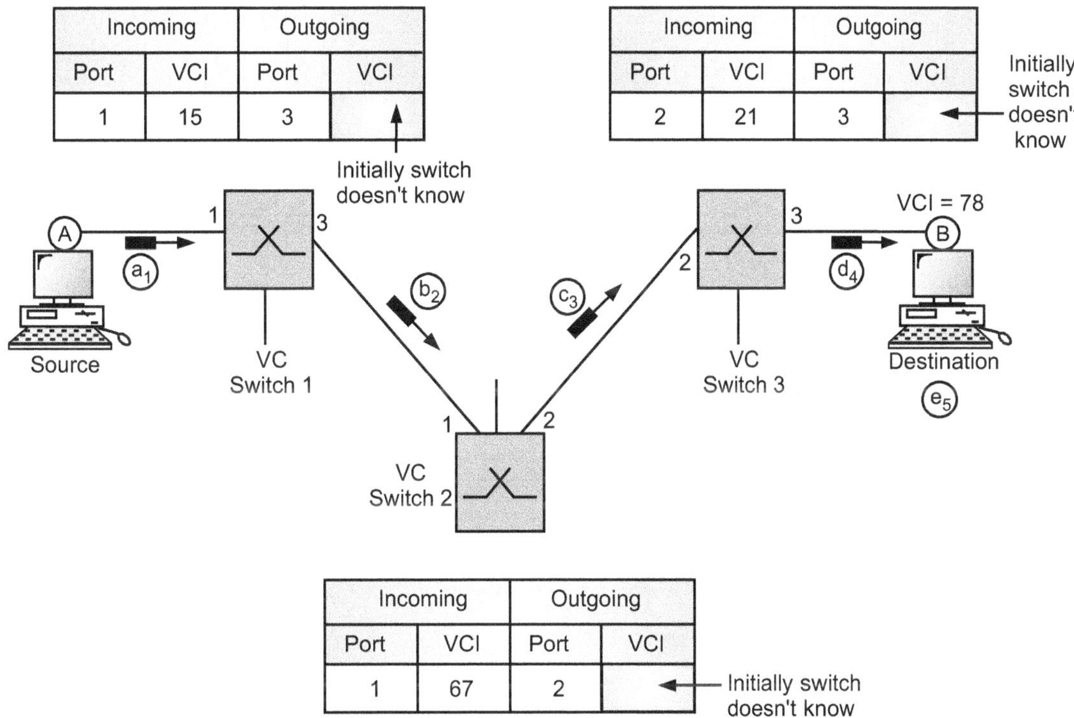

(b) Set-up Request Process from Source 'A' to Destination 'B'

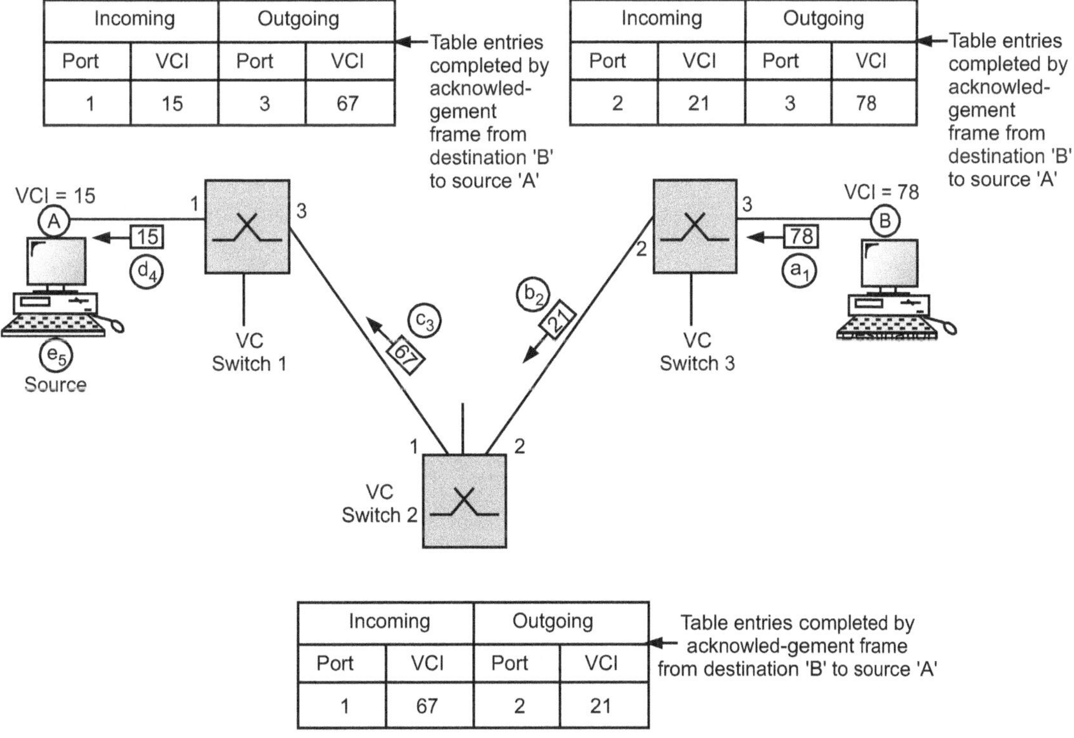

(c) Set-up Acknowledgement Process from Destination 'B' to Source 'A'

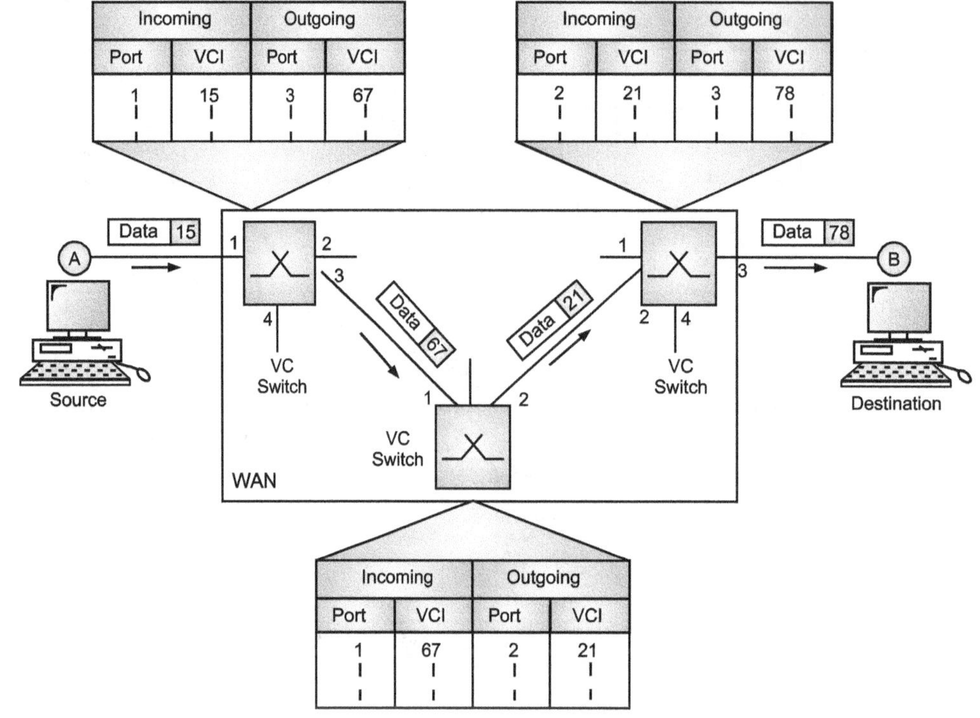

(d) Source to Destination Data Transfer from

'A' to 'B' end system according to Updated Table Entries

Fig. 1.67 : Communication between Source 'A' and Destination 'B' in VC Switched Network

- **Thus, Fig. 1.67 (a) says :**
- When incoming frame 1 with VCI = 15 and frame 2 with VCI = 78 arrives to VC switch, depending upon switching table entries mentioned, frame 1-VCI is changed from 15 → 21 and sent on port 3. Also frame 2-VCI is changed from 78 → 42 and sent on port 2.

- **Fig. 1.67 (b) says :**
  - ➢ Set-up request process from source 'A' to destination 'B'.
  - ➢ It clearly shows random incoming VCI number and undefined outgoing VCI numbers.
  - ➢ Set-up request starts from source 'A' → VC switch 1 → VC switch 2 → VC switch 3 → destination 'B'.

- **Fig. 1.67 (c) says :**
  - ➢ Special acknowledgement frame is sent from destination 'B' to source 'A'.
  - ➢ This special acknowledgement frame completes the table entries like

  VC switch 3 → Outgoing VCI = 78
      Incoming VCI = 21
  VC switch 2 → Outgoing VCI = 21
      Incoming VCI = 67
  VC switch 1 → Outgoing VCI = 67
      Incoming VCI = 15

- Thus, source 'A' gets the VCI = 15 as source frame VCI number and hence source 'A' sends | data || 15 | frame like to immediate VC switch 1 in data transfer phase.

- **Thus, Fig. 1.67 (d) says :**
  - ➢ Once table entries are confirmed the first frame generated | data | 15 | from source 'A' is with VCI = 15.
  - ➢ Thus, data transfer takes place from source 'A' → VC switch 1 → VC switch 2 → VC switch 3 → destination 'B'.

- Hence, Fig. 1.67 shows communication between source 'A' and destination 'B' in VC switched network.

  Where, set-up phase request process is given by Fig. 1.67 (b).

  Set-up phase acknowledgement process is given by Fig. 1.67 (c).

  Data transfer between 'A' and 'B' is given by Fig. 1.67 (d).

- Finally, when data transfer is completed between 'A' and 'B' systems then the remaining process is teardown process or connection release process. In this process or in this phase source 'A' sends a special frame called a teardown request to destination 'B'. Destination 'B' also responds with teardown confirmation frame and sends to source 'A'. Thus, all corresponding switching table entries are deleted from VC switches.

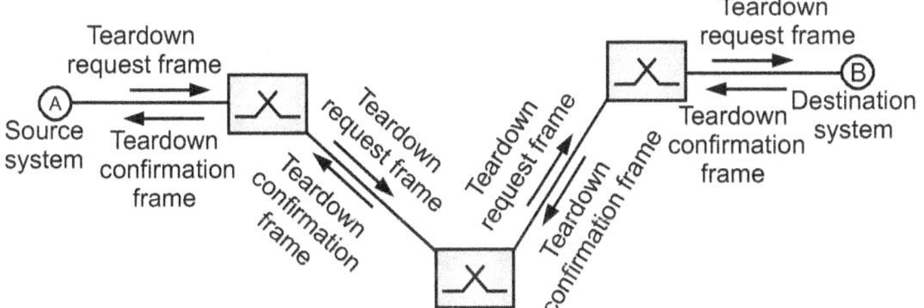

Fig. 1.68 : Typical Teardown Request and Confirmation set-up between Source 'A' and Destination 'B' after Data Transfer is Over

- Thus, the efficiency of the VC switched network is greater as compared to circuit switched network and datagram switched network.

- This happens because the main advantage of VC switched network is even if resource allocation is on demand, the source can check availability of the resources without actually reserving it.

- Thus, though path between source 'A' and destination 'B' is same, packets may arrive at the destination with different delays if resource allocation is done on demand as we discussed.

- The total delay in communication between source 'A' and destination 'B' is given by,

> Total delay = Transmission delay + Propagation delay
> + Set-up delay + Teardown delay

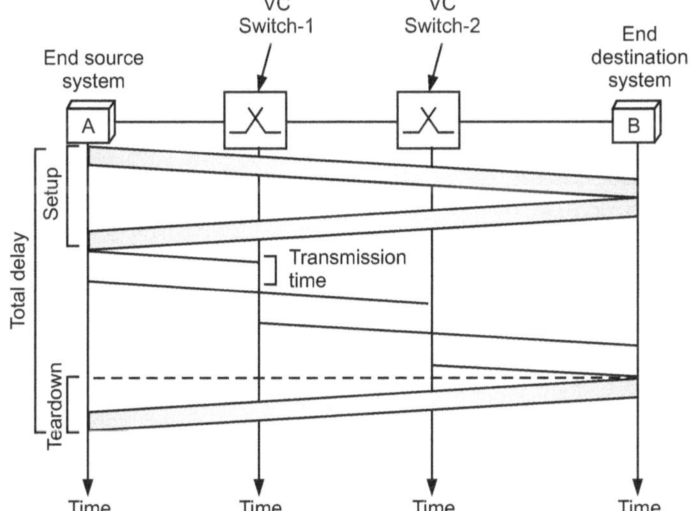

Fig. 1.69 : Delay Present in VC Switched Network

- Thus, total delay in above typical VC switched network is given as,

$$T_D = 3T_t + 3T_p + \text{Set-up delay} + \text{Teardown delay}$$

where,

$3T_t$ = Three transmission times available (hence $3T_t$)

$3T_p$ = Three propagation delays available (hence $3T_p$)

Set-up delay consists of request + Acknowledged process delay.

Teardown delay consists of request + Acknowledged process delay.

- Following are the typical VC switched networks :

➢ X.25 VC switched networks.

➢ Frame relay VC switched networks.

➢ ATM (Asynchronous Transfer Mode) networks.

➢ MPLS (Multiprotocol Label Switching) networks.

1.15.4 Concept of Virtual Circuit

- A *virtual circuit* is a logical connection created to ensure reliable communication between two network devices.

- A virtual circuit denotes the existence of a logical, bidirectional path from sender device to another receiver device across an VC switched network.

- Physically, the connection can pass through any number of intermediate nodes, such as VC switches.

- Multiple virtual circuits (logical connections) can be multiplexed onto a single physical circuit (a physical connection).

- Virtual circuits are demultiplexed at the remote end, and data is sent to the appropriate destinations. Fig. 1.70 illustrates four separate virtual circuits being multiplexed onto a single physical circuit.

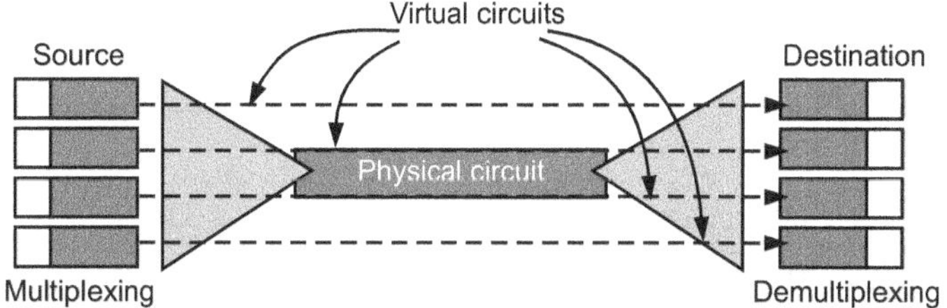

Fig. 1.70 : Virtual Circuits and Physical Circuit between Source and Destination

- Thus, we have seen the trade-off between connection establishment and forwarding time costs that exist in circuit switching and datagram packet switching.

- In VC switching, routing is performed at circuit establishment time to keep fast packet forwarding.

- Other advantages of VC switching include the traffic engineering capability of circuit switching, and the resources usage efficiency of datagram packet switching.

- Nevertheless, a main issue of VC switched networks is the behavior on a topology change.

- As opposed to datagram packet switched networks which automatically recompute routing tables on a topology change like a link failure, in VC switching all virtual circuits that pass through a failed link are interrupted.

- Hence, rerouting in VC switching relies on traffic engineering techniques.

1.15.5 Types of Virtual Circuits

- Switched Virtual Circuits (Physical connection is always present temporary logical connection is to be established every time for the data transfer) Switched Virtual Circuits (SVCs) are temporary connections used in situations requiring only sporadic data transfer between DTE devices across the Frame Relay network.

- Permanent Virtual Circuits (Physical connection is always present and permanent logical connection is already established so that any time data can be transferred) Permanent Virtual Circuits (PVCs) are permanently established connections that are used for frequent and consistent data transfers between DTE devices across the network. Communication across a PVC does not require the call setup and termination states that are used with SVCs. PVCs do not require that sessions be established and terminated. Therefore, DTEs can begin transferring data whenever necessary because the session is always active.

1.15.6 Typical Applications of VC Switched Networks

- File Transfer
  - ➤ Character-interactive traffic (e.g. text editing)
  - ➤ High Resolution graphics
- Access to Internet and Intranet
- Multimedia, Real Time Voice, Video, Fax

- LAN Peer-to-Peer, WAN Interconnection

- Multi-protocol networking applications
  - ➤ ATM, SNA, TCP/IP

- Private backbone networks.

1.15.7 Comparison of Different Switching Technique

- Now, we have seen the following switching techniques in detail.
  - ➤ Circuit switching.
  - ➤ Datagram switching.
  - ➤ Virtual circuit switching.

- Let's compare the following :
  - ➤ Circuit switching V/s Packet switching.
  - ➤ Datagram switching V/s Virtual circuit switching.
  - ➤ Circuit switching V/s Datagram switching V/s VC switching V/s Message switching.

1.15.8 Circuit Switching V/s Packet Switching

Table 1.14

Parameter	Circuit Switched	Packet Switched
Call setup requirement	Required	Not required
Dedicated physical path requirement	Yes, it is required	No, it is not required
Whether each packet follows the same path (route)	Yes, it follows the same path (route)	No, it does not follow the same path (route)
Bandwidth Available for Transmission	It is fixed	It is dynamic
Congestion can occur at	Setup time	On every packet
Bandwidth Wastage	Yes	No
Store and Forward transmission	Not available	Yes, it is available
Transparency in System	Yes, it is present	Not present
Charge applied	Per unit time	Per unit packet

1.15.9 Datagram (DG) V/s Virtual Circuit (VC) Switching

Table 1.15

Parameter	Datagram (DG) Network	Virtual Circuit (VC) Network
Requirement of Circuit Setup	Requirement of circuit setup is not needed	Requirement of circuit setup is needed.
Routing of Packet	Each packet is routed independently	Route is chosen when VC setup is over and all packet follows the same
Router failure	None	Less probability
If router fails	Packets are lost during the crash	All VCs that passed through the failed router are terminated
Achievement of Quality of service	It is difficult here	It is easy if enough resources can be allocated in advance for each VC
Congestion control	It is difficult over here	It is easy if enough resources can be allocated in advance for each VC
Examples of network	TCP/IP internet network	X.25, Frame Relay and ATM networks

1.16 STRUCTURE OF DIFFERENT SWITCHES

Typical switches classification is as given below.

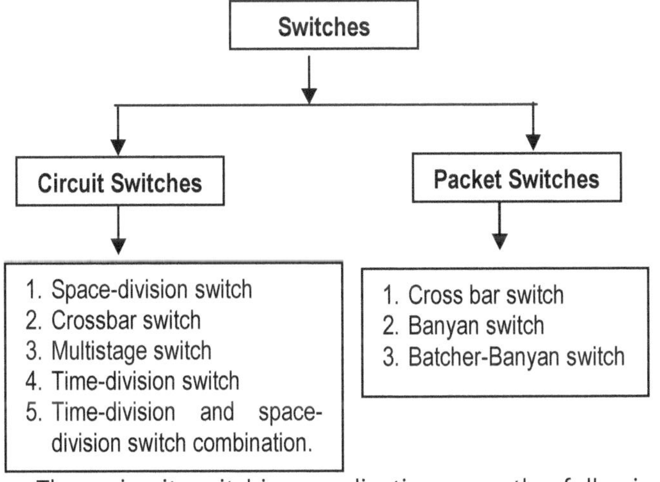

- Thus, circuit switching application uses the following switches as switching fabrics in their device architecture :
  - ➢ Space-division switch.
  - ➢ Crossbar switch.
  - ➢ Multistage switch.
  - ➢ Time-division switch.
  - ➢ Time-division and space-division switch combination.

- The packet switching application uses the following switches as switching fabrics in their device architecture :
  - ➢ Crossbar switch.
  - ➢ Banyan switch.
  - ➢ Batcher-Banyan switch.

1.16.1 Circuit Switches

- We have already classified the types of switches like :
  1. Circuit switches.
  2. Packet switches.

- Basically, there are two major types of switching used in telecommunication networks :
  1. Circuit switching.
  2. Packet switching.

- Basically the circuit switching is
  - Developed for voice.
  - Used nowadays also for data.
  - Has well-specified delays.
  - Faces echo problems.
  - Uses TDM and FDM multiplexing.

- The circuit switching is used in PSTN, ISDN and PCM networks.

- In the circuit switching, following types of switches are used :
  1. Space-division switches
  2. Time-division switches
  3. Combination of both

- Basically, the packet switching is
  - ➢ Developed for data
  - ➢ Used nowadays also for voice
  - ➢ Faces traditionally variable delays
  - ➢ Uses statistical multiplexing

- Packet switching is used in IP (connectionless service), Frame relay (connection oriented service), ATM (connection oriented service), MPLS (connection oriented service) and X.25 (connectionless and connection oriented service) networks.

- In the packet switching, following types of switches are used :
  1. Banyan switches
  2. Batcher-Banyan switches

- We are interested in circuit switching and circuit switches now.

- Hardware and/or software devices allowing temporary connections between two or more devices are known as switches.

- Basic circuit switch device is as shown in Fig. 1.71.

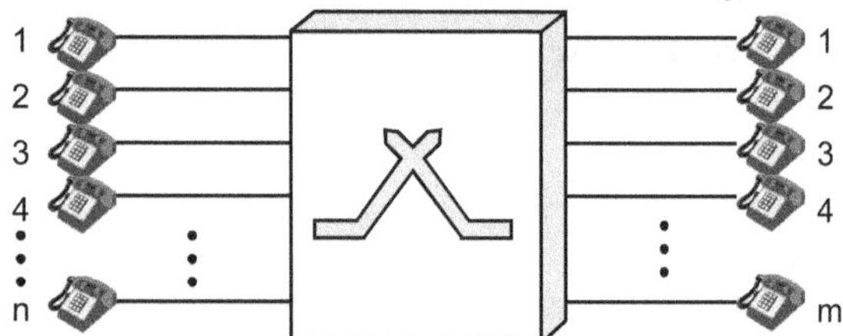

Fig. 1.71 : Basic Circuit Switch Device

- Basic circuit switch device contains :

  ➢ n inputs and m outputs; these do not have to be equal.

  ➢ These inputs and outputs create a temporary connection between an input link and output link.

- Another circuit switch known as folded switch is shown in Fig. 1.72.

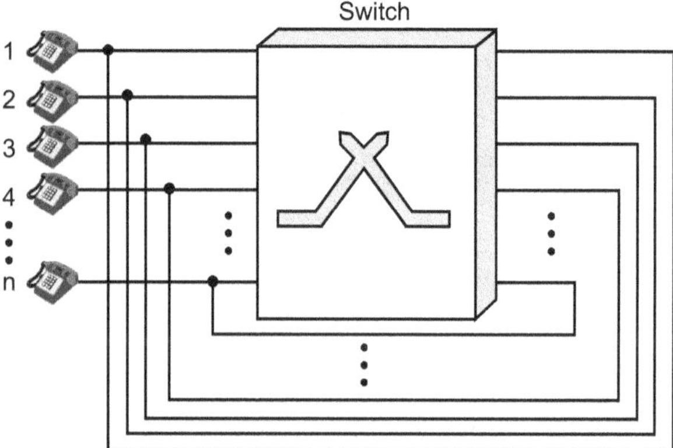

Fig. 1.72 : An n-by-n Folded Switch allows every device to connect to every other device in Full-Duplex Mode

1.16.2 Space-Division switches and Switching

- In space-division switching paths between devices are separated from each other spatially.

- Space-division switching was originally used in analog networks; now it is used in both digital and analog networks.

- Crossbar switches and multistage switches are commonly used in space-division switching.

1.16.3 Crossbar Switches

- Typical crossbar switch structure is as shown in Fig. 1.73.

- Crossbar switch connects n inputs to m outputs in a grid using transistors at each crosspoint.

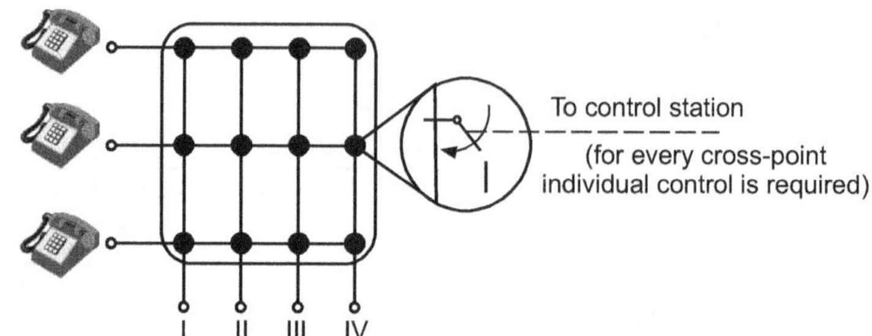

Fig. 1.73: Typical Crossbar Switch

- Crossbar switch requires a significant number of crosspoints(for example 1000 inputs and 1000 outputs require the switch with 1000000 crosspoints), and crossbar with such number of crosspoints are not possible practically.

- Crossbar switch is inefficient; statistics says fewer than 25% of switches are in used at a given time and rest are idle.

1.16.4 Multistage Switches

- Typical multistage switch structure is as shown in Fig. 1.74.

- Multistage switch reduces the number of crosspoints in switch.

- Multistage switch combines crossbar switches in several stages.

- Multistage switch devices are linked to switches that are linked to a hierarchy of other switches.

- Middle stages of multistage switch usually have fewer switches than first and last stages.

- In multistage switch, crosspoints are fewer but still allowing multiple paths through the network which increases reliability.

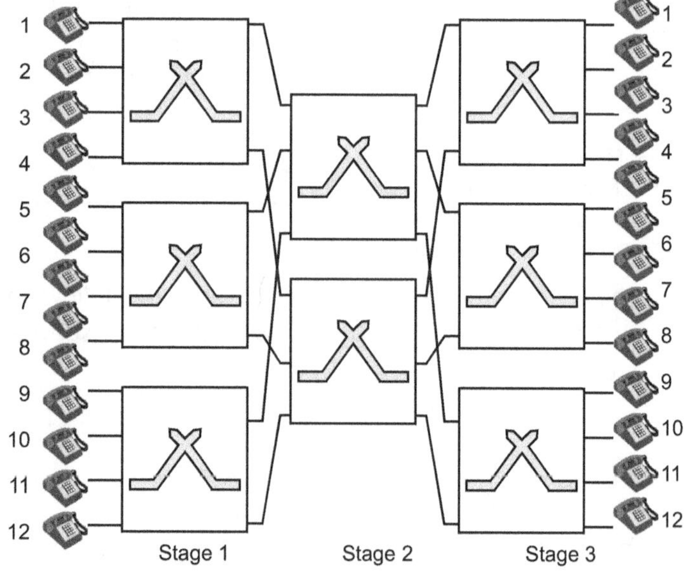

Fig. 1.74 : Typical Multistage Switch Structure

- Multiple paths are established using multistage switch as shown in Fig. 1.75.

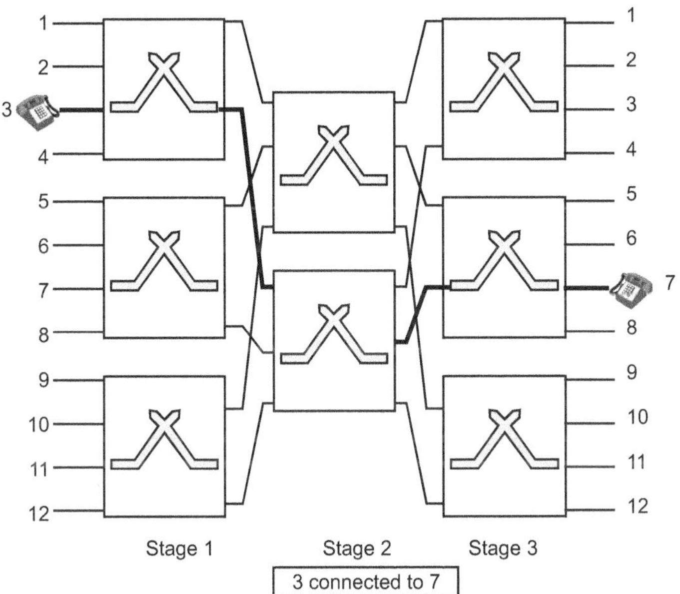

Fig. 1.75 : Multiple Paths are established using Multistage Switch

- Here in multistage switch, the concept of blocking is very important.

- Reduced number of crosspoints may mean that at times of heavy traffic, an input may not be able to connect to an output if there is no path available (i.e. all switches are occupied).

- Blocking does not occur in single-stage switch; a non-blocking path is always available.

EXERCISE

1. What are the different advantages of installing a network ?

2. What are the different disadvantages of installing a network ?

3. Write a short note on network topology.

4. Compare all network topologies.

5. Classify the different networks.

6. Write a short note on :
   - (a) LAN　　　　(b) WAN
   - (c) MAN　　　　(d) Internet

7. What is Network Layered Architecture ?

8. What are the different benefits of layered design ?

9. What are service primitives and relationship of service to protocols ?

10. Compare connection oriented V/s connectionless service.

11. Draw and explain ISO-OSI reference model.

12. Explain the function of each layer of OSI-ISO reference model.

13. List out the different network architectures.

14. Draw and explain TCP/IP network architecture.

15. Compare OSI model and TCP/IP model.

16. What is MAC address ?

17. What are the types of IP addresses ?

18. Explain private IP addresses.

19. Write a short note on port addresses.

20. Classify the different transmission media.

21. Write notes on :
   - (a) UTP cable
   - (b) STP cable
   - (c) Coaxial cable

22. What are the Cat 3, Cat 4, Cat 5, Cat 5e cables ?

23. Explain Cat 6 and Cat 7 wires in detail.

24. What is the difference between thin Ethernet cable and thick Ethernet cable ?

25. Draw and explain the construction of fiber cable.

26. Explain the typical fiber optic communication system.

27. Classify the fiber optic cable depending upon material and principle of operation.

28. Explain the advantages and disadvantages of fiber optic cable.

29. Compare Coaxial cable V/s Fiber optic cable.

30. Draw and explain typical electromagnetic spectrum.

31. Classify the different switching techniques.

32. Write short notes on :
   - (a) Circuit switching networks
   - (b) Phases in circuit switching
   - (c) Delay in circuit switching

33. Write short notes on :
   - (a) Datagram switching networks
   - (b) Delay in datagram networks
   - (c) Types of datagram packet switches.

34. Write short notes on :

    (a) Similarities of VC switching with circuit switching and datagram networks

    (b) VC switching between source and destination

    (c) Virtual Circuit Identifier (VCI)

    (d) Phases in VC switching

    (e) Delays in VC switching networks.

35. What is Virtual Circuit ? What is SVC and PVC ?

36. Explain the different applications of VC switching.

37. Compare the following :

    (a) Circuit switching V/s Packet switching

    (b) Datagram switching V/s Virtual circuit switching

    (c) Circuit switching V/s Datagram switching V/s VC switching V/s Message switching

38. Classify the different switches.

39. Write short notes on :

    (a) Circuit switches

    (b) Packet switches.

DATA LINK LAYER

2.1 INTRODUCTION TO DATA LINK LAYER

- Physical layer takes care of transmitting information over a communication channel.

- Information transmitted may be affected by noise or distortion caused in the channel.

- Hence, the transmission over communication channel is not reliable.

- The data transfer is also affected by delay and has finite rate of transmission. This reduces the efficiency of transmission.

- Data link layer is designed to take care of these problems i.e. data link layer improves reliability and efficiency of channel.

- We can also say that the services provided by physical layer are not reliable.

- Hence, we require some layer above physical layer which can take care of these problems. The layer above physical layer is Data Link Layer (DLL).

2.2 DLC SERVICES

Following are some of the functions of a data link layer.

- **Error Control :** Physical layer is error prone. The errors introduced in the channel need to be corrected.

- **Flow Control :** There might be mismatch in the transmission rate of sender and the rate at which receiver receives. This mismatch must be taken care of.

- **Addressing :** In the network where there are multiple terminals, whom to send the data has to be specified.

- **Frame Synchronization :** In physical layer, information is in the form of bits. These bits are grouped in blocks of frames at data link layer. In order to identify beginning and end of frames, some identification mark is put before and/or after each frame.

- **Link Management :** In order to manage co-ordination and co-operation among terminals in the network, initiation, maintenance and termination of link is required to be done properly. These procedures are handled by data link layer. The control signals required for this purpose use the same channels on which data is exchanged. Hence, identification of control and data information is another task of data link layer.

- **Services Provided to Network Layer :** Data link layer provides services to the layer above it viz. network layer. The basic service is transferring packets from network layer on source machine to network layer on destination machine as shown in Fig. 2.1.

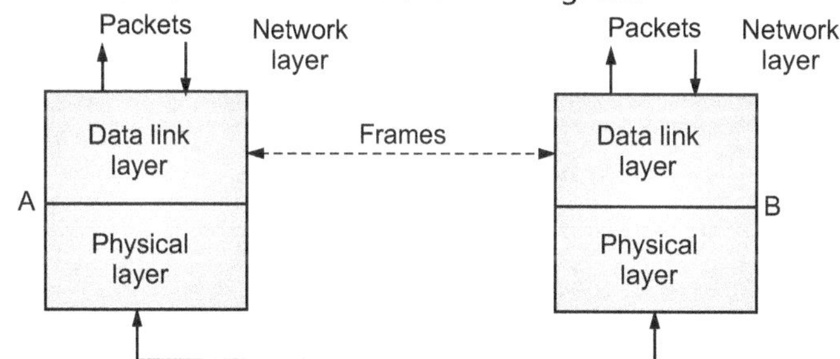

Fig. 2.1 : Service provided to Network Layer

The service model describes the service provided by a protocol.

There are two categories of service models :

1. Connection-oriented service.
2. Connectionless service.

- In connection-oriented service, connection is established between the peer entities first and then data transfer begins. There will be connection setup, data transfer and connection release procedure required to be carried out.

- Connectionless services do not require a connection setup procedure. Information blocks are transmitted using address information in each Protocol Data Unit (PDU).

- Acknowledged connectionless services provide acknowledgement for each PDU so that data transfer is reliable.

- Unacknowledged connectionless services do not provide acknowledgement for each PDU. This is also called best effort service. In such case, network layer has to provide reliable service i.e. acknowledged service.

- The service model specifies the Quality of Service (QoS). It includes expected performance level in transfer of information. Examples of some QoS parameters are :

  ➤ Probability of error.
  ➤ Probability of loss.
  ➤ Transfer delay.

2.3 DLL PROTOCOLS

High Level Data Link Control (HDLC)

- It is the most widely used DLL protocol.
- It has a set of functions which provides communication service to network layer.
- HDLC supports variety of applications for which it has three types of stations, two link configurations and three data transfer modes.

Types of Stations :

1. **Primary Station :** It controls operation of link. It issues commands.
2. **Secondary Station :** Primary station controls it by issuing command frames transmitted by secondary station.
3. **Combined Station :** It has features of both primary and secondary stations i.e. it issues both commands and response.

Types of Configuration :

1. **Unbalanced Configuration :** It has one primary and one or more secondary stations. It supports both full duplex and half duplex configuration.
2. **Balanced Configuration :** It consists of two combined stations supporting half duplex and full duplex transmission.

These configurations are shown in Fig. 2.2 (a), (b) and (c).

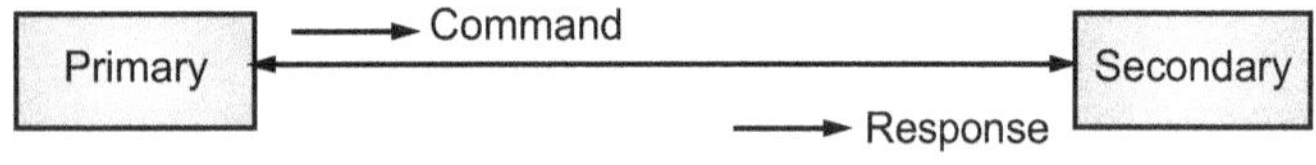

(a) Unbalanced Point-to-Point Link

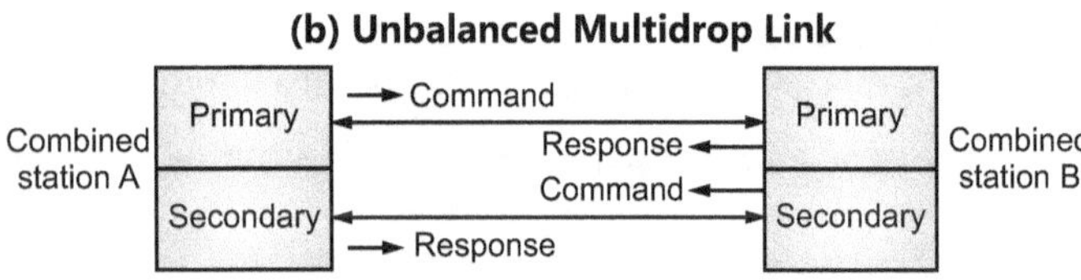

(b) Unbalanced Multidrop Link

(c) Balanced Point-to-Point Link

Fig. 2.2

Types of Data Transfer Modes :

1. **Normal Response Mode (NRM) :** Used with unbalanced configuration. Primary can initiate data transfer to a secondary. Secondary can only transfer data in response to command from primary.

2. **Asynchronous Balanced Mode (ABM) :** Used with balanced configuration. Any one of the combined station can initiate transmission without the permission of other station.

3. **Asynchronous Response Mode (ARM) :** Used with unbalanced configuration secondary can initiate transmission without permission from primary. But primary has control of the link.

NRM can be used on multidrop lines and point-to-point links. ABM is most widely used. ARM is rarely used.

2.4 HDLC FRAME FORMAT

- The functionality of a protocol depends on the control fields that are defined in the header and trailer.
- The various data transfer modes are determined by the frame structure.
- Fig. 2.3 shows HDLC frame format.

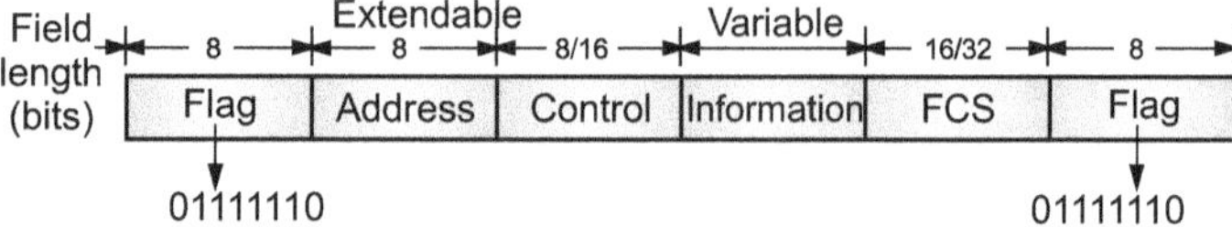

Fig. 2.3 : HDLC frame format

- Information is attached with an header consisting of flag, address and control fields and a trailer consisting of checksum and flag.
- Flag fields which are 8 bit long delimit the frame at both ends with a pattern 01111110. Bit stuffing is used to achieve data transparency.
- The addressing function of DLL is identifying the station that transmitted the frame and the station that will be receiving the frame. The address field specifies this. It is extendable over more than 8 bits in multiples. If this field is all 1's, the frame is broadcast to all secondaries.
- There are three types of control fields to identify three types of frames.
  1. **Information Frame (I-Frame)** has 0 in the first bit of control field.
  2. **Supervisory Frame (S-Frame)** has 10 in first two bits of control field.
  3. **Unnumbered Frame (U-Frame)** has 11 in first two bits of control field.

Error control and flow control functions of DLL are provided by I-frame and S-frame. The three frame fields are shown in Fig. 2.4.

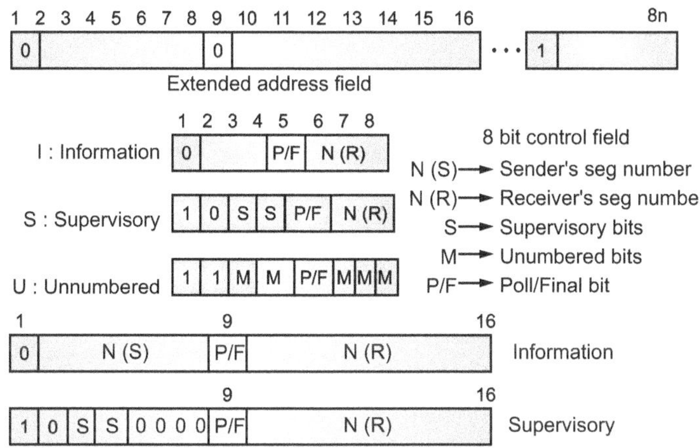

Fig. 2.4 : HDLC frame types

- Information field consists of 3 bit sequence number N(S) and N(R) of sender and acknowledgement number of receiver respectively (piggybacked). All frames have a P/F bit. Its uses depend on situation. In command frames it is called P bit and is set to 1 whenever a response frame is expected (Polled) from peer entity. In response frames it is called F bit and is set to 1 to indicate that the frame is response to a command frame (Poll). The length of information frame can be variable.

- There are four types of supervisory frames decided by S-field.

  1. If SS = 00. It is receive ready (RR) which acknowledges received frames in absence of piggybacking.

  2. If SS = 01. It is reject frame indicating negative acknowledgement and transmitter should go back and transmit frames N(R) onward.

  3. SS = 10 means receive not ready (RNR). Buffer full condition.

  4. SS = 11 indicates selective reject, where N(R) is frame to be retransmitted.

- Combination of I-frame and S-frame allows HDLC to implement ARQ techniques.

- The unnumbered frames implement number of control functions.

  The M bits decide the function.

  They are as below.

  (i) Set Asynchronous Balanced Mode (SABM) : To set up asynchronous balanced mode connection.

  (ii) Set Normal Response Mode (SNRM) : To set up normal response mode.

  (iii) Disconnect (DISC) : Indicates station wishes to disconnect connection.

  (iv) Unnumbered Acknowledgement (UA) : Acknowledges frames during call set up.

  (v) Frame Reject (FRMR) : Reject unacceptable frame.

- The information field contains sequence of bits in multiples of octets. Length of F-field is variable.

- Frame Check Sequence (FCS) field consists of error detecting code calculated from frame bits except flag fields. It has 16 bit CRC CCITT code.

Operation of HDLC

Let us now see how HDLC operates.

Connection Establishment and Release :

- Station A sends SAMB (Set Asynchronous Balanced Mode) frame indicating that it wants to establish a new connection.

- Station B sends unnumbered acknowledgement if it is ready to proceed. Otherwise it will REJECT the request by sending RNR frame.

- Whenever station wants to release connection it sends DISC frame and other station sends unnumbered acknowledgement. It is shown in Fig. 2.5.

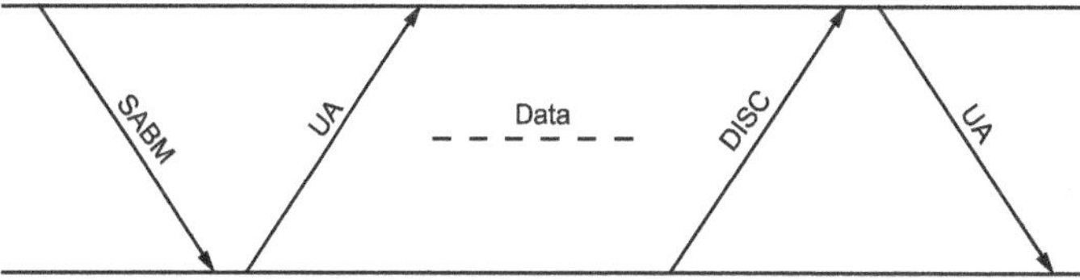

Fig. 2.5 : HDLC Connection Establishment and Release

Exchange of Frames using Normal Response Mode :

Assuming that connection is established between station A as primary and stations B and C as secondary, exchange of frames is shown in Fig. 2.6.

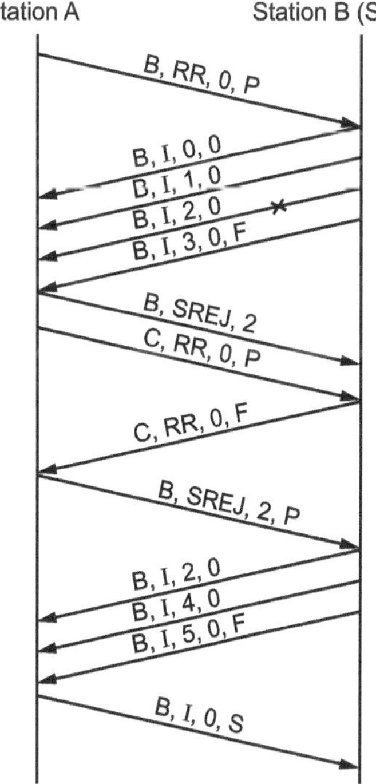

Fig. 2.6 : Exchange of Frames using NRM

- Station A polls B with N(R) = 0. Station B responds by sending frames 0, 1, 2, 3, with F bits set in last frame.

- Station A sends rejection of frame 2 and polls station C, which responds with receive ready frame.

- Station A again sends request to transmit frame 2 to B with poll bit set. Station B responds by sending frames 2, 4, 5 with F bit set in last frame.

- Station A sends information frame piggybacking acknowledgement of 5.

Exchange of frames using Asynchronous Balanced Mode (Shown Fig. 2.6).

Address field consists of address of receiving station, if it is information frame and address of transmitting station, if it is command frame or a response frame. Whenever frame is in error a REJ frame is send indicating number of bad frame so that transmitting station resends all the frames starting from that frame (Go_back_N) as is seen in case frame 1 rejected by station B.

Information frame consists of N(S) and N(R), where N(S) is transmitted frame number and N(R) is expected frame number i.e. piggyback acknowledgement. In case station does not have information frame to send it sends RR frame with acknowledgement of previously received frame.

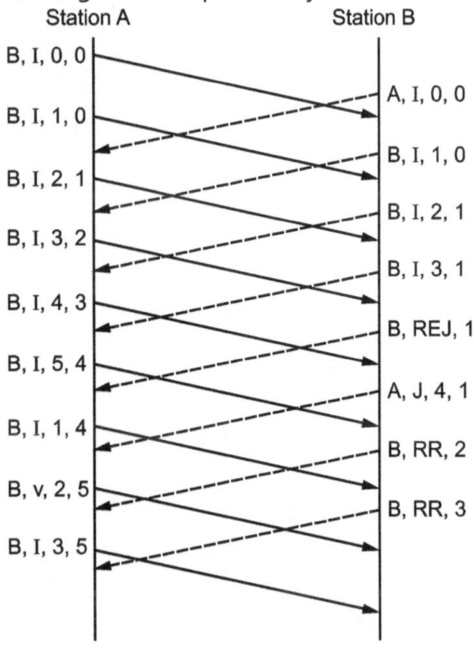

Fig. 2.7 : Exchange of Frames using ABM

2.5 POINT-TO-POINT PROTOCOL (PPP)

The *Point-to-Point Protocol (PPP)* originally emerged as an encapsulation protocol for transporting IP traffic over point-to-point links. PPP also established a standard for the assignment and management of IP addresses, asynchronous (start/stop) and bit-oriented synchronous encapsulation, network protocol multiplexing, link configuration, link quality testing, error detection, and option negotiation for such capabilities as network layer address negotiation and data-compression negotiation. PPP supports these functions by providing an extensible Link Control Protocol (LCP) and a family of Network Control Protocols (NCPs) to negotiate optional configuration parameters and facilities. In addition to IP, PPP supports other protocols, including Novell's Internetwork Packet Exchange (IPX) and DECnet.

PPP Components :

PPP provides a method for transmitting datagrams over serial point-to-point links. PPP contains three main components :

1. A method for encapsulating datagrams over serial links. PPP uses the High-Level Data Link Control (HDLC) protocol as a basis for encapsulating datagrams over point-to-point links.

2. An extensible LCP to establish, configure, and test the data link connection.

3. A family of NCPs for establishing and configuring different network layer protocols. PPP is designed to allow the simultaneous use of multiple network layer protocols.

General Operation :

To establish communications over a point-to-point link, the originating PPP first sends LCP frames to configure and (optionally) test the data link. After the link has been established and optional facilities have been negotiated as needed by the LCP, the originating PPP sends NCP frames to choose and configure one or more network layer protocols. When each of the chosen network layer protocols has been configured, packets from each network layer protocol can be sent over the link. The link will remain configured for communications until explicit LCP or NCP frames close the link, or until some external event occurs (for example, an inactivity timer expires or a user intervenes).

Physical Layer Requirements :

PPP is capable of operating across any DTE/DCE interface. Examples include EIA/TIA-232-C (formerly RS-232-C), EIA/TIA-422 (formerly RS-422), EIA/TIA-423 (formerly RS-423), and International Telecommunication Union Telecommunication Standardization Sector (ITU-T) (formerly CCITT) V.35. The only absolute requirement imposed by PPP is the provision of a duplex circuit, either dedicated or switched, that can operate in either an asynchronous or synchronous bit-serial mode, transparent to PPP link layer frames. PPP does not impose any restrictions regarding transmission rate other than those imposed by the particular DTE/DCE interface in use.

PPP Link Layer :

PPP uses the principles, terminology, and frame structure of the International Organization for Standardization (ISO) HDLC procedures (ISO 3309-1979), as modified by ISO 3309:1984/PDAD1 "Addendum 1: Start/Stop Transmission." ISO 3309-1979 specifies the HDLC frame structure for use in synchronous environments. ISO 3309 : 1984/PDAD1 specifies proposed modifications to ISO 3309-1979 to allow its use in asynchronous environments. The PPP control procedures use the definitions and control field encodings standardized in ISO 4335-1979 and ISO 4335-1979/Addendum 1-1979. The PPP frame format appears as shown in Fig. 2.8.

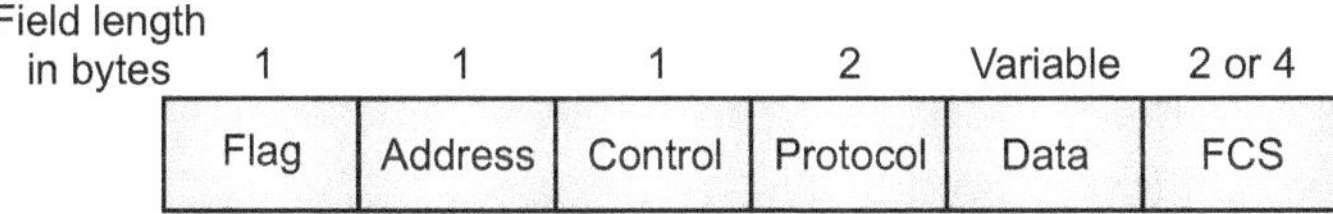

Fig. 2.8 : Six Fields make up the PPP Frame

The following descriptions summarize the PPP frame fields illustrated in Fig. 2.8.

- **Flag:** A single byte that indicates the beginning or end of a frame. The flag field consists of the binary sequence 01111110.

- **Address:** A single byte that contains the binary sequence 11111111, the standard broadcast address. PPP does not assign individual station addresses.

- **Control:** A single byte that contains the binary sequence 00000011, which calls for transmission of user data in an unsequenced frame. A connectionless link service similar to that of Logical Link Control (LLC) Type 1 is provided.

- **Protocol:** Two bytes that identify the protocol encapsulated in the information field of the frame. The most up-to-date values of the protocol field are specified in the most recent Assigned Numbers Request For Comments (RFC).

- **Data:** Zero or more bytes that contain the datagram for the protocol specified in the protocol field. The end of the information field is found by locating the closing flag sequence and allowing 2 bytes for the FCS field. The default maximum length of the information field is 1,500 bytes. By prior agreement, consenting PPP implementations can use other values for the maximum information field length.

- **Frame Check Sequence (FCS) :** Normally 16 bits (2 bytes). By prior agreement, consenting PPP implementations can use a 32-bit (4-byte) FCS for improved error detection.

The LCP can negotiate modifications to the standard PPP frame structure. Modified frames, however, always will be clearly distinguishable from standard frames.

PPP Link-Control Protocol :

- The PPP LCP provides a method of establishing, configuring, maintaining, and terminating the point-to-point connection. LCP goes through four distinct phases.

- First, link establishment and configuration negotiation occur. Before any network layer datagrams (for example, IP) can be exchanged, LCP first must open the connection and negotiate configuration parameters. This phase is complete when a configuration-acknowledgment frame has been both sent and received.

- This is followed by link quality determination. LCP allows an optional link quality determination phase following the link-establishment and configuration-negotiation phase. In this phase, the link is tested to determine whether the link quality is sufficient to bring up network layer protocols. This phase is optional. LCP can delay transmission of network layer protocol information until this phase is complete.

- At this point, network layer protocol configuration negotiation occurs. After LCP has finished the link quality determination phase, network layer protocols can be configured separately by the appropriate NCP and can be brought up and taken down at any time. If LCP closes the link, it informs the network layer protocols so that they can take appropriate action.

- Finally, link termination occurs. LCP can terminate the link at any time. This usually is done at the request of a user but can happen because of a physical event, such as the loss of carrier or the expiration of an idle-period timer.

- Three classes of LCP frames exist. Link-establishment frames are used to establish and configure a link. Link-termination frames are used to terminate a link, and link-maintenance frames are used to manage and debug a link.

- These frames are used to accomplish the work of each of the LCP phases.

2.6 MEDIA ACCESS CONTROL

There are two categories of networks based on the manner in which they provide interconnection.

1. **Switched Networks :** They provide interconnection by means of transmission lines, multiplexers and switches. Addressing is hierarchical and routing is used.

2. **Broadcast Network :** Information is broadcasted over a channel and is available to all users. Non-hierarchical addressing will identify correct user. Routing is not used.

Broadcast networks require what is called as medium access control protocol to co-ordinate the access of channels. A single medium is shared by number of users, hence these networks are called multiple access networks.

2.7 MULTIPLE ACCESS COMMUNICATIONS

When number of users share the same medium for transmission as shown in Fig. 2.9, all the stations sharing the medium can hear transmission from any given station.

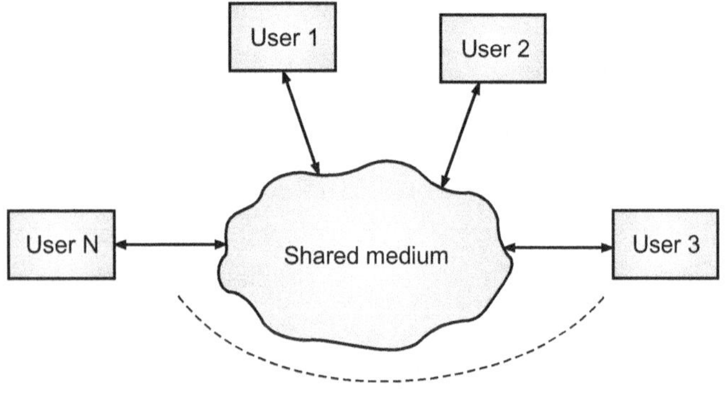

Fig. 2.9 : Multiple Access Communication

If two or more stations try transmitting simultaneously their transmission will collide.

There are two schemes for allocating channels for transmission to a particular user.

1. **Static Channel Allocation :**

- It is static and collision free sharing of medium. If there are N users, the bandwidth of channel is divided into N subchannels and each user uses them separately. It is called Frequency Division Multiplexing (FDM).

- FDM is efficient in case when all the channels have continuous traffic and inefficient when traffic is bursty.

- Another static channel allocation method is Time Division Multiplexing (TDM) where each user is allocated a fixed time slot for transmission.

2. **Dynamic Channel Allocation :**

- In situations where user traffic is bursty the channel is allocated on a per frame basis. This is called Medium Access Control Scheme.

- There are two approaches of implementing this scheme :
  - ➤ Scheduling
  - ➤ Random Access

While allocating the channels dynamically, following assumptions are made :

- There are N independent stations which generate frames for transmission randomly. Once a frame is generated station is blocked i.e. it does nothing until frame is successfully transmitted.

- There is only one channel available.

- If two frames are transmitted simultaneously, it results into collision.

- Frame transmission can begin at any instant. Frames are transmitted in a fixed time slot.

- Stations can sense if channel is in use. If it is sensed busy no station will attempt to use it.

Stations cannot serve channel before use. Only after transmission they decide whether transmission was successful or not.

There are number of protocols devised to handle the multiple access over a shared channel. These protocols can be classified as :

1. Random Access Protocols.

   e.g. Aloha, CSMA, CSMA/CD, CSMA/CA

2. Controlled Access Protocols

   e.g. Reservation, Polling, Token Passing.

3. Channelization Protocols

   e.g. FDMA, TDMA and CDMA.

2.8 RANDOM ACCESS

2.8.1 ALOHA

- It is a random access scheme for transmitting information for terminals sharing the same channel.

- It is simple in operation.

- Information is transmitted over the shared channel as soon as it becomes available.

- If there is collision because of more stations transmitting simultaneously, they will wait for random amount of time before transmitting the information again. It is called back-off.

- Fig. 2.10 shows frame transmission using ALOHA.

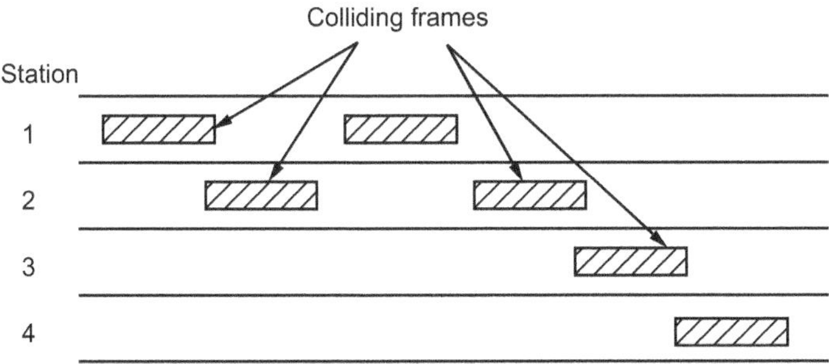

Fig. 2.10 : ALOHA System

Efficiency of ALOHA :

Let L be the length of frame (bits) (constant).

R be rate of transmission.

$$\therefore \quad \text{Frame time } = X = \frac{L}{R}$$

Let some frame arrive at time t_o and end at $t_o + X$.

This frame will collide if there is transmission from other stations between $t_o - X$ and $t_o + X$ as shown in Fig. 2.11.

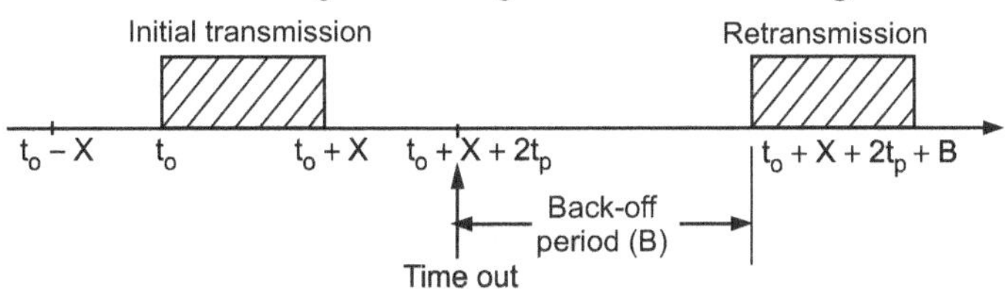

Fig. 2.11 : ALOHA System

$$\therefore \quad \text{Vulnerable time} = (t_o + X) - (t_o - X)$$

$$= 2X$$

Let G be the total arrival rate of the system in frames/X seconds. G is also throughput of the system. Let G be total arrival rate of the system in frames/X seconds. G is also called total load of the system.

With the assumption that the back-off spreads retransmissions such that new and repeated frame transmission are equally likely to occur, the number of frames transmitted in a time interval has Poisson distribution with average number of arrivals of 2G arrivals/2X seconds.

Hence, probability that k frames are generated during a given frame time are

$$P[k \text{ transmissions in 2X seconds}] = \frac{(2G)^k}{k!} e^{-2G}$$

Hence, throughput S is equal to total arrival rate G times probability of successful transmission.

$$\therefore \qquad S = P[\text{no collision}]$$

$$= P[0 \text{ transmissions in 2X seconds}]$$

$$= G \frac{(2G)^0}{0!} e^{-2G}$$

$$= Ge^{-2G}$$

The plot of S versus G is shown in Fig. 2.12.

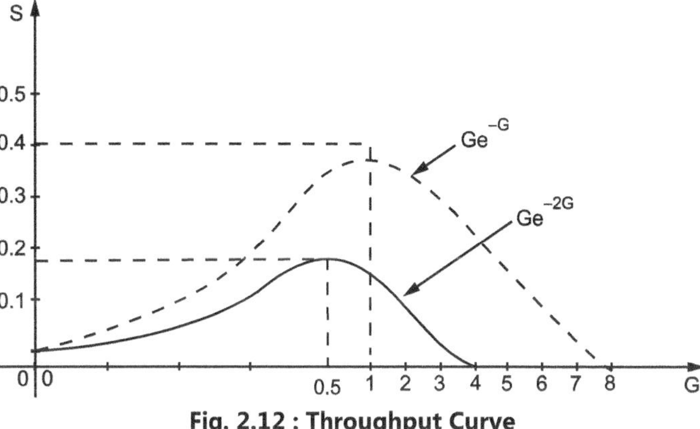

Fig. 2.12 : Throughput Curve

It can be seen that the maximum value of $S = \frac{1}{2e}$ at $G = 0.5$. That is system can achieve throughput of 18.4% only.

2.8.2 Slotted ALOHA

Performance of ALOHA can be improved by putting a restriction on time of transmission i.e. stations will transmit only at a fixed time (Synchronize fashion). Thus, reducing the probability of collisions.

All stations keep track of transmission time slots and are allowed to initiate transmission only at beginning of slot.

Vulnerable time i.e. time of collision reduces to $t_o - X$ to X i.e. X second as shown in Fig. 2.13.

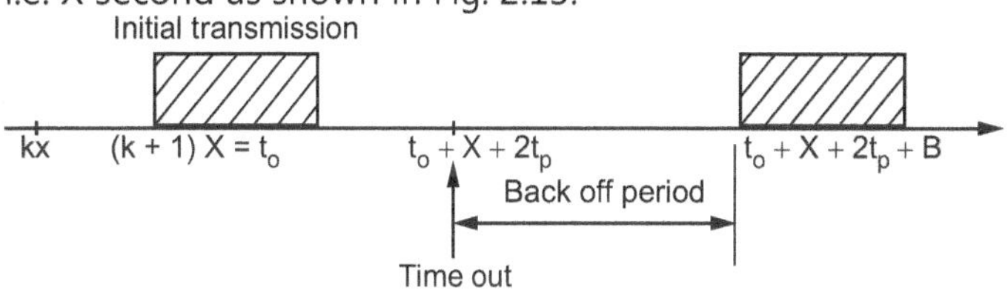

Fig. 2.13

There are G arrivals in X seconds, (G is total arrival rate)

$$\therefore \qquad S = G \times P \text{ [no collision]}$$

$$= G \times P \text{ [0 transmission in X seconds]}$$

$$= G \times \frac{(G)^0}{0!} e^{-G}$$

$$= Ge^{-G}$$

Above equation is plotted in Fig. 2.11. Slotted ALOHA has maximum throughput of $\frac{1}{e} = 36.8\%$ at G = 1.

2.8.3 Carrier Sense Multiple Access Protocols

Protocols in which stations listen for carriers and take suitable action are called Carrier Sense Protocols.

Following are some carrier sense protocols :

1. 1-Persistent CSMA :
- When a station has some data to send it listens to the channel.
- If channel is busy it waits until it becomes free or idle continuously sensing the channel.

- If channel is idle it transmits the frame.
- It is called 1-persistent because whenever channel is idle the station transmits with probability 1.

2. Non-Persistent CSMA :

- When station has some data to send, it listens to the channel and if channel is idle it sends data.
- If channel is busy, it waits until it becomes free/idle.
- But then it does not sense the channel continuously as in 1-persistent CSMA.
- It waits for random period and then again senses the channel.

3. p-Persistent CSMA :

- It applies to slotted channels.
- When a station is ready to transit data and channel is idle it transmits with probability p and decides not to transmit with probability q = 1 − p until next slot. If that slot is also idle it decides to transmit or defer with probability p and q. This process is repeated until either frame is transmitted or another station has started transmission.
- If the channel is busy it waits until next slot and repeats above step.

2.8.4 Carrier Sense Multiple Access with Collision Detection (CSMA/CD)

1. A Shared Medium :

- The Ethernet network may be used to provide shared access by a group of attached nodes to the physical medium which connects the nodes. These nodes are said to form a Collision Domain. All frames sent on the medium are physically received by all receivers, however the Medium Access Control (MAC) header contains a MAC destination address which ensure only the specified destination actually forwards the received frame (the other computers all discard the frames which are not addressed to them).

- Consider a LAN with four computers each with a Network Interface Card (NIC) connected by a common Ethernet cable.

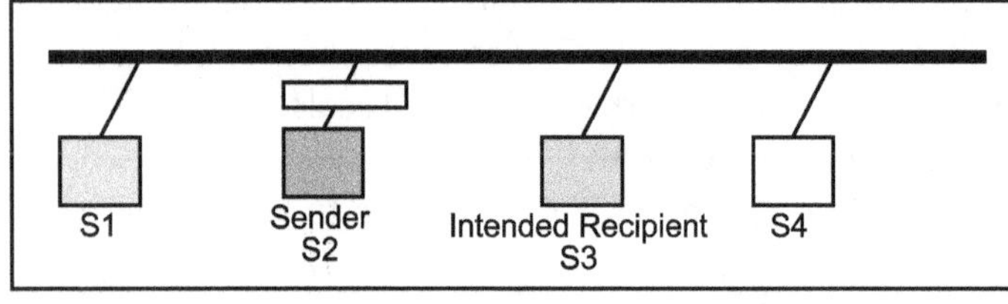

Fig. 2.14 (a)

- One computer (S_2) uses a NIC to send a frame to the shared medium, which has a destination address corresponding to the source address of the NIC in the S_3 computer.

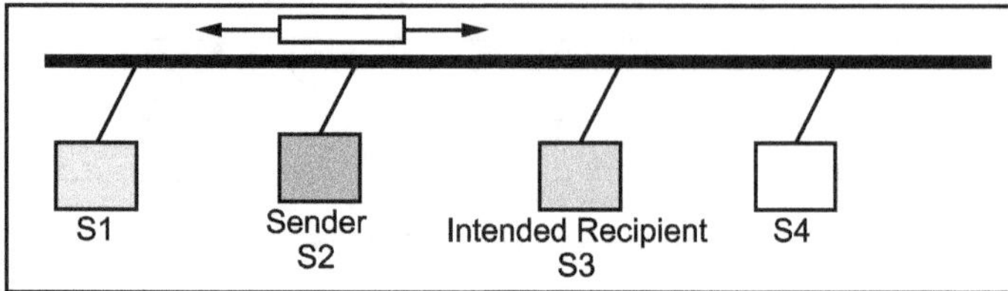

Fig. 2.14 (b)

- The cable propagates the signal in both directions, so that the signal (eventually) reaches the NICs in all four of the computers. Termination resistors at the ends of the cable absorb the frame energy, preventing reflection of the signal back along the cable.

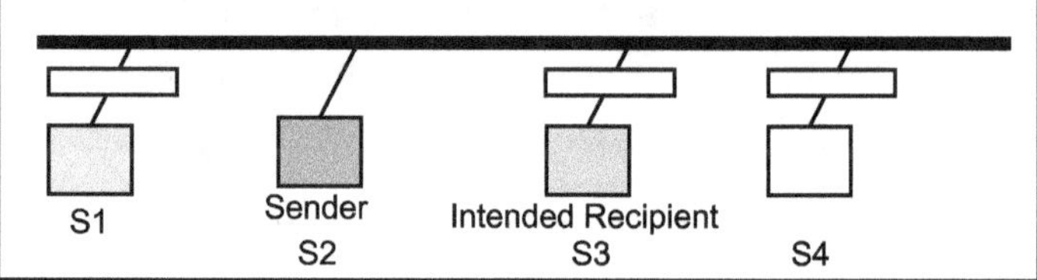

Fig. 2.14 (c)

- All the NICs receive the frame and each examines it to check its length and checksum. The header destination MAC address is next examined, to see if the frame should be accepted, and forwarded to the network-layer software in the computer.

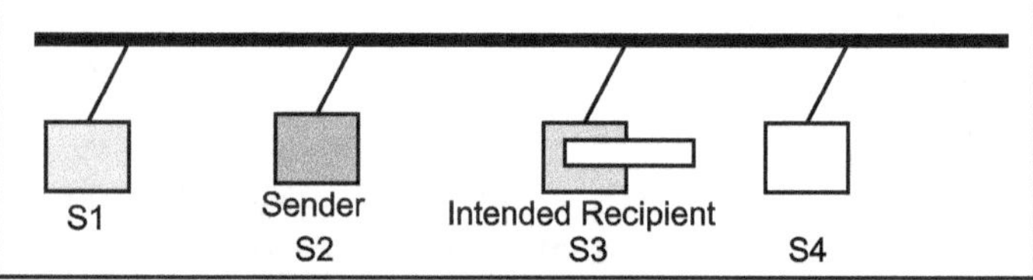

Fig. 2.14 (d)

- Only the NIC in the computer S_3 recognises the frame destination address as valid, and therefore this NIC alone forwards the contents of the frame to the network layer. The NICs in the other computers discard the unwanted frame.

- The shared cable allows any NIC to send whenever it wishes, but if two NICs happen to transmit at the same time, a collision will occur, resulting in the data being corrupted.

2. ALOHA and Collisions

- To control which NICs are allowed to transmit at any given time, a protocol is required. As seen earlier, the simplest protocol is known as ALOHA (this is actually an Hawaiian word, meaning "hello"). ALOHA allows any NIC to transmit at any time, but states that each NIC must add a checksum/CRC at the end of its transmission to allow the receiver(s) to identify whether the frame was correctly received.

- ALOHA is therefore a best effort service, and does not guarantee that the frame of data will actually reach the remote recipient without corruption. It therefore relies on ARQ protocols to retransmit any data which is corrupted. An ALOHA network only works well when the medium has a low utilization, since this leads to a low probability of the transmission colliding with that of another computer, and hence a reasonable chance that the data is not corrupted.

3. Carrier Sense Multiple Access (CSMA)

- Ethernet uses a refinement of ALOHA, known as Carrier Sense Multiple Access (CSMA), which improves performance when there is a higher medium utilization. When a NIC has data to transmit, the NIC *first* listens to the cable (using a transceiver) to see if a carrier (signal) is being transmitted by another node.

- This may be achieved by monitoring whether a current is flowing in the cable (each bit corresponds to 18-20 milliAmps (mA)). The individual bits are sent by encoding them with a 10 (or 100 MHz for Fast Ethernet) clock using Manchester encoding. Data is only sent when no carrier is observed (i.e. no current present) and the physical medium is therefore idle. Any NIC which does not need to transmit, listens to see if other NICs have started to transmit information to it.

- However, this alone is unable to prevent two NICs transmitting at the same time. If two NICs *simultaneously* try transmit, then both could see an idle physical medium (i.e. neither will see the other's carrier signal), and both will conclude that no other NIC is currently using the medium. In this case, both will then decide to transmit and a *collision* will occur.

- The collision will result in the corruption of the frame being sent, which will subsequently be discarded by the receiver since a corrupted Ethernet frame will (with a very high probability) not have a valid 32-bit MAC CRC at the end.

4. Collision Detection (CD)

- A second element to the Ethernet access protocol is used to detect when a collision occurs. When there is data waiting to be sent, each transmitting NIC also monitors its own transmission. If it observes a collision (excess current above what it is generating, i.e. > 24 mA for coaxial Ethernet), it stops transmission immediately and instead transmits a 32-bit jam sequence. The purpose of this sequence is to ensure

that any other node which may currently be receiving this frame will receive the jam signal in place of the correct 32-bit MAC CRC, this causes the other receivers to discard the frame due to a CRC error.

- To ensure that all NICs start to receive a frame before the transmitting NIC has finished sending it, Ethernet defines a minimum frame size (i.e. no frame may have less than 46 bytes of payload). The minimum frame size is related to the distance which the network spans, the type of media being used and the number of repeaters which the signal may have to pass through to reach the furthest part of the LAN. Together these define a value known as the *Ethernet Slot Time*, corresponding to 512 bit times at 10 Mbps.

- When two or more transmitting NICs each detect a corruption of their own data (i.e. a collision), each responds in the same way by transmitting the jam sequence. The following sequence depicts a collision :

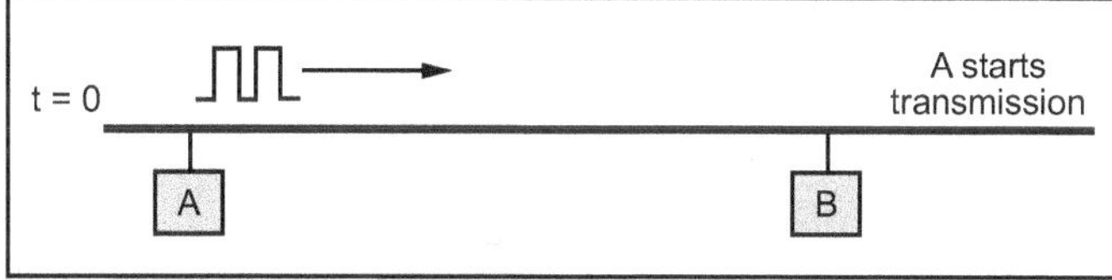

Fig. 2.15 (a)

At time t = 0, a frame is sent on the idle medium by NIC A.

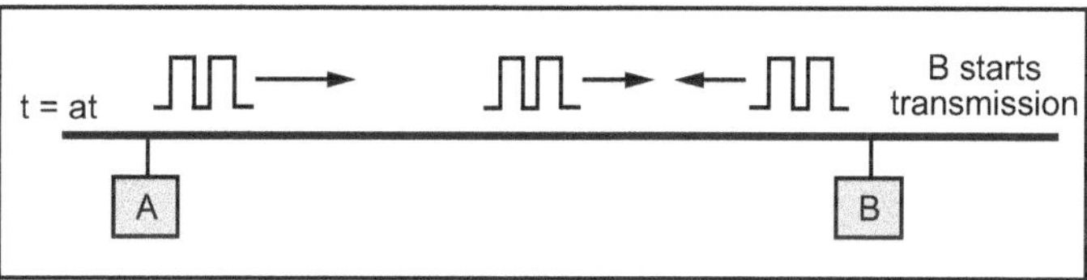

Fig. 2.15 (b)

A short time later, NIC B also transmits. (In this case, the medium, as observed by the NIC at B happens to be idle too.)

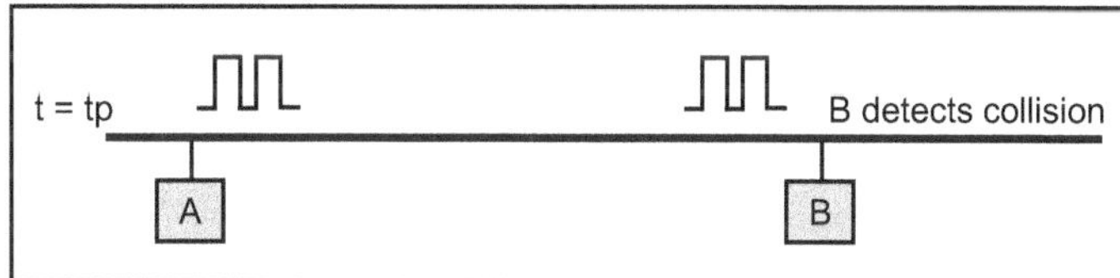

Fig. 2.15 (c)

After a period, equal to the propagation delay of the network, the NIC at B detects the other transmission from A, and is aware of a collision, but NIC A has not yet observed that NIC B was also transmitting. B continues to transmit, sending the Ethernet Jam sequence (32 bits).

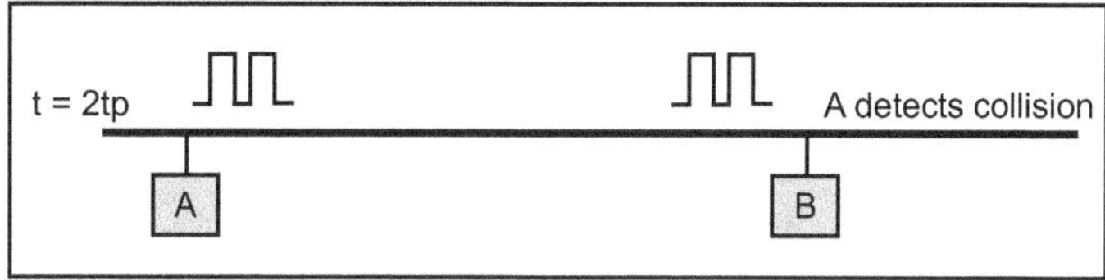

Fig. 2.15 (d)

After one complete round trip propagation time (twice the one way propagation delay), both NICs are aware of the collision. B will shortly cease transmission of the jam sequence, however A will continue to transmit a complete jam sequence. Finally the cable becomes idle.

5. Retransmission Back-Off

- An overview of the transmit procedure is shown below. The transmitter initializes the number of transmissions of the current frame (n) to zero, and starts listening to the cable (using the carrier sense logic (CS) - e.g. by observing the R_x signal at transceiver to see if any bits are being sent). If the cable is not idle, it waits (defers) until the cable is idle. It then waits for a small Inter-Frame Gap (IFG) (e.g. 2.6 microseconds) to allow to time for all receiving nodes to return to prepare themselves for the next transmission.

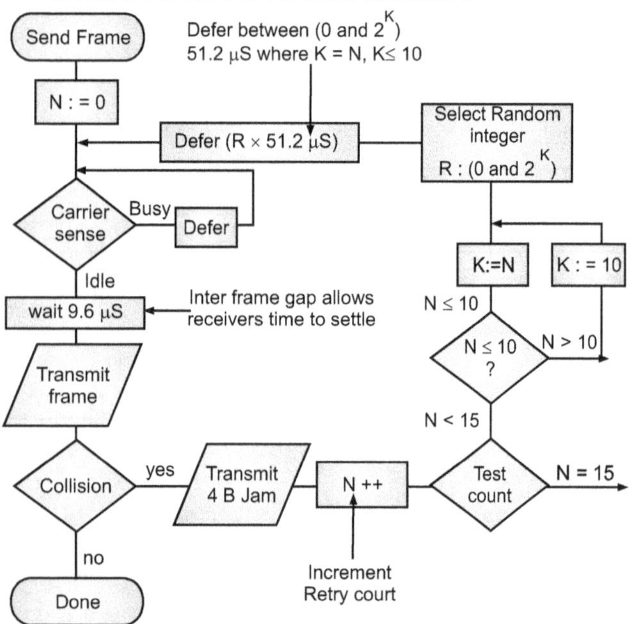

Fig. 2.16

- Transmission then starts with the preamble, followed by the frame data and finally the CRC-32. After this time, the transceiver Tx logic is turned-off and the transceiver returns to passively monitoring the cable for other transmissions.

- During this process, a transmitter must also continuously monitor the collision detection logic (CD) in the transceiver to detect if a collision occurs. If it does, the transmitter aborts the transmission (stops sending bits) within a few bit periods, and starts the collision procedure, by sending a jam signal to the transceiver Tx logic. It then calculates a retransmission time.

- If all NICs attempted to retransmit immediately following a collision, then this would certainly result in another collision. Therefore a procedure is required to ensure that there is only a low probability of simultaneous retransmission. The scheme adopted by Ethernet uses a random back-off period, where each node selects a random number, multiplies this by the slot time (minimum frame period, 51.2 µs) and waits for this random period before attempting retransmission. The small Inter-Frame Gap (IFG) (e.g., 2.6 µs) is also added.

- On a busy network, a retransmission may still collide with another retransmission (or possibly new frames being sent for the first time by another NIC). The protocol therefore counts the number of retransmission attempts (using a variable N in Fig. 2.17) and attempts to retransmit the same frame up to 15 times.

For each retransmission, the transmitter constructs a set of numbers :

{0, 1, 2, 3, 4, 5, ... L} where L is ([2 to the power (K)]−1) and where K = N; K<= 10;

A random value R is picked from this set, and the transmitter waits (defers) for a period.

R × (slot time) i.e. R × 51.2 Micro Seconds

- For example, after two collisions, N = 2, therefore K = 2, and the set is {0, 1, 2, 3} giving a one in four chance of collision. This corresponds to a wait selected from {0, 51.2, 102.4, 153.6} µs.

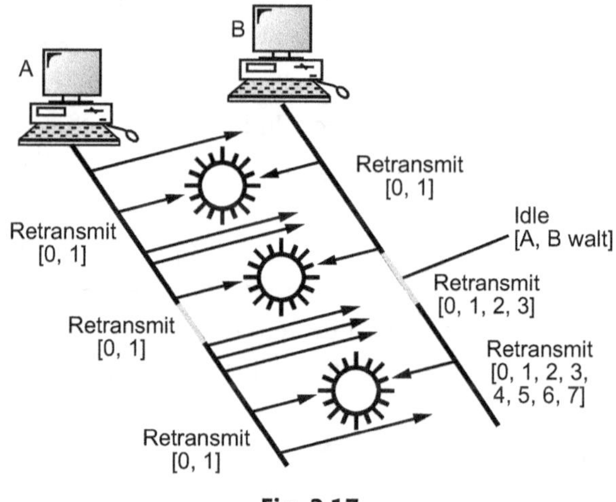

Fig. 2.17

After 3 collisions, N = 3, and the set is {0, 1, 2, 3, 4, 5, 6, 7}, that is a one in eight chance of collision.

But after 4 collisions, N = 4, the set becomes {0, 1, 2, 3, 4, 5, 6, 7, 8, 9, 10, 11, 12, 13, 14, 15}, that is a one in 16 chance of collision.

- The scaling is performed by multiplication and is known as exponential back-off. This is what lets CSMA/CD scale to large numbers of NICs - even when collisions may occur. The first ten times, the back-off waiting time for the transmitter suffering collision is scaled to a larger value. The algorithm includes a threshold of 1024. The reasoning is that the more attempts that are required, the more greater the number of NICs which are trying to send at the same time, and therefore longer the period which needs to be deferred. Since a set of numbers {0, 1, ..., 1023} is a large set of numbers, there is very little advantage from further increasing the set size.

- Each transmitter also limits the maximum number of retransmissions of a single frame to 16 attempts (N=15). After this number of attempts, the transmitter gives up transmission and discards the frame, logging an error. In practice, a network that is not overloaded should never discard frames in this way.

6. Late Collisions

- In a proper functioning Ethernet network, a NIC may experience collision within the first slot time after it starts transmission. This is the reason why an Ethernet NIC monitors the CD signal during this time and use CSMA/CD. A faulty CD circuit, or misbehaving NIC or transceiver may lead to a late collision (i.e. after one slot time). Most Ethernet NICs therefore continue to monitor the CD signal during the entire transmission. If they observe a late collision, they will normally inform the sender of the error condition.

7. Performance of CSMA/CD

- It is simple to calculate the performance of a CSMA/CD network where only one node attempts to transmit at any time. In this case, the NIC may saturate the medium and near about 100% utilization of the link may be achieved, providing almost 10 Mbps of throughput on a 10 Mbps LAN.

- However, when two or more NICs attempt to transmit at the same time, the performance of Ethernet is less predictable. The fall in utilization and throughput occurs because some bandwidth is wasted by collisions and back-off delays. In practice, a busy shared 10 Mbps Ethernet network will typically supply 2-4 Mbps of throughput to the NICs connected to it.

- As the level of utilization of the network increases, particularly if there are many NICs competing to share the bandwidth, an overload condition may occur. In this case, the throughput of Ethernet LANs reduces very considerably, and much of the capacity is wasted by the CSMA/CD algorithm, and very little is available for sending useful data. This is the reason why a shared Ethernet LAN should not connect more than 1024 computers. Many engineers use a threshold of 40% utilization to determine if a LAN is overloaded. A LAN with a higher utilization will observe a high collision rate, and likely a very variable transmission time (due to back off). Separating the LAN into two or more collision domains using bridges or switches would likely provide a significant benefit (assuming appropriate positioning of the bridges or switches).

- Shared networks may also be constructed using Fast Ethernet, operating at 100 Mbps. Since Fast Ethernet always uses fibre or twisted pair, a hub or switch is always required.

8. Ethernet Capture

- A drawback of sharing a medium using CSMA/CD, is that the sharing is not necessarily fair. When each computer connected to the LAN has little data to send, the network exhibits almost equal access time for each NIC. However, if one NIC starts sending an excessive number of frames, it may dominate the LAN. Such conditions may occur, for instance, when one NIC in a LAN acts as a source of high quality packetized video. The effect is known as "Ethernet Capture".

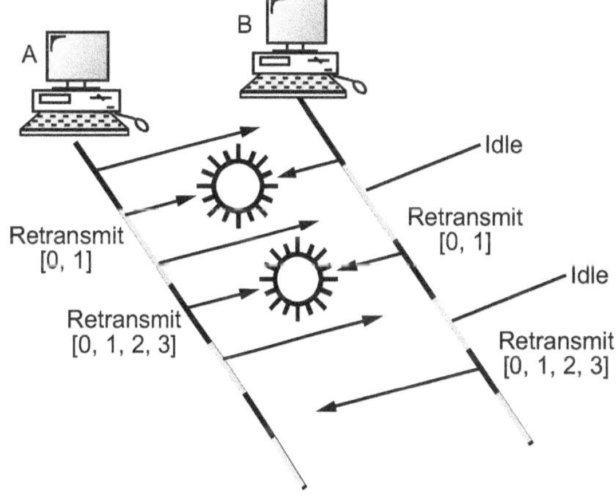

Fig. 2.18

(i) Ethernet Capture by Node A.

- Fig. 2.18 illustrates Ethernet Capture. Computer A dominates computer B. Originally both computers have data to transmit. A transmits first. A and B then both simultaneously try to transmit. B picks a larger retransmission interval than A and defers. A sends, then sends again.

- There is a short pause, and then both A and B attempt to resume transmission. A and B both back-off, however, since B was already in back-off (it failed to retransmit), it chooses from a larger range of back-off times (using the exponential back-off algorithm). A is therefore more likely to succeed, which it does in the example. A and B both attempt to send, however, since this fails in this case, B further increases its back-off and is now unable to fairly compete with A.

- Ethernet Capture may also arise when many sources compete with one source which has much more data to send. Under these situations some nodes may be "locked out" of using the medium for a period of time. The use of higher speed transmission (e.g. 100 Mbps) significantly reduces the probability of Capture, and the use of full duplex cabling eliminates the effect.

- Ethernet LANs may be implemented using a variety of media (not just the coaxial cable described above). The types of media segments supported by Ethernet are :

  ➢ 10 Base 5 Low loss coaxial cable (also known as "thick" Ethernet).

  ➢ 10 Base 2 Low cost coaxial cable (also known as "thin" Ethernet).

  ➢ 10 Base T Low cost twisted pair copper cable (also known as Unshielded Twisted Pair (UTP)).

  ➢ 10 Base F Fibre optic cable.

The network design rules for using these types of media are summarized below :

Table 2.1

Segment Type	Max Number of Systems per Cable Segment	Max Distance of a Cable Segment
10 Base 5 (Thick Coax)	100	500 m
10 Base 2 (Thin Coax)	30	185 m
10 Base T (Twisted Pair)	2	100 m
10 Base F (Fibre Optic)	2	2000 m

(ii) Network Design Rules for Different Types of Cable

- There is also a version of Ethernet which operates using twisted pair cabling or fibre optic links at 100 Mbps and at 1 Gbps. 100 Mbps networks may operate full duplex (using a Fast Ethernet Switch) or half duplex (using a Fast Ethernet Hub). 1 Gbps networks usually operate between a pair of Ethernet Switches. Many LANs combine the various speeds of operation using dual-speed switches which allow the same switch to connect some ports to one speed of network, and other ports at another speed. The higher speed ports are usually used to connect switches to one another.

2.8.5 Carrier Sense Multiple Access with Collision Avoidance (CSMA/CA)

- In case of CSMA/CD, the transmitting station detects collision based on signal energy level. When there is no collision signal energy level will be minimum because only one station is transmitting. When there is collision energy level will be more.

- In case of wireless transmission it is not possible to detect collision based on signal energy level. It is because there is more loss. Hence, the solution is avoid collision as it cannot be detected.

- There are three strategies used to avoid collisions :

  1. Interframe Space (IFS)

  2. Contention Window

  3. Acknowledgements.

1. Interframe Space (IFS) :

When a station wants to transmit and finds the channel is idle it waits for a period of time called interframe space (IFS). After the IFS time for a station is over and it finds that the channel is idle it can send, but not immediately it still waits for contention time. IFS can be variable for each station.

2. Contention Window :

A station that is ready to send waits for a time equal to random number of slots, the algorithm used to generate the random number of slots is binary back-off.

3. Acknowledgement :

If there is collision even with implementation of above, we can have the mechanism of acknowledgement where the transmitting station waits for acknowledgement from receiving station for a particular time. If time out occurs it can retransmit. Fig. 2.19 depicts the three strategies.

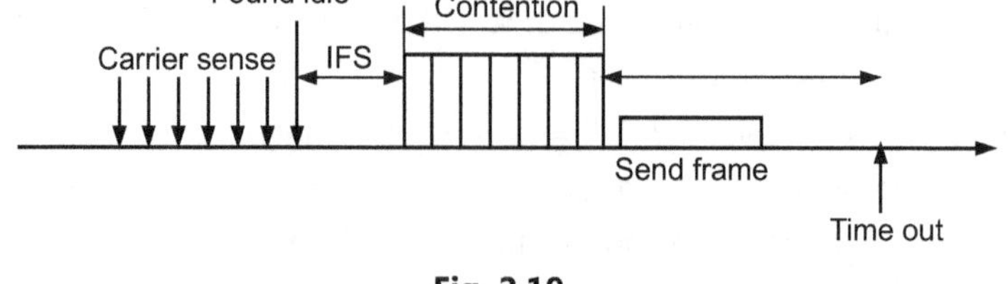

Fig. 2.19

The procedure is shown in flowchart below.

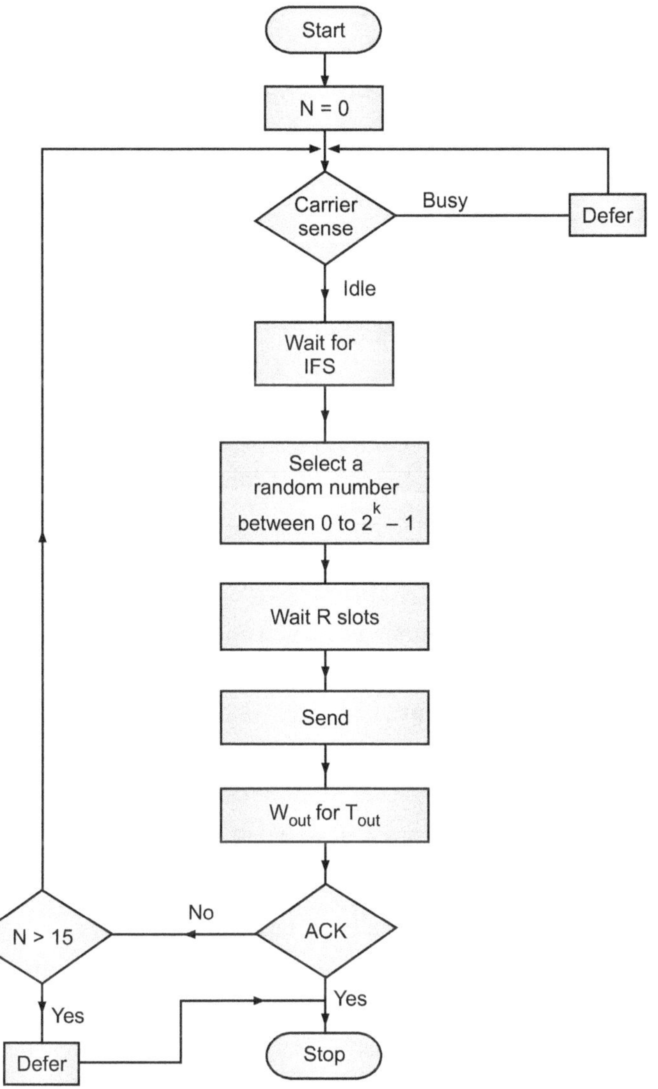

2.9 CONTROLLED ACCESS

- In controlled access, the station consults one another to find which station has the right to send. We discuss three popular methods :

- **Reservation :** In reservation method, a station needs to make reservation before sending data. Time is divided into intervals. In each interval, a reservation frame preceds the data frames sent in that interval.

- If there are N stations in the system, there are exactly N reservation minislots in reservation frame. Each minislot belongs to a station. When a station needs to send data frame, it makes reservation in its own minislot. The stations that have made reservations can send their data frames after the reservation frame.

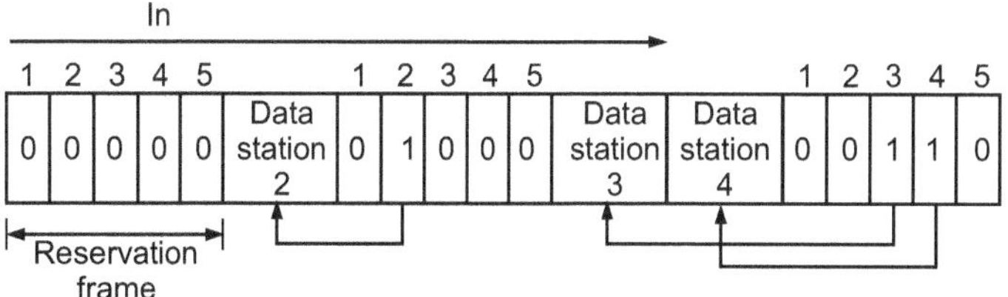

Fig. 2.20 : Reservation Access

- In Fig. 2.20, it shows a situation with five stations and five minislot reservation frames. In the first interval, only stations 1, 3 and 4 have made reservations. In the second interval, only station 1 has made a reservation.

- **Polling :** Polling works with topologies in which one device is designated as a primary station and the other devices are secondary stations. All data exchanges must be made through primary device even when the ultimate destination is a secondary device. The primary device controls the link. The secondary devices follow its instructions. The primary device is always the initiator of session.

- If the primary wants to receive data, it asks the secondaries. If they have anything to send, this is called poll function.

- If the primary wants to send data, it tells secondary to get ready to receive. This is called select function.

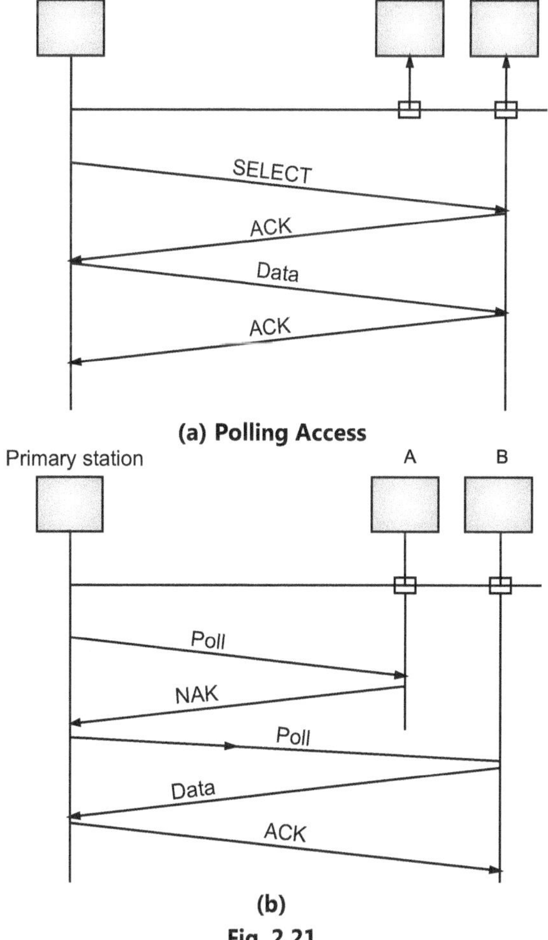

Fig. 2.21

- **Poll :** The poll function is used for primary device to receive transmissions from the secondary devices. When the primary is ready for that, it must ask (poll) each device in turn if it has anything to send. When the first secondary is approached, it responds either with NAK frame if it has nothing to send or with data (in the form of data frame) if it does. If the response is negative (a NAK frame) then the primary polls the next secondary in the same manner until it finds one with data to send. When the response is positive (a data frame) the primary reads the frame and returns an acknowledgement (ACK frame) verifying its receipt.

- **Select :** The select function is used whenever the primary device has something to send. The primary device controls the link. If the primary is neither sending nor receiving data, it knows the link is available.

- If it has something to send, the primary device sends it. But however, it does not know, whether the target device is prepared to receive. So the primary must alert the secondary to the upcoming transmission and wait for an acknowledgement of the secondary ready status. Before sending data the primary creates and transmits a select (SEL) frame, one field of which includes the address of intended secondary.

- **Token Passing :** In the token passing method, the stations in a network are organized in a logical ring. For each station, there is a predecessor and successor. The predecessor is the station which is logically before the station in the ring, the successor is the station which is after the station in the ring. The current station is one that is accessing to the channel now. The right to this access has been passed from the predecessor to the current station. The right will be passed to the successor when the current station has no more data to send.

- A special packet called 'token' is used to pass the right to access from one station to another. The token keeps on circulating in the ring whichever station has the token it has right to access the channel. Whenever a station has data to send it waits for token. Whenever token is received it will send data. Whenever it has finished sending data, the token is passed to next station.

- Token management is an important issue in this type of network. The network should ensure that the token is not lost because of some problem in the network. e.g. faulty station. Sometimes the data transfer may

require priority assignments. The token can be used by stations as per their priority.

Logical Ring :

- This standard evolved from the needs of companies like General Motors implementing terminals for factory automation.

- 802.3 was not suitable for them as a station in ethernet has to wait for a long time to send a frame in worst case. 802.3 frames do not have priorities which makes important frames waiting for unimportant frames.

- A token bus system is a medium access control technique for bus/tree stations form a logical ring around which a token is passed. A station receiving the token may transmit data and then must pass the token on to next station in the ring.

- Though it is similar to token ring it does not implement the ring physically as it has drawback of getting entire network down in case one terminal is down.

- If there are n stations in the token bus network and T_P seconds are required to send a frame, it will take not more than nT_P seconds to get a turn for a station.

- A token bus network is shown in Fig. 2.22.

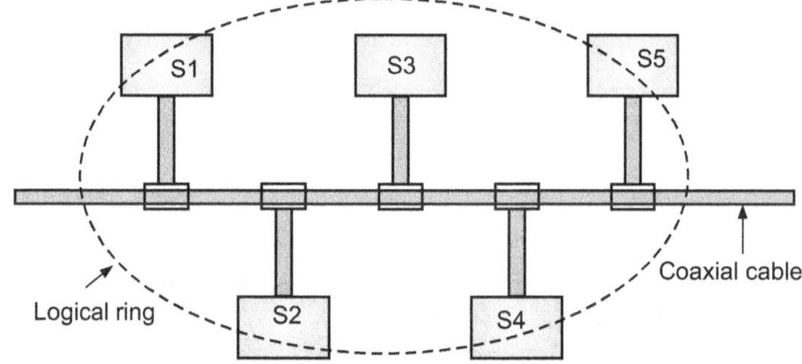

Fig. 2.22 : Token Bus

- The stations can be configured so that they can be part of the logical ring or may opt out of the ring.

- Initially highest number station gets the token. Then it passes the token to its neighbour (right or left).

- Token thus moves around the ring with token holder permitted to send the frame.

- If station has no data it must pass the token to next station.

- A 85 ohm broadband cable is used at physical layer.

2.10 MULTIPLE ACCESS TECHNIQUES

A transponder channel aboard may be fully loaded by single transmission from an earth station in Satellite communications. This is referred to as single access mode of operation. It is also possible for a transponder to be loaded by a number of carriers. These may originate from a number of earth stations geographically separate and each earth station may transmit one or more of the carriers. This mode of operation is termed multiple access. The most commonly used methods of multiple access are FDMA, TDMA and CDMA.

FDMA :

- Frequency Division Multiple Access (FDMA) is the most common analog system.

- It is a technique whereby spectrum is divided up into frequencies and then assigned to users.

- With FDMA, only one subscriber at any given time is assigned to a channel.

- The channel therefore is closed to other conversations until the initial call is finished, or until it is handed-off to a different channel.

- A "full-duplex" FDMA transmission requires two channels, one for transmitting and the other for receiving. FDMA has been used for first generation analog systems.

TDMA :

- Time Division Multiple Access (TDMA) improves spectrum capacity by splitting each frequency into time slots.

- TDMA allows each user to access the entire radio frequency channel for the short period of a call.

- Other users share this same frequency channel at different time slots.

- The base station continually switches from user to user on the channel.

- TDMA is the dominant technology for the second generation mobile cellular networks.

CDMA :

- Code Division Multiple Access is based on "spread" spectrum technology.

- Since it is suitable for encrypted transmissions, it has long been used for military purposes.

- CDMA increases spectrum capacity by allowing all users to occupy all channels at the same time.

- Transmissions are spread over the whole radio band, and each voice or data call are assigned a unique code to differentiate from the other calls carried over the same spectrum.

- CDMA allows for a soft hand-off, which means that terminals can communicate with several base stations at the same time.

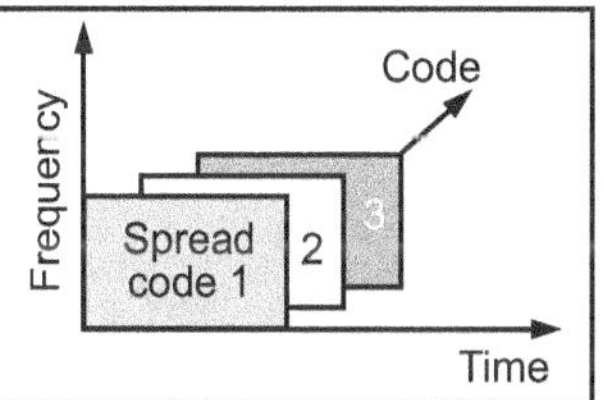

Fig. 2.23 : FDMA-TDMA-CDMA graphical representation

In addition to above multiple access techniques, we can have

(1) Fixed Assigned Multiple Access (FAMA) mode using either FDMA, TDMA, CDMA. In this mode, the format does not change even if traffic load changes.

(2) Demand Assigned Multiple Access (DAMA) mode using either FDMA, TDMA, CDMA. In this mode, the formats are changed depending on traffic demand. It is more efficient but costlier to implement and maintain.

Advantages of Digital Technology

- All multiple access techniques depend on the adoption of digital technology. Digital technology is now the standard for the public telephone system where all analog calls are converted to digital form for transmission over the backbone. Digital transmission has a number of advantages over analog transmission :

  ➢ It economizes on bandwidth.

  ➢ It allows easy integration with personal communication systems (PCS) devices.

  ➢ It maintains superior quality of voice transmission over long distances.

  ➢ It is difficult to decode.

  ➢ It can use lower average transmitter power.

  ➢ It enables smaller and less expensive individual receivers and transmitters.

  ➢ It offers voice privacy.

How FDMA Works ?

- TDMA is basically analog's FDMA with a time-sharing component built into the system.

- FDMA allocates a single channel to one user at a time (see Fig. 2.24).

- If the transmission path deteriorates, the controller switches the system to another channel.

- Although technically simple to implement, FDMA is wasteful of bandwidth : the channel is assigned to a single conversation whether or not somebody is speaking. Moreover, it cannot handle alternate forms of data, only voice transmissions.

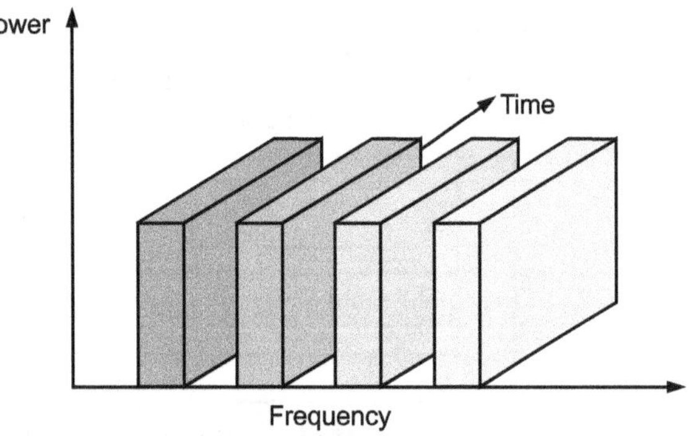

Fig. 2.24 : FDMA

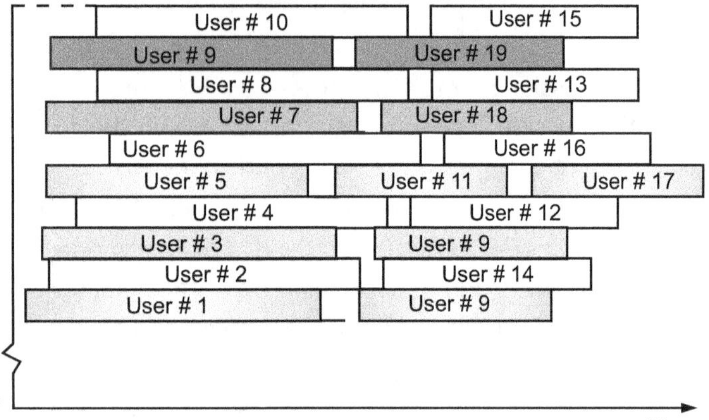

Fig. 2.25 : Schematic allocation of subscriber channels within an assigned frequency band (range)

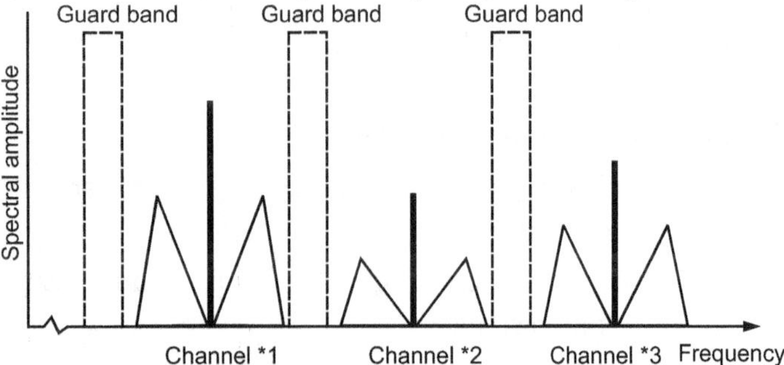

Fig. 2.26 : Schematic frequency spectrum of several subscriber channels

- Analog transmission is considered an "older" cellular phone technology.

- Analog technology was built in the early 1980's. Analog allows a cellular phone to transmit signals by sending voice, video, and data that are always changing, and so does the network systems.

- Analog is considered an older method of modulating voice or data information radio signals.

- Analog transmissions use FDMA technology. FDMA stands for "Frequency Division Multiple Access". FDMA is used exclusively for analog cellular systems, even though in theory FDMA can also be used with digital.

- Essentially, FDMA splits the allocated spectrum into many channels. In current analog cell systems, each channel is 30 kHz.

- When a FDMA cell phone establishes a call, it reserves the frequency channel for the entire duration of the call.

- The voice data is modulated into this channel's frequency band (using frequency modulation) and sent over the airwaves.

- At the receiver, the information is recovered using a band-pass filter. The phone then uses a common digital control channel to acquire channels.

- FDMA analog transmissions are the least efficient cellular networks since each analog channel can only be used one user at a time. Analog channels don't take full advantage of bandwidth.

- Not only are these FDMA channels larger than necessary given modern digital voice compression, but they are also wasted whenever there is silence during a cell phone conversation.

- Analog signals are especially susceptible to noise and the extra noise cannot get filtered out. Given the nature of the signal, analog cell phones must use higher power (between 1 and 3 watts) to get acceptable call quality.

- Given these analog features, it is easy to see why FDMA is being replaced by newer digital networks such as TDMA and CDMA.

Advantages of FDMA :

- If channel is not in use, it sits idle.

- Channel bandwidth is relatively narrow (30 kHz).

- Simple algorithmically, and from a hardware standpoint.

- Fairly efficient when the number of stations is small and the traffic is uniformly constant.

- Capacity increase can be obtained by reducing the information bit rate and using efficient digital code.

- No need for network timing.

- No restriction regarding the type of baseband or type of modulation.

Disadvantages of FDMA :

- The presence of guard bands.
- Requires right RF filtering to minimize adjacent channel interference.
- Maximum bit rate per channel is fixed.
- Small inhibiting flexibility in bit rate capability.
- Does not differ significantly from analog system.

How TDMA Works ?

- TDMA relies upon the fact that the audio signal has been digitized; that is, divided into a number of milliseconds-long packets.
- It allocates a single frequency channel for a short time and then moves to another channel.
- The digital samples from a single transmitter occupy different time slots in several bands at the same time as shown in Fig. 2.27.

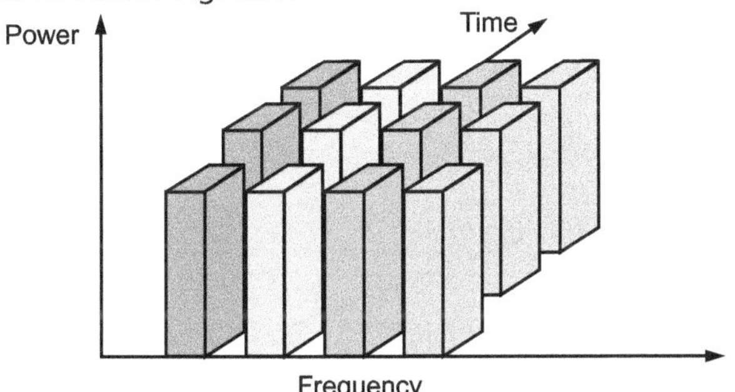

Fig. 2.27 : TDMA

- The access technique used in TDMA has three users sharing a 30 kHz carrier frequency.
- TDMA is also the access technique used in the European digital standard, GSM, and the Japanese digital standard, Personal Digital Cellular (PDC).
- The reason for choosing TDMA for all these standards was that it enables some vital features for system operation in an advanced cellular environment.
- Today, TDMA is an available, well-proven technique in commercial operation in many systems.
- To illustrate the process, consider the following situation. Fig. 2.28 shows four different, simultaneous conversations occurring.

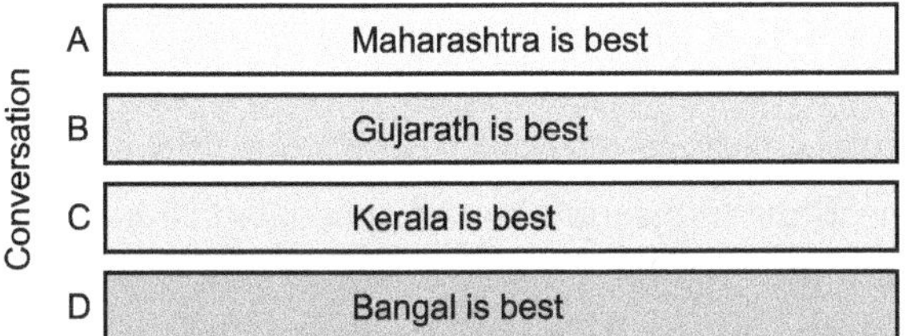

Fig. 2.28 : Four Conversations-Four Channels

- A single channel can carry all four conversations if each conversation is divided into relatively short fragments, is assigned a time slot, and is transmitted in synchronized timed bursts as shown in Fig. 2.29.
- After the conversation in time-slot four is transmitted, the process is repeated.

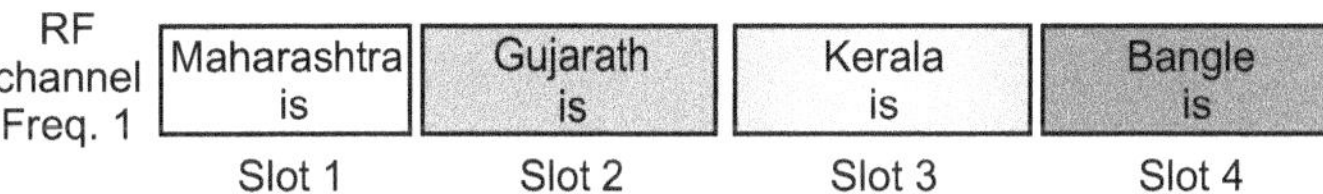

Fig. 2.29 : Four Conversations-One Channel

- Effectively, the implementations of TDMA immediately tripled the capacity of cellular frequencies by dividing a 30 kHz channel into three time slots, enabling three different users to occupy it at the same time.
- Currently, systems are in place that allow six times capacity. In the future, with the utilization of hierarchical cells, intelligent antennas, and adaptive channel allocation, the capacity should approach 40 times analog capacity.

Advanced TDMA:

- TDMA substantially improved upon the efficiency of analog cellular.
- However, like FDMA, it had the weakness that it wasted bandwidth : the time slot was allocated to a specific conversation whether or not anyone was speaking at that moment.
- Enhanced version of TDMA Extended Time Division Multiple Access (ETDMA) attempts to correct this problem. Instead of waiting to determine whether a subscriber is transmitting, ETDMA assigns subscribers dynamically.
- ETDMA sends data through those pauses which normal speech contains.
- When subscribers have something to transmit, they put one bit in the buffer queue.
- The system scans the buffer, notices that the user has something to transmit, and allocates bandwidth accordingly.
- If a subscriber has nothing to transmit, the queue simply goes to the next subscriber. So, instead of being arbitrarily assigned, time is allocated according to need.
- If partners in a phone conversation do not speak over one another, this technique can almost double the spectral efficiency of TDMA, making it almost 10 times as efficient as analog transmission.

Benefits of TDMA

- In addition to increasing the efficiency of transmission, TDMA offers a number of other advantages over standard cellular technologies.

- First and foremost, it can be easily adapted to the transmission of data as well as voice communication.

- TDMA offers the ability to carry data rates of 64 kbps to 120 Mbps (expandable in multiples of 64 kbps).

- This enables operators to offer personal communication-like services including fax, voice band data, and short message services (SMSs) as well as bandwidth-intensive applications such as multimedia and videoconferencing.

- Unlike spread-spectrum techniques which can suffer from interference among the users all of whom are on the same frequency band and transmitting at the same time, TDMA's technology, which separates users in time, ensures that they will not experience interference from other simultaneous transmissions.

- TDMA also provides the user with extended battery life and talk time since the mobile is only transmitting a portion of the time (from 1/3 to 1/10) of the time during conversations.

- TDMA installations offer substantial savings in base-station equipment, space and maintenance, an important factor as cell sizes grow ever smaller.

- TDMA is the most cost-effective technology for upgrading a current analog system to digital.

- TDMA is the only technology that offers an efficient utilization of hierarchical cell structures (HCSs) offering pico, micro, and macrocells. HCSs allow coverage for the system to be tailored to support specific traffic and service needs.

- Because of its inherent compatibility with FDMA analog systems, TDMA allows service compatibility with the use of dual-mode handsets.

Advantages of TDMA :

- Flexible bit rate.

- No frequency guard band required.

- No need for precise narrowband filters.

- Easy for mobile or base stations to initiate and execute hands off.

- Extended battery life.

- TDMA installations offer savings in base station equipment, space and maintenance.

- The most cost-effective technology for upgrading a current analog system to digital.

Drawbacks of TDMA :

- One of the disadvantages of TDMA is that each user has a predefined time slot. However, users roaming from one cell to another are not allotted a time slot.

- Thus, if all the time slots in the next cell are already occupied, a call might well be disconnected. Likewise, if all the time slots in the cell in which a user happens to be in are already occupied, a user will not receive a dial tone.

- Another problem with TDMA is that it is subjected to multipath distortion. A signal coming from a tower to a handset might come from any one of several directions.

- It might have bounced off several different buildings before arriving (see Fig. 2.30) which can cause interference.

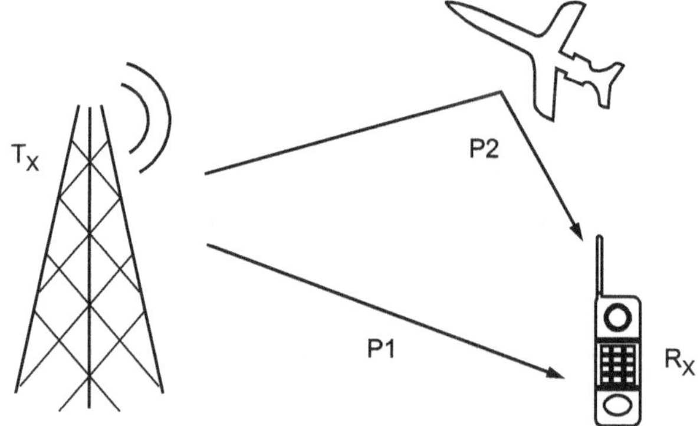

Fig. 2.30 : Multipath Interference

- One way of getting around this interference is to put a time limit on the system.

- The system will be designed to receive, treat, and process a signal within a certain time limit. After the time limit has expired, the system ignores signals.

- The sensitivity of the system depends on how far it processes the multipath frequencies. Even at thousandths of seconds, these multipath signals cause problems.

Disadvantages of TDMA :

- Cross-link transmissions must occur one cross-link at a time.

- Time synchronization needed between all distributed users.

- Propagation delay corrections must be applied when cross-link signal path lengths vary in order to avoid signal collisions.

- The greater the number of spacecraft, the longer the duty interval for cross-link transmissions by a given spacecraft resulting in lowering the overall data throughput for the distribution.

- Changing the user distances of separation requires dynamic assessments of time slot allocations to compensate for variable signal delays.

How CDMA Works ?

- CDMA is a digital wireless technology.

- It is a general type of technology, implemented in many specific technologies.

- But the term "CDMA" is also commonly used to refer to one specific implementation : IS-95 - a mobile-phone technology that competes with technologies such as GSM. CDMA is a "spread spectrum" technology, which means that it spreads the information contained in a particular signal of interest over a much greater bandwidth than the original signal.

- *Code division multiple access,* CDMA, is a spread spectrum system in which two or more spread spectrum signals communicate simultaneously, each operating over the same frequency band.

- In a CDMA system, each user is given a unique sequence (pseudo-random code).

- This sequence identifies the user. For example, if user-A has sequence-a, and user-B has sequence-b, a receiver wanting to listen to user-A would use sequence-a to decode the wanted intelligence. It would then receive all the energy being transmitted by user-A and disregard the power transmitted by user-B. Fig. 2.31 shows CDMA in use in a cellular system.

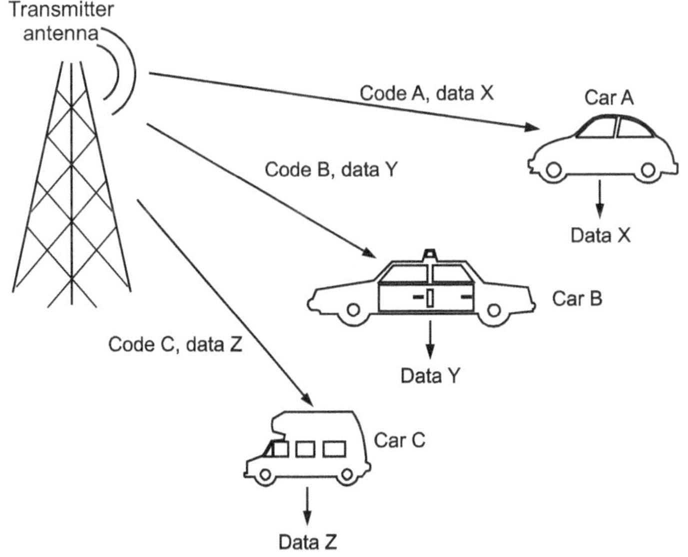

Fig. 2.31 : Spreading at Transmitter and De-spreading at Receivers

Benefits of CDMA :

- It has many advantages over TDMA.

- CDMA is less prone to deep multipath fading caused by transmissions arriving at the receiver that have followed different paths. i.e. one signal direct, and another reflecting off a large object.

- In fact, one approach in common use with CDMA systems, a rake receiver, takes advantage of multipath, normally a major source of interference and signal degradation in other systems. In a rake receiver, each receiver coherently combines the three strongest multipath signals to provide an enhanced signal with better voice quality.

- Rake receiver : A rake receiver is a radio receiver designed to counter the effects of multipath fading. It does this by using several "sub-receivers" each delayed slightly in order to tune in to the individual multipath components. Each component is decoded independently, but at a later stage combined in order to make the most use of the different transmission characteristics of each transmission path. This could very well result in higher SNR in a multipath environment than in a "clean" environment.

- CDMA can operate with much lower transmit powers leading to smaller handsets and smaller batteries and longer life. For example, some field trials in the different countries claimed to demonstrate that the average transmits power of CDMA phones averaged 6 mW, or roughly 10% of analogue phones for similar coverage.

- CDMA can reduce interference between cells in cellular networks and improve 'hand-over' by summing and correlating transmissions from adjacent cells.

- CDMA systems have the ability to co-exist with conventional narrow-band transmissions.

- CDMA can simplify cell planning by removing the need to specify rigid frequency allocations to individual cells.

- CDMA is claimed to provide higher spectrum efficiency compared to TDMA, although real verification of these claims is still lacking.

Advantages of CDMA :

- Multipath fading may be substantially reduced because of large signal bandwidth.

- No absolute limit on the number of users.

- Easy addition of more users.

- Impossible for hackers to decipher the code sent.

- Better signal quality.

- No sense of handoff when changing cells.

Disadvantages of CDMA :

- As the number of users increases, the overall quality of service decreases.
- Self-jamming.
- Near- Far- problem arises.

However, it seems clear that for the near future at least, TDMA will remain the dominant technology in the wireless market.

Table 2.2 : Comparison of Multiple Access Techniques

TDMA	FDMA	CDMA
Users transmit inturn in their own unique time slots.	All users transmit at the same time but in unique frequency band.	Many users simultaneously transmit spread spectrum signals that occupy same frequency band.
Guard times are kept between two users.	Guard bands are kept between two users.	Both guard band and guard times are kept.
Power efficiency is high.	Power efficiency is low.	Power efficiency is highest.
Time synchronization is required.	Synchronization is not necessary.	Synchronization is not required.
Interference from adjacent users can degrade performance.	Interference from adjacent users can degrade performance.	Interference from adjacent users is negligible.
Implementation is complex.	Implementation is simple.	Implementation is complex.

2.11 ETHERNET

- Local Area Network (LAN) is a data communications network connecting terminals, computers and printers within a building or other geographically limited areas. These devices could be connected through wired cables or wireless links. Ethernet, Token Ring and Wireless LAN using IEEE 802.11 are examples of standard LAN technologies.

- Ethernet is by far the most commonly used LAN technology. Token Ring technology is still used by some companies. FDDI is sometimes used as a backbone LAN interconnecting Ethernet or Token Ring LANs. WLAN using IEEE 802.11 technologies is rapidly becoming the new leading LAN technology for its mobility and easy to use features.

- Local Area Network could be interconnected using Wide Area Network (WAN) or Metropolitan Area Network (MAN) technologies. The common WAN technologies include TCP/IP, ATM, Frame Relay, etc. The common MAN technologies include SMDS and 10 Gigabit Ethernet.

- LANs are traditionally used to connect a group of people who are in the same local area. However, the working group are becoming more geographically distributed in today's working environment. The virtual LAN (VLAN) technologies are defined for people in different places to share the same networking resource.

- Local Area Network protocols are mostly used at data link layer (layer 2). IEEE is the leading organization defining most of the LAN protocols.

Protocol Structure - Local Area Network and LAN Protocols

The key LAN protocols are listed as follows :

Table 2.3

LAN - Local Area Network Protocols	
Ethernet	Ethernet LAN protocols as defined in IEEE 802.3 suite
	Fast Ethernet : Ethernet LAN at data rate 100Mbps (IEEE 802.3u)
	Gigabit Ethernet : Ethernet at data rate 1000Mbps (IEEE 802.3z, 802.3ab)
	10Gigabit Ethernet : Ethernet at data rate 10 Gbps (IEEE 802.3ae)
WLAN	Wireless LAN in IEEE 802.11, 802,11a, 802.11b, 802.11g and 802.11n
	IEEE 802.11i : WLAN Security Standards
	IEEE 802.1X : WLAN Authentication & Key Management
	IEEE 802.15 : Bluetooth for Wireless Personal Area Network (WPAN)
VLAN	IEEE 802.1Q : Virtual LAN Bridging Switching Protocol
	GARP : Generic Attribute Registration Protocol (802.1P)
	GMRP : GARP Multicast Registration Protocol (802.1P)
	GVRP : GARP VLAN Registration Protocol (802.1P, 802.1Q)
Token Bus	IEEE 802.4 : LAN Protocol
Token Ring	IEEE 802.5 LAN protocol
FDDI	Fiber Distributed Data Interface
Others	LLC : Logic Link Control (IEEE 802.2)
	SNAP : SubNetwork Access Protocol
	STP : Spanning Tree Protocol (IEEE 802.1D)
	IEEE 802.1p : LAN Layer 2 QoS/CoS Protocol

2.11.1 Ethernet : IEEE 802.3 Local Area Network (LAN) Protocols

- Ethernet protocols refer to the family of local-area network (LAN) covered by the IEEE 802.3. In the Ethernet standard, there are two modes of operation : half-duplex and full-duplex modes. In the half duplex mode, data are transmitted using the popular Carrier-Sense Multiple Access/Collision Detection (CSMA/CD) protocol on a shared medium.

- The main disadvantages of the half-duplex are the efficiency and distance limitation, in which the link distance is limited by the minimum MAC frame size. This restriction reduces the efficiency drastically for high-rate transmission. Therefore, the carrier extension technique is used to ensure the minimum frame size of 512 bytes in Gigabit Ethernet to achieve a reasonable link distance.

- Four data rates are currently defined for operation over optical fiber and twisted-pair cables :
  - ➢ 10 Mbps - 10Base-T Ethernet (IEEE 802.3)
  - ➢ 100 Mbps - Fast Ethernet (IEEE 802.3u)
  - ➢ 1000 Mbps - Gigabit Ethernet (IEEE 802.3z)
  - ➢ 10-Gigabit - 10 Gbps Ethernet (IEEE 802.3ae).

- The Ethernet system consists of three basic elements :
  1. The physical medium used to carry Ethernet signals between computers,
  2. A set of medium access control rules embedded in each Ethernet interface that allow multiple computers to fairly arbitrate access to the shared Ethernet channel, and
  3. An Ethernet frame that consists of a standardized set of bits used to carry data over the system.

- As with all IEEE 802 protocols, the ISO data link layer is divided into two IEEE 802 sublayers, the Media Access Control (MAC) sublayer and the MAC-client sublayer. The IEEE 802.3 physical layer corresponds to the ISO physical layer.

- The MAC sub-layer has two primary responsibilities :
  1. Data encapsulation, including frame assembly before transmission, and frame parsing/error detection during and after reception.
  2. Media access control, including initiation of frame transmission and recovery from transmission failure.

The MAC-client sub-layer may be one of the following :

- Logical Link Control (LLC), which provides the interface between the Ethernet MAC and the upper layers in the protocol stack of the end station. The LLC sublayer is defined by IEEE 802.2 standards.

- Bridge entity, which provides LAN-to-LAN interfaces between LANs that use the same protocol (for example, Ethernet to Ethernet) and also between different protocols (for example, Ethernet to Token Ring). Bridge entities are defined by IEEE 802.1 standards.

- Each Ethernet-equipped computer operates independently of all other stations on the network : there is no central controller. All stations attached to an Ethernet are connected to a shared signaling system, also called the medium. To send data a station first listens to the channel, and when the channel is idle the station transmits its data in the form of an Ethernet frame, or packet.

- After each frame transmission, all stations on the network must contend equally for the next frame transmission opportunity. Access to the shared channel is determined by the Medium Access Control (MAC) mechanism embedded in the Ethernet interface located in each station. The medium access control mechanism is based on a system called Carrier Sense Multiple Access with Collision Detection (CSMA/CD).

- As each Ethernet frame is sent onto the shared signal channel, all Ethernet interfaces look at the destination address. If the destination address of the frame matches with the interface address, the frame will be read entirely and be delivered to the networking software running on that computer. All other network interfaces will stop reading the frame when they discover that the destination address does not match their own address.

- When it comes to how signals flow over the set of media segments that make up an Ethernet system, it helps to understand the topology of the system. The signal topology of the Ethernet is also known as the logical topology, to distinguish it from the actual physical layout of the media cables. The logical topology of an Ethernet provides a single channel (or bus) that carries Ethernet signals to all stations.

- Multiple Ethernet segments can be linked together to form a larger Ethernet LAN using a signal amplifying and retiming device called a repeater. Through the use of repeaters, a given Ethernet system of multiple segments can grow as a "non-rooted branching tree". "Non-rooted" means that the resulting system of linked segments may grow in any direction, and does not have a specific root segment. Most importantly, segments must never be connected in a loop. Every segment in the system must have two ends, since the Ethernet system will not operate correctly in the presence of loop paths.

- Even though the media segments may be physically connected in a star pattern, with multiple segments attached to a repeater, the logical topology is still that of a single Ethernet channel that carries signals to all stations.

2.11.2 Standard Ethernet

- MAC Layer : In standard Ethernet MAC layer performs two functions :
  1. Controls the access.
  2. Data received from network layer is used for preparation of frame to pass it to physical layer.

Table 2.4 : Ethernet MAC Data Frame for 10/100 Mbps Ethernet

Number of Bytes	7	1	2/6	2/6	2	46-1500bytes	4
Name of Field	Pre	SFD	DA	SA	Length/Type	Data unit + pad	FCS

- **Preamble (PRE) - 7 Bytes :** The PRE is an alternating pattern of ones and zeros that tells receiving stations that a frame is coming, and that provides a means to synchronize the frame-reception portions of receiving physical layers with the incoming bit stream.

- **Start-of-Frame Delimiter (SFD) - 1 Bytes :** The SFD is an alternating pattern of ones and zeros, ending with two consecutive 1-bits indicating that the next bit is the leftmost bit in the leftmost byte of the destination address.

- **Destination Address (DA) - 6 Bytes :** The DA field identifies which station(s) should receive the frame.

- **Source Addresses (SA) - 6 Bytes.** The SA field identifies the sending station.

- **Length/Type - 2 Bytes :** This field indicates either the number of MAC-client data bytes that are contained in the data field of the frame, or the frame type ID if the frame is assembled using an optional format.

- **Data :** Is a sequence of n bytes (46 ≤ n ≤ 1500) of any value. (The total frame minimum is 64 bytes).

- **Frame Check Sequence (FCS) - 4 Bytes :** This sequence contains a 32-bit Cyclic Redundancy Check (CRC) value, which is created by the sending MAC and is recalculated by the receiving MAC to check for damaged frames.

Physical Layer :

- There are several physical layer implementations. Some of them are :
  1. 10Base5 : Bus, thick coaxial.
  2. 10Base2 : Bus thin coaxial.
  3. 10Base-T : Star UTP.
  4. 10Base-F : Star, fibre.

1. 10Base5 : Thick Ethernet :

- The first implementation is called 10Base5, thick Ethernet, or Thicknet. The nick name derives from the size of the cable, which is roughly the size of a garden hose and too stiff to bend with your hands. 10Base5 was the first Ethernet specification to use a bus topology with an external transceiver (transmitter/receiver) connected via to tap to a thick coaxial cable. Fig. 2.32 shows a schematic diagram of a 10Base5 implementation.

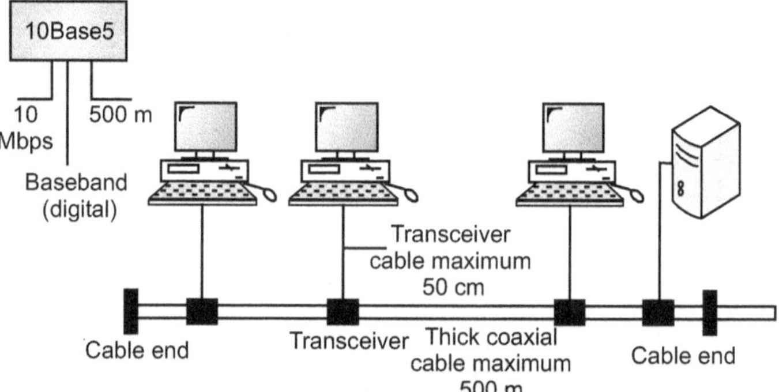

Fig. 2.32 : 10Base5 Implementation

- The transceiver is responsible for transmitting, receiving and detecting collisions. The transceiver is connected to the station via a transceiver cable that provides separate path for sending and receiving. This means that collision can only happen in the coaxial cable.

- The maximum length of the coaxial cable must not exceed 500 m, otherwise, there is excessive degradation of the signal. If a length of more than 500 m is needed, upto five segments, each a maximum of 500 meter, can be connected using repeaters.

2. 10Base2 : Thin Ethernet :

- The second implementation is called 10Base2, thin Ethernet, or Cheapernet, 10Base2 also uses a bus topology, but the cable is much thinner and more flexible. The cable can be bent to pass very close to the stations. In this case, the transceiver is normally part of the Network Interface Card (NIC), which is installed inside the station. Fig. 2.33 shows the schematic diagram of a 10Base2 implementation.

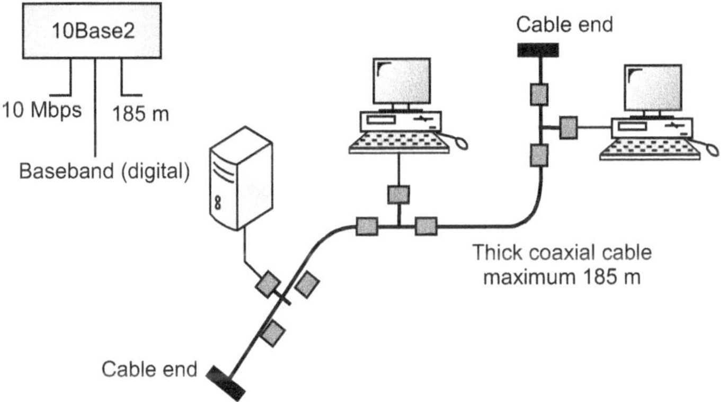

Fig. 2.33 : 10Base2 Implementation

- Note that the collision here occurs in the thin coaxial cable. This implementation is more cost effective than 10Base5 because thin coaxial cable is less expensive than thick coaxial and the tee connections are much cheaper than taps. Installation is simpler because the thin coaxial cable is very flexible. However, the length of each segment cannot exceed 185 m (close to 200 m) due to the high level of attenuation in thin coaxial cable.

3. 10Base-T : Twisted-Pair Ethernet :

- The third implementation is called 10Base-T or twisted-pair Ethernet. 10Base-T uses a physical star topology. The stations are connected to a hub via two pairs of twisted cable, as shown in Fig. 2.34.

- Note that two pairs of twisted cable create two paths (one for sending and one for receiving) between the station and the hub. Any collision here happens in the hub. Compared to 10Base5 or 10Base2, we can see that the hub actually replaces the coaxial cable as far as collision is concerned. The maximum length of the twisted cable here is defined as 100 m, to minimize the effect of attenuation in the twisted cable.

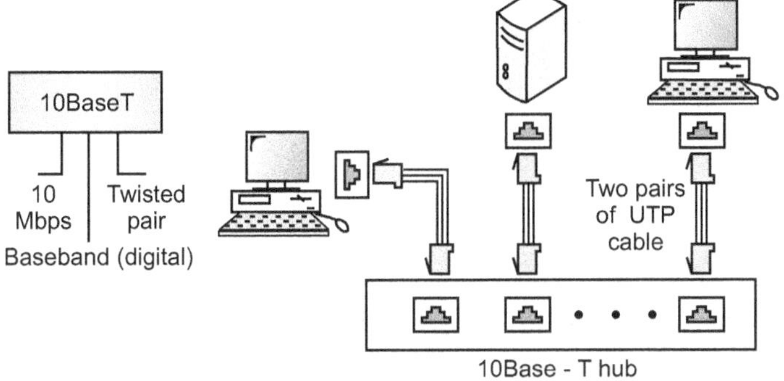

Fig. 2.34 : 10Base-T Implementation

4. 10Base-F : Fiber Ethernet :

- Although there are several types of optical fiber 10 Mbps Ethernet, the most common is called 10Base-F. 10Base-F uses a star topology to connect stations to a hub. The stations are connected to the hub using two fiber-optic cables, as shown in Fig. 2.35.

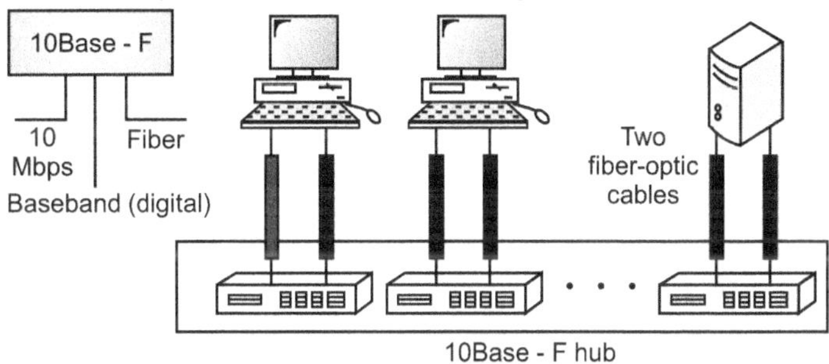

Fig. 2.35 : 10Base-F Implementation

- Following table gives Comparison of Physical Layer Implementation of Ethernet.

Table 2.5 : Summary of Standard Ethernet Implementations

Characteristics	10Base5	10Base2	10Base-T	10Base-F
Media used	Thick coaxial cable	Thin coaxial cable	2 UTP	2 Fiber
length	< 500 m	< 185 m	< 100 m	< 2000 m
Line coding technique	Split phase Manchester	Split phase Manchester	Split phase Manchester	Split phase Manchester

2.11.3 Bridged Ethernet

- In order to have compatibility between 10 Mbps and 100 Mbps LANs, some changes were required. They are :
1. Bridged Ethernet.
2. Switched Ethernet.
3. Full Duplex Ethernet.

- The bridged ethernet divides LAN by bridges because of which there is improvement in bandwidth and separation of collision domains.

- If we have 10 Mbps LAN and 10 nodes are there, the bandwidth will be divided among these nodes depending on need. e.g. if only one station wants to transmit entire bandwidth will be available to it. But if all of them want to transmit each one will have 1 Mbps bandwidth.

- We can improve the bandwidth efficiency by using a bridge. If 10 nodes are divided into 2 groups of 5 each, each group will have an average bandwidth of 10/5 = 2 Mbps instead of 1 Mbps.

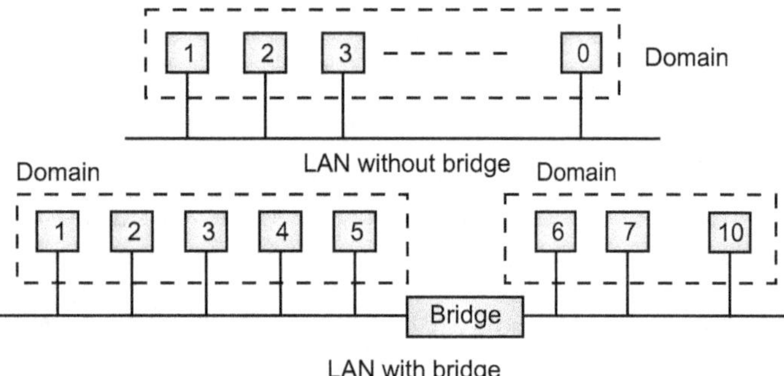

Fig. 2.36

- Another advantage is number of nodes in collision domain are reduced and hence probability of collision reduces by 50%.

2.11.4 Switched Ethernet

- If we divide the number of nodes in the LAN, there is improvement in bandwidth efficiency. We can have only single node in each network. If there are N nodes in LAN, there will be N networks. It is called switched LAN as shown below. The collision domain is also divided into N domains.

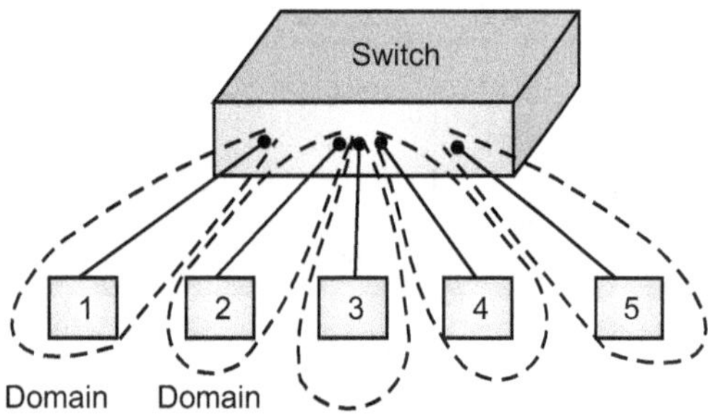

Fig. 2.37

- The bandwidth will be shared between station and the switch (i.e. 5 Mbps each).

Full Duplex Ethernet :

- In half duplex a station can either send or receive. In full duplex mode Ethernet, send and receive operations can be done simultaneously. The capacity of each domain is doubled because of this. Two links will be used in such configuration.

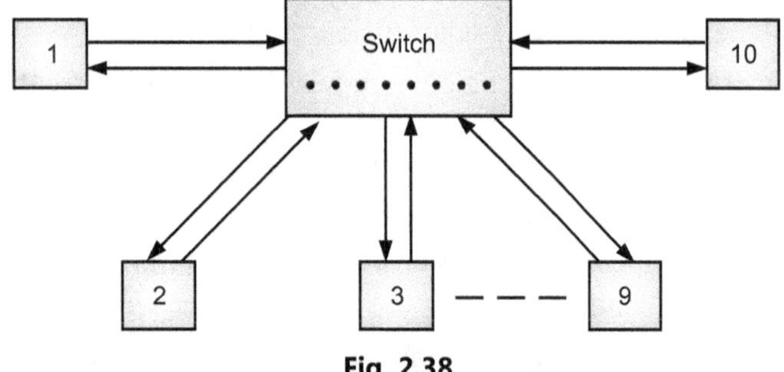

Fig. 2.38

- In this mode, there is node of CSMA/CD, since each station is independent.

2.11.5 Fast Ethernet : 100 Mbps Ethernet

- Fast Ethernet (100BASE-T) offers a speed increase ten times that of the 10BaseT Ethernet specification, while preserving such qualities as frame format, MAC mechanisms, and MTU. Such similarities allow the use of existing 10BaseT applications and network management tools on Fast Ethernet networks. Officially, the 100BASE-T standard is IEEE 802.3u.

- Like Ethernet, 100BASE-T is based on the CSMA/CD LAN access method. There are several different cabling schemes that can be used with 100BASE-T, including :

  ➢ **100BASE-TX :** Two pairs of high-quality twisted-pair wires.

  ➢ **100BASE-T4 :** Four pairs of normal-quality twisted-pair wires.

  ➢ **100BASE-FX :** Fiber optic cables.

- The Fast Ethernet specifications include mechanisms for Auto-Negotiation of the media speed. This makes it possible for vendors to provide dual-speed Ethernet interfaces that can be installed and run at either 10-Mbps or 100-Mbps automatically.

- The IEEE identifiers include three pieces of information. The first item, "100", stands for the media speed of 100-Mbps. The "BASE" stands for "baseband," which is a type of signaling. Baseband signaling simply means that Ethernet signals are the only signals carried over the media system.

- The third part of the identifier provides an indication of the segment type. The "T4" segment type is a twisted-pair segment that uses four pairs of telephone-grade twisted-pair wire. The "TX" segment type is a twisted-pair segment that uses two pairs of wires and is based on the data grade twisted-pair physical medium standard developed by ANSI. The "FX" segment type is a fiber optic link segment based on the fiber optic physical medium standard developed by ANSI and that uses two strands of fiber cable. The TX and FX medium standards are collectively known as 100BASE-X.

- The 100BASE-TX and 100BASE-FX media standards used in Fast Ethernet are both adopted from physical media standards first developed by ANSI, the American National Standards Institute. The ANSI physical media standards were originally developed for the Fiber Distributed Data Interface (FDDI) LAN standard (ANSI standard X3T9.5), and are widely used in FDDI LANs.

Protocol Structure - Fast Ethernet : 100 Mbps Ethernet (IEEE 802.3u)The basic IEEE 802.3 Ethernet MAC Data Frame for 10/100 Mbps Ethernet

Table 2.6

Number of Bytes	7	1	2/6	2/6	2	612 <= n <= 1500	4 bytes
Name of Field	Pre	SFD	DA	SA	Length / Type	Data unit + pad	FCS

- **Preamble (PRE) - 7 Bytes :** The PRE is an alternating pattern of ones and zeros that tells receiving stations that a frame is coming, and that provides a means to synchronize the frame-reception portions of receiving physical layers with the incoming bit stream.

- **Start-of-Frame Delimiter (SFD) - 1 Byte :** The SFD is an alternating pattern of ones and zeros, ending with two consecutive 1-bits indicating that the next bit is the left-most bit in the left-most byte of the destination address.

- **Destination Address (DA) - 6 Bytes :** The DA field identifies which station(s) should receive the frame.

- **Source Addresses (SA) - 6 Bytes :** The SA field identifies the sending station.

- **Length/Type - 2 Bytes :** This field indicates either the number of MAC-client data bytes that are contained in the data field of the frame, or the frame type ID if the frame is assembled using an optional format.

- **Data :** Is a sequence of n bytes ($612 \leq n \leq 1500$) of any value. Note that since transmission speed has increased from 10 Mbps to 100 Mbps, frame transmission time reduces by factor of 10. Hence, minimum frame size increases by a factor of 10 to 640.

- **Frame Check Sequence (FCS) - 4 Bytes :** This sequence contains a 32-bit Cyclic Redundancy Check (CRC) value, which is created by the sending MAC and is recalculated by the receiving MAC to check for damaged frames.

Gigabit (1000 Mbps) Ethernet

- Ethernet protocols refer to the family of Local-Area Network (LAN) covered by the IEEE 802.3 standard. The Gigabit Ethernet is based on the Ethernet protocol, but increased speed tenfold over the fast Ethernet, using shorter frames with carrier extension. It is published as the IEEE 802.3z and 802.3ab, supplement to the IEEE 802.3 base standards.

- The Gigabit Ethernet standards are fully compatible with Ethernet and Fast Ethernet installations. It retains Carrier Sense Multiple Access/Collision Detection (CSMA/CD) as the access method. It supports full-duplex as well as half duplex modes of operation. Single-mode and multi mode fiber and short-haul coaxial cable, and twisted pair cables are supported. The Gigabit Ethernet architecture is displayed in Fig. 2.39.

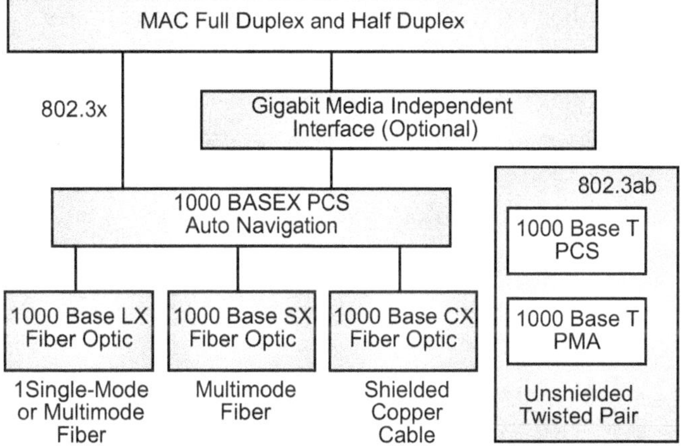

Fig. 2.39

- The IEEE 802.3z defines the Gigabit Ethernet over fiber and cable, which has a physical media standard 1000Base-X (1000BaseSX - short wave covers up to 500 m, and 1000BaseLX - long wave covers up to 5 km). The IEEE 802.3ab defines the Gigabit Ethernet over the unshielded twisted pair wire (1000Base-T covers up to 75m).

- The Gigabit interface converter (GBIC) allows network managers to configure+ each gigabit port on a port-by-port basis for short-wave (SX), long-wave (LX), long-haul (LH), and copper physical interfaces (CX). LH GBICs extended the single-mode fiber distance from the standard 5 km to 10 km.

Protocol Structure - Gigabit (1000 Mbps) Ethernet

- 1000Base-X has a minimum frame size of 416 bytes, and 1000Base-T has a minimum frame size of 520 bytes. An extension field is used to fill the frames that are shorter than the minimum length.

Table 2.7

Number of Bytes	7	1	6	6	2	494 <= n <=1500	4	Variable
Name of Field	Pre	SFD	DA	SA	Length/ Type	Data unit + pad	FCS	Ext

- **Preamble (PRE) - 7 Bytes :** The PRE is an alternating pattern of ones and zeros that tells receiving stations that a frame is coming, and that provides a means to synchronize the frame-reception portions of receiving physical layers with the incoming bit stream.

- **Start-of-Frame Delimiter (SFD) - 1 Byte :** The SFD is an alternating pattern of ones and zeros, ending with two consecutive 1-bits indicating that the next bit is the left-most bit in the left-most byte of the destination address.

- **Destination Address (DA) - 6 Bytes :** The DA field identifies which station(s) should receive the frame.
- **Source Addresses (SA) - 6 Bytes :** The SA field identifies the sending station.
- **Length/Type - 2 Bytes :** This field indicates either the number of MAC-Client Data Bytes that are contained in the data field of the frame, or the frame type ID if the frame is assembled using an optional format.
- **Data :** Is a sequence of n bytes (494 <= n <=1500) of any value.
- **Frame Check Sequence (FCS) - 4 Bytes :** This sequence contains a 32-bit cyclic redundancy check (CRC) value, which is created by the sending MAC and is recalculated by the receiving MAC to check for damaged frames.
- **Ext :** Extension, which is an non-data variable extension field for frames that are shorter than the minimum length.

2.11.6 10-Gigabit Ethernet

- It is the fastest Ethernet which use optical fibre cable. It is specified as IEEE 802.3ae standard.

The goals of Ten-Gigabit Ethernet are :

1. Upgradation of data rate to 10 Gbps.
2. Make it compatible with other Ethernet standards.
3. Use 48 bit address.
4. Use same frame format.
5. Make it compatible with other technologies such as ATM and frame relay.
6. Keep same frame lengths (maximum and minimum).
7. Allow existing LANs in WAN and MAN.

- The specifications of MAC sublayer are full duplex mode of operation. Hence no contention. No need of CSMA/CD.

The physical layer specifications are :

- Fibre optic cables over long distance.
- Three different layers are :

1. **10GBase-S :** Uses (300 m) short wave 850 nm multimode fibre.
2. **10GBase-L :** Uses (10 km) long wave 1310 nm singlemodefibre.
3. **10GBase-E :** Uses (40 km) extended 1550 nm single mode fibre.

SOLVED EXAMPLES

Example 2.1 : *Consider a 64 kbps geostationary satellite channel is used to send 512 byte data frames in one direction, with very short acknowledgement coming back the other way. What is maximum throughput for window size of 1, 7, 15 and 127 ? (Sliding window protocol is used).*

Solution :

Given :
$$R = 64 \text{ kbps}$$
$$= 64000 \text{ bits/sec}$$
$$\text{Frame size} = 512 \text{ byte}$$
$$= 4096 \text{ bits}$$

$$\text{Frame time, } T_F = \frac{\text{Frame size}}{\text{Rate}}$$

$$= \frac{4096}{64000} = 64 \times 10^{-3} \text{ sec} = 64 \text{ ms}$$

Round trip propagation delay.
$$2T_P = 540 \text{ ms}$$

Throughput is given by,

$$\eta = \frac{W_S T_F}{T_F + 2T_P}$$

(i) $W_S = 1$

$$\eta = \frac{1 \times 64}{64 + 540} = 0.1059$$

i.e. throughput is 10.59%

(ii) $$\eta = \frac{7 \times 64}{64 + 540} = 0.7413$$

i.e. throughput is 74.13%

(iii) $W_S = 15$

Here, $W_S \geq \dfrac{2T_P}{T_F} + 1$

$\therefore$ $\eta = 1$

i.e. throughput is 100%.

(iv) $W_S = 127$

Here also,

$$W_S \geq \frac{2T_P}{T_F} + 1$$

$\therefore$ $\eta = 1$

i.e. throughput is 100%.

[i.e. when window size $W_S \geq \dfrac{2T_P}{T_F} + 1$, sender keeps on sending the frames and does not remain idle at any time].

Example 2.2 : *A 4 Mbps token ring has a token holding timer value of 10 m/sec. What is the longest frame that can be sent on this ring ?*

Solution :

Given : $R = 4$ Mbps $= 4 \times 10^6$ bits/sec.

Token holding time $= 10 \times 10^{-3}$ sec.

$\therefore$ Number of bits that can be transmitted in 10 msec.

$$= 4 \times 10^6 \times 10 \times 10^{-3}$$

$$= 40000 \text{ bits}$$

$$= 5000 \text{ bytes}$$

Hence, longest frame is 5000 bytes including overheads. Data portion will be slightly less than this.

Example 2.3 : *Consider the use of 1000 bit frames on a 1 Mbps satellite channel. What is the maximum links utilization for*

(i) *Stop and Wait ARQ*

(ii) *Continuous ARQ with Windowsize 7.*

(iii) *Continuous ARQ with Windowsize 127.*

(iv) *Continuous ARQ with Windowsize 255.*

Solution :

Given : Frame size $= 1000$ bits

Bit rate $= 1 \times 10^6$ bits/sec.

$\therefore$ $T_F = \dfrac{1000}{1 \times 10^6} = 1 \times 10^{-3} = 1$ msec.

For satellite channel,

$T_P = 270$ ms.

(i) Stop-and-Wait ARQ :

$$\text{Utilization} = U = \dfrac{1}{1 + 2\dfrac{T_P}{T_F}}$$

$$= \dfrac{1}{1 + 2 \times \dfrac{270}{1}}$$

$$= 0.0018 \text{ i.e. } 0.18\%$$

(ii) For Continuous ARQ (Sliding Window) :

$$\text{Utilization} = U = \dfrac{W_S}{1 + 2 \times \dfrac{T_P}{T_F}}$$

$$W_S = 7$$

$\therefore$ $U = 7 \times \dfrac{1}{541}$

$$= 0.0129$$

i.e. 1.29%

(iii) $W_S = 127$

$$U = \dfrac{127}{541}$$

$$= 0.2343$$

i.e. 23.43%

(iv) $W_S = 255$

$$U = \dfrac{255}{541}$$

$$= 0.4713 \text{ i.e. } 47.13\%$$

Example 2.4 : *A channel has a bit rate of 4 kbps and propagation delay 20 ms. For what range of frame size does stop-and-wait ARQ gives throughput $\geq 50\%$?*

Solution :

Given : $R = 4$ kbps

$$= 4 \times 10^3 \text{ bits/sec}$$

$$T_P = ?$$

$$\eta \geq 50\%$$

For stop-and-wait ARQ,

$$\eta = \dfrac{1}{1 + \dfrac{2T_P}{T_F}}$$

$\therefore$ $\dfrac{1}{1 + \dfrac{2T_P}{T_F}} \geq \dfrac{1}{2}$

$\therefore$ $\dfrac{1}{1 + \dfrac{2 \times 20}{T_F}} \geq \dfrac{1}{2}$

$$\dfrac{T_F}{T_F + 40} \geq \dfrac{1}{2}$$

$$2T_F \geq T_F + 40$$

$$T_F \geq 40$$

$$T_F \geq 40 \text{ ms}$$

$\therefore$ Frame size $= R \times T_F$

$$= \dfrac{4 \times 10^3}{10^3} \times 40 \times 10^{-3}$$

$$= 160 \text{ bits}$$

$\therefore$ Frame size ≥ 160 bits

$$\geq 20 \text{ bytes}$$

Example 2.5 : *A group of N users share 56 kbps pure ALOHA channel. Each station outputs 1000 bit frame on an average of once every 100 sec., even if the previous has not yet been sent (buffered). What is maximum value of N ?*

Solution :

For pure ALOHA,

Maximum throughput = 0.184

$\therefore$ Maximum usable channel bandwidth

$$R = 0.184 \times 56 \text{ kbps}$$
$$R = 10.3 \text{ kbps}$$

Rate of transmission of stations $= \dfrac{1000}{100} = 10$ bps

$\therefore$ Number of stations that can use the channel will be,

$$N = \dfrac{10.3 \times 10^3}{10} = 1030$$

Example 2.6 : *10,000 Airline reservation stations are competing for the use of a single slotted ALOHA channel. The average station makes 18 requests/hr. A slot is 125 µsec. What is approximate total channel load ?*

Solution : Total channel load is number of transmissions per slot.

Given : One station makes 18 requests/hr.

i.e. 18/3000 requests/sec.

1/200 requests/sec.

$\therefore$ Number of requests made by 10000 stations in 1 sec

$$= 10000 \times \dfrac{1}{200} = 50$$

i.e. 50 requests in 1 sec.

Hence, number of requests per time slot of 125 µsec = $125 \times 10^{-6} \times 50$

i.e.   Total load G $= 0.006250$

Example 2.7 : *A large population of ALOHA users manages to generate 50 requests/sec. including both originals and retransmissions. Time is slotted in units of 40 msec.*

(a) What is the chance of success on first attempt ?

(b) What is the probability of exactly k collisions and then a success ?

(c) What is expected number of transmission attempts needed ?

Solution :

Given : Stations generate 50 requests/sec.

$\therefore$ Number of requests per time slot of 40 msec

$$= 40 \times 10^{-3} \times 50 = 2$$

$\therefore$ $G = 2$

(a) For slotted ALOHA,

Chance of success in first attempt will be

$$= e^{-G} \text{ (Poisson's Distribution)}$$
$$= e^{-2}$$
$$= 0.135$$

(b) The probability of exactly k collisions and a success

$$= (1 - e^{-G})^k \, e^{-G}$$
$$= 0.135 \times 0.865^{-k}$$

(c) The expected number of transmissions is,

$$e^G = 7.4$$

Example 2.8 : *Calculate maximum throughput possible for ALOHA and pure ALOHA for a radio system with 9600 bps channel used for call setup request to base station. Let frame length be 200 bits.*

Solution :

(i) The maximum throughput for pure ALOHA = 0.184.

Given :

Rate of transmission = 9600 bps

Frame length $= 120$ bits

$\therefore$ Number of frames per sec.

$$= \dfrac{9600 \text{ bits/sec.}}{120 \text{ bits/frame}}$$
$$= 80 \text{ frames/sec}$$

$\therefore$ Throughput $= 80 \times 0.184 = 15$ frames/sec.

(ii) For slotted ALOHA

Maximum throughput = 0.368

$\therefore$ Throughput $= 0.368 \times 80$

$$= 30 \text{ frames/sec.}$$

Example 2.9 : *Measurement of slotted ALOHA channel with infinite number of users. Show that 10% of slots are idle.*

(a) What is channel load G ?

(b) What is throughput ?

(c) Is the channel under loaded or overloaded ?

Solution :

(i) Given : $N = \infty$

Idle slots $= 10\%$

$\therefore$ Probability of frame not generated $= 0.1$

Using Poisson's Law :

Probability that k transmissions are done in a slot is

$$= \dfrac{G^k \times e^{-G}}{k\,!}$$

$\therefore$ Probability that no frame generated

$$= \frac{G^0\,e^{-G}}{k\,!}$$

$$= e^{-G}$$

$\therefore \qquad 0.1 = e^{-G}$

$\therefore \qquad ln\,0.1 = -G$

$\therefore$ Channel load, $G = 2.3$

(ii) Throughput s $= Ge^{-G}$

$$= 2.3\,e^{-2.3}$$

$$= 0.23$$

(iii) A channel is overloaded, if $G > 1$.

Since $G = 2.3$. It is overloaded.

Example 2.10 : *A CSMA/CD network running at 1 Gbps over 1 km cable with no repeaters. The signal speed in the cable is 200000 km/sec. What is minimum frame size ?*

Solution :

Given : $\qquad d = 1$ km

$\qquad\qquad v = 2 \times 10^8$ m/sec

$\therefore$ Propagation time $(T_P) = \dfrac{d}{v} = \dfrac{1000}{2 \times 10^8} = 5 \times 10^{-6}$ sec

For CSMA/CSD the frame must not be transmitted in time $2T_P$.

$\therefore \qquad 2T_P = 10 \times 10^{-6} = 10$ μsec.

Rate of transmission = 1 Gbps

$\therefore$ 1×10^9 bits are transmitted in 1 sec.

$\therefore$ Number of bits transmitted in 10 μsec.

$$= 1 \times 10^9 \times 10 \times 10^{-6}$$

$$= 10 \times 10^3$$

$$= 10000 \text{ bits}$$

$\therefore$ The frame size must be greater than 10000 bits which is minimum frame size.

Example 2.11 : *At a transmission rate of 5 Mbps and propagation speed of 200 m/μsec to how many metres of cable is the 1 bit delay in token ring interface is equivalent ?*

Solution :

Bit rate R $= 5$ Mbps $= 5 \times 10^6$

Bit duration $= \dfrac{1}{R} = 2 \times 10^{-7}$

Propagation speed $= 200 \times 10^6$ m/s

$$= 2 \times 10^8 \text{ m/s}$$

Distance equivalent/bit

$$d = v \times R$$

$$= 2 \times 10^8 \times 2 \times 10^{-7}$$

$$= 40 \text{ m}$$

[i.e. if we add one more station it will introduce delay equivalent to 40 m cable].

Example 2.12 : *Draw sender and receiver windows for a system using Go-back N ARQ and selective repeat ARQ.*

(a) Frame 0 is sent; frame 0 is acknowledged.

(b) Frame 1 and 2 are sent; frame 1 and 2 are acknowledged.

(c) Frames 3, 4 and 5 are sent NAK 4 is received.

(d) Frames 4, 5, 6, 7 are sent frames 4 through 7 are acknowledged.

Solution :

(i) Go-back N ARQ

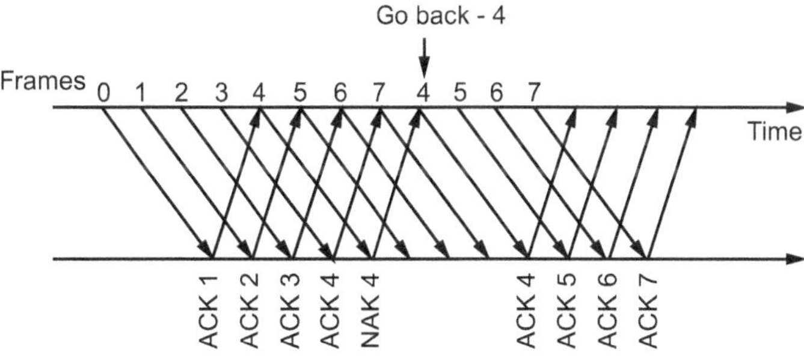

Fig. 2.40 (a)

(ii) Selective Repeat ARQ

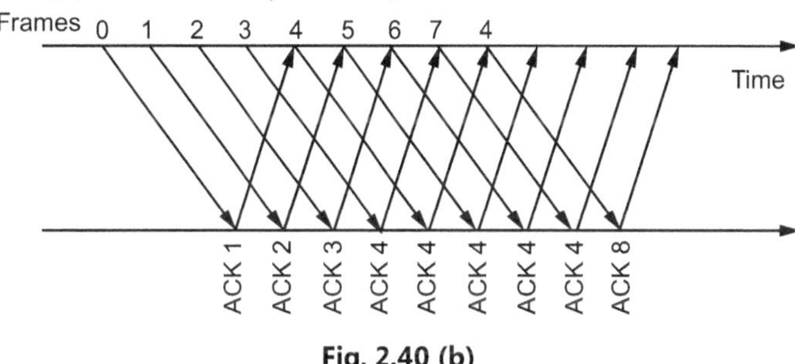

Fig. 2.40 (b)

Example 2.13 : *Computer A uses stop-and-wait ARQ protocol to send packets to computer B. Distance between A and B is 4000 km. How long does it take for computer A to send out a packet of size 1000 byte if throughput is 100000 kbps ? How much time the computer is idle ?*

(Assume propagation speed to be speed of light).

Solution :

Given :

$$d = 4000 \text{ km}$$

$$N_F = 1000 \text{ byte} = 8000 \text{ bits}$$

$$v = 3 \times 10^8 \text{ m/s}$$

$$T_F = ?$$

$\therefore$ Propagation delay $T_P = \dfrac{4000 \times 10^3}{3 \times 10^8} = \dfrac{4}{3} \times 10^{-2}$

$$s = 0.0133 \text{ s}$$

Throughput for stop and wait ARQ is

(i) Here throughput means rate at which data is coming out of station A.

$\therefore$ Time token to transmit one packet is,

$$T_F = \frac{8000}{100000 \times 10^3} = 8 \times 10^{-6}$$

(ii) Idle time $= 2T_P - T_F$

$$= 0.01333 \times 2 - 8 \times 10^{-5}$$

$$= 0.026658663$$

Example 2.14 : *Calculate the maximum link utilization efficiency for stop-and-wait flow control mechanism if the frame size is 2400 bits, bit rate is 4800 bps and distance between the devices is 2000 km. Speed of propagation be 2×10^8 m/s.*

Solution :

Frame transmission time $T_F = \dfrac{2400}{4800} = 0.5$

Propagation time $T_P = \dfrac{d}{v} = \dfrac{2000 \times 10^3}{2 \times 10^8} = 0.01$ s

Link utilization is throughput

$$\eta = \frac{T_F}{T_F + 2T_P} = \frac{0.5}{0.5 + 0.01 \times 2} = 96.15\%$$

EXERCISE

1. State the protocols devised to handle multiple access communication.
2. Explain pure aloha.
3. Explain slotted aloha.
4. What are carrier sense multiple access protocols ?
5. Explain CSMA/CD technique.
6. How collision is detected in Ethernet ?
7. What is retransmission Back-off ?
8. Explain CSMA/CA technique.
9. What is controlled access ? How it is achieved ?
10. What is logical ring ? Explain.
11. State and explain various multiple access techniques.
12. Compare FDMA, TDMA and CDMA.
13. What are advantages and disadvantages of FDMA ?
14. What are advantages and disadvantages of TDMA ?
15. What are advantages and disadvantages of CDMA ?
16. What is Ethernet ?
17. State various forms of Ethernet based on data rate.
18. Explain standard Ethernet and frame format of standard ethernet.
19. What is bridged ethernet ?
20. What is switched ethernet ?
21. Explain various physical layer implementations of 10 Mbps Ethernet.
22. What is Fast Ethernet ? Give its frame format.
23. What is Gigabit Ethernet ? Give its frame format.
24. Explain Ten Gigabit ethernet.
19. Give frame format of HDLC protocol. Explain each field.
20. Give frame format of PPP protocol. Explain each field.

WIRELESS LANS & VIRTUAL CIRCUIT NETWORKS

3.1 WIRELESS LAN

Need of Wireless LAN

- An increasing number of LAN users are becoming mobile.

- These mobile users require that they should be connected to the network regardless of where they are because they want simultaneous access to the network.

- This makes the use of cables, or wired LANs, impractical if not impossible.

- Wireless LANs are very easy to install.

- There is no requirement for wiring every workstation and every room.

- This ease of installation makes wireless LANs inherently flexible.

- If a workstation must be moved, it can be done easily and without additional wiring, cable drops or reconfiguration of the network.

- Another advantage is its portability. If a company moves to a new location, the wireless system is much easier to move than ripping up all of the cables that a wired system would have snaked throughout the building.

- Most of these advantages also translate into monetary savings.

- AdHoc networks (discussed later) are easily set up in a wireless environment.

How Wireless LAN Work ?

- Wireless LAN (WLAN) uses electromagnetic airwaves (radio and infrared) to communicate information from one point to another without relying on any physical connection.

- Radio waves are often referred to as radio carriers because they simply perform the function of delivering energy to a remote receiver.

- The data being transmitted is superimposed on the radio carrier so that it can be accurately extracted at the receiving end.

- This is generally referred to as modulation of the carrier by the information being transmitted.

- Once data is superimposed (modulated) onto the radio carrier, the radio signal occupies more than a single frequency, since the frequency or bit rate of the modulating information adds to the carrier.

- Multiple radio carriers can exist in the same space at the same time without interfering with each other if the radio waves are transmitted on different radio frequencies.

- In a typical WLAN configuration, a transmitter/receiver (transceiver) device, called an access point (AP), connects to the wired network from a fixed location using standard Ethernet cable.

- At a minimum, the access point receives, buffers, and transmits data between the WLAN and the wired network infrastructure.

- A single access point can support a small group of users and can function within a range of less than one hundred to several hundred feet.

- The access point (or the antenna attached to the access point) is usually mounted high but may be mounted essentially anywhere that is practical as long as the desired radio coverage is obtained.

- End users access the WLAN through wireless LAN adapters, which are implemented as USB adapters, PC cards in notebook computers, ISA or PCI cards in desktop computers, or fully integrated devices within handheld computers. WLAN adapters provide an interface between the client network operating system (NOS) and the airwaves (via an antenna).

- The nature of the wireless connection is transparent to the network operating system.

Advantages of Wireless LANs :

Wireless networks offer the following productivity, service, convenience, and cost advantages over traditional wired networks :

- **Mobility Improves Productivity and Service :** Wireless LAN systems can provide LAN users with access to real-time information anywhere in their organization. This mobility supports productivity and service opportunities not possible with wired networks.

- **Installation Speed and Simplicity :** Installing a wireless LAN system can be fast and easy and can eliminate the need to pull cable through walls and ceilings.

- **Installation Flexibility :** Wireless technology allows the network to go where wire cannot go.

- **Reduced Cost-of-Ownership :** While the initial investment required for wireless LAN hardware can be higher than the cost of wired LAN hardware, overall installation expenses and life-cycle costs can be significantly lower. Long-term cost benefits are greatest in dynamic environments requiring frequent moves, adds, and changes.

- **Scalability :** Wireless LAN systems can be configured in a variety of topologies to meet the needs of specific applications and installations. Configurations are easily changed and range from independent networks suitable for a small number of users to full infrastructure networks of thousands of users that allow roaming over a broad area.

Applications of Wireless LANs :

- LAN extensions

- Ad-hoc networks

- Nomadic access and so many

3.2 ARCHITECTURE OF WIRELESS NETWORK

- Each computer, mobile which is portable or fixed, is referred to as a station in 802.11 wireless networks. When two or more stations come together to communicate with each other, they form a Basic Service Set (BSS).The minimum BSS consists of two stations. 802.11 LANs use the BSS as the standard building block. The BSS can be either without AP (Access Point) or with AP (Access Point) which is as shown in Fig. 3.1.The BSS without AP can not send data to another BSS. So it is called as standalone or ad-hoc network.

- Two or more BSSs are interconnected using a Distribution System or DS. This concept of DS increases network coverage, which can be either wired or wireless. Entry to the DS is accomplished with the use of access points. An access point is a station, thus addressable. So data moves between the BSS and the DS with the help of these access points. Creating large and complex networks using BSSs and DSs leads us to the next level of hierarchy, the Extended Service Set or ESS.

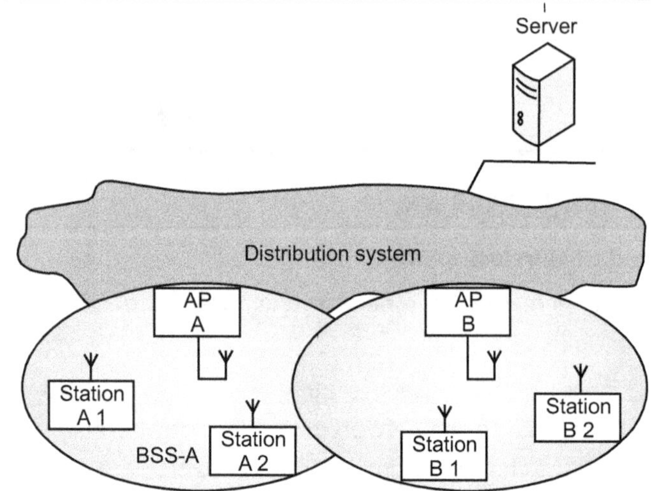

Fig. 3.1 : Components of Wireless Local Area Network

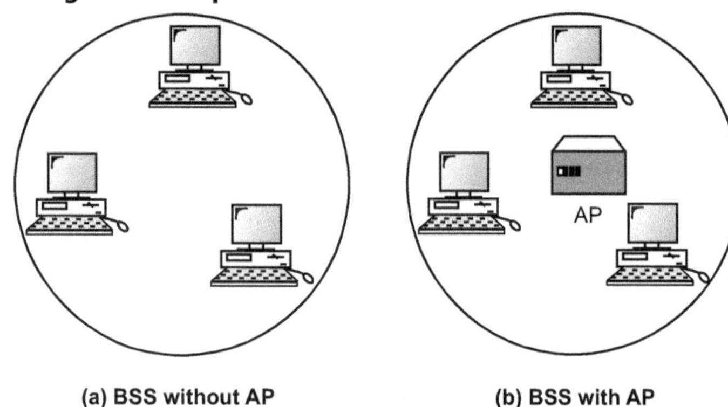

(a) BSS without AP   (b) BSS with AP

Fig. 3.2 : Types of BSS

- An Extended Service set contains two or more BSS with APs. The BSSs in the system are connected to each other via a distribution system which is generally a wired LAN as shown in Fig. 3.2.The beauty of the ESS is the entire network looks like an independent basic service set.

- This means that stations within the ESS can communicate or even more between BSSs transparently. The implementation of the DS is not specified by 802.11. So a distribution system may be created from existing or new technologies. As the implementation for the DS is not specified, 802.11 does specify the services, which the DS must support. Services are divided into two sections, Station Services (SS) and Distribution System Services (DSS).

3.3 DSS SERVICES

There are five services provided by the DSS : Association, Re-association, Disassociation, Distribution, and Integration. The first three services deal with station mobility.

1. **No-Transition :** If a station is moving within its own BSS or is not moving, the stations mobility is termed as No-transition.

2. **BSS Transition :** If a station moves between BSSs within the same ESS, its mobility is termed as BSS transition.

3. **ESS Transition :** If the station moves between BSSs of differing ESSs, it is ESS transition.

The DSS Services are Explained Below :

1. **Association :**

   - A station must affiliate itself with the BSS infrastructure if it wants to use the LAN.

   - This is done by associating itself with an access point.

   - Associations are dynamic in nature because stations move, turn on or turn off.

   - A station can only be associated with one AP. This ensures that the DS always knows where the station is.

   - Association supports no-transition mobility but is not enough to support BSS transition.

2. **Re-association :**

   - This service allows the station to switch its association from one AP to another.

   - Both association and re-association are initiated by the station.

3. **Disassociation :**

   - It is when the association between the station and the AP is terminated.

   - This can be initiated by either party.

   - A disassociated station cannot send or receive data.

4. **Distribution :**

   - This service determines how to route frames sent to the recipient.

   - The message is sent to the local AP (input AP), then distributed through the DS to the AP (output AP) that the recipient is associated with.

   - If the sender and receiver are in the same BSS, the input and output APs are the same. So the distribution service is logically invoked whether the data is going through the DS or not.

5. **Integration :**

   - This service handles situation when a frame needs to be sent through a non 802.11 network with a different addressing scheme or frame format, this service handles required translation. Thus 802.x LANs are integrated into the 802.11 DS.

- Notice that I have not mentioned ESS-transition. That is because it is not supported. A station can move to a new ESS but will have to reinitiate connections.

3.4 SS SERVICES

Station services are Authentication, De-authentication, Privacy, and MAC Service Data Unit (MSDU) Delivery.

1. **Authentication :**

   - With a wireless system, the medium is not exactly bounded as with a wired system.

   - In order to control access to the network, stations must first establish their identity.

   - This is much like trying to enter a radio net in the military. Before you are acknowledged and allowed to converse, you must first pass a series of tests to ensure that you are who you say you are. That is really all authentication is.

There are two types of authentication services offered by 802.11.

(i) **Open System Authentication :** This means that anyone who attempts to authenticate will receive authentication.

(ii) **Shared Key Authentication :** In order to become authenticated the users must be in possession of a shared secret. The shared secret is implemented with the use of the Wired Equivalent Privacy (WEP) algorithm. The shared secret is delivered to all stations ahead of time in some secure method (such as someone walking around and loading the secret onto each station).

2. **De-authentication :**

   - It is when either the station or AP wishes to terminate a stations authentication.

   - When this happens the station is automatically disassociated.

3. **Privacy :**

   - It is an encryption algorithm, which is used so that other 802.11 users cannot eavesdrop (maliciously get the data) on your LAN traffic.

   - IEEE 802.11 specifies Wired Equivalent Privacy (WEP) as an optional algorithm to satisfy privacy. If WEP is not used then stations are "in the clear" or "in the red", meaning that their traffic is not encrypted.

   - Data transmitted in the clear are called plaintext. Data transmissions, which are encrypted, are called ciphertext.

- All stations start "in the red" until they are authenticated.

4.   MSDU Delivery :

- Data transmission over 802.11 is not guaranteed to be completely reliable.

- MSDU delivery ensures that the information in the MAC service data unit is delivered between the medium access control service access points with detecting and correcting errors.

- Authentication is basically a network wide password. Privacy is whether or not encryption is used.

3.5 THE 802.11 PROTOCOL STACK AND ARCHITECTURE

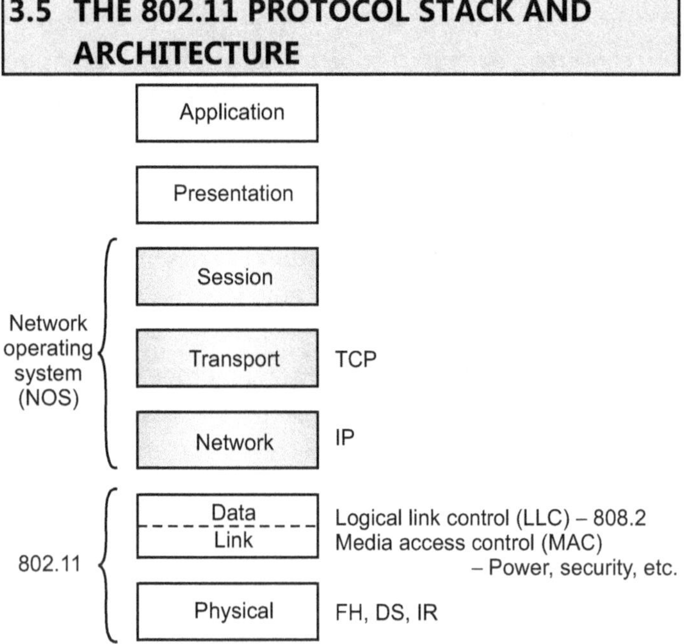

Fig. 3.3 : Protocol stack of 802.11

3.5.1 Physical Layer

There are three media that can be used for transmission over wireless LANs.

1. Infrared,

2. Microwave.

3. Radio frequency

- In 1985, the United States released the industrial, scientific, and medical (ISM) frequency bands.

- These bands are 902-928MHz, 2.4 - 2.4853 GHz, and 5.725 - 5.85 GHz and do not require licensing by the Federal Communications Commission (FCC).

- This prompted most of the wireless LAN products to operate within ISM bands.

1.   Infrared

- Infrared (IR) systems use very high frequencies to carry data.

- Like light, IR cannot penetrate opaque objects; it is either directed (line-of-sight) or diffuse technology.

- Inexpensive directed systems provide very limited range (3ft) and typically are used for personal area networks but occasionally are used in specific wireless LAN applications.

- Diffuse (or reflective) IR wireless LAN systems do not require line-of-sight, but cells (BSS) are limited to individual rooms.

- Infrared systems are simple in design and therefore inexpensive.

- They use the same signal frequencies used on fiber optic links.

- IR systems detect only the amplitude of the signal and so interference is greatly reduced.

- These systems are not bandwidth limited and thus can achieve transmission speeds greater than the other systems.

- Infrared transmission operates in the light spectrum and does not require a license from the FCC to operate.

- The drawbacks to IR systems are that the transmission spectrum is shared with the sun and other things such as fluorescent lights. If there is enough interference from other sources it can make the LAN useless.

- IR systems require a clear line of sight (LOS). IR signals cannot penetrate opaque objects.

2.   Microwave

- Microwave (MW) systems operate at less than 500 milliwatts of power in compliance with FCC regulations.

- They use narrow-band transmission with single frequency modulation and are set up mostly in the 5.8 GHz band. The big advantage to Microwave systems is higher throughput achieved.

3.   Radio Frequency

- Radio frequency technique can be implemented in two ways narrowband technology or spread spectrum technology. Radio frequency systems must use spread spectrum technology since it doesn't require license from FCC.

- This spread spectrum technology currently comes in two types : direct sequence spread spectrum (DSSS) and frequency hopping spread spectrum (FHSS).

(i) Direct-Sequence Spread Spectrum (DSSS)

- Direct-sequence spread spectrum (DSSS) generates a redundant bit pattern for each bit to be transmitted. This bit pattern is called a chip (or chipping code). The longer the chip, the greater the probability that the original data can be recovered and, of course, the more bandwidth required. Each bit is transmitted as 11 chips using a Barker Sequence. Even if one or more bits in the chip are damaged during transmission, statistical techniques embedded in the radio can recover the original data without the need for retransmission.

- To an unintended receiver, DSSS appears as low-power wideband noise and is rejected (ignored) by most narrowband receivers. With direct sequence spread spectrum the transmission signal is spread over an allowed band.

- A random binary string is used to modulate the transmitted signal. This random string is called the spreading code. The data bits are mapped to into a pattern of "chips" and mapped back into a bit at the destination. The number of chips that represent a bit is the spreading ratio.

- The higher the spreading ratio, the more the signal is resistant to interference. The lower the spreading ratio, the more bandwidth is available to the user. The FCC dictates that the spreading ratio must be more than 10. IEEE 802.11 standard requires a spreading ratio of 11. The transmitter and the receiver must be synchronized with the same spreading code.

(ii) Frequency-Hopping Spread Spectrum (FHSS)

- Frequency-hopping spread spectrum (FHSS) uses a narrowband carrier that changes frequency in a pattern known to both transmitter and receiver. To an unintended receiver, FHSS appears to be short-duration impulse noise. This technique splits the band into small 79 sub channels, each 1-MHz wide. The signal then hops from sub channel to sub channel transmitting short bursts of data on each channel for a set period of time, called dwell time. The hopping sequence must be synchronized at the sender and the receiver or information is lost. The FCC requires that the band is split into at least 75 subchannels and that the dwell time is no longer than 400 ms.

- Frequency hopping is less susceptible to interference because the frequency is constantly shifting. This feature gives FH systems a high degree of security. In order to jam a frequency hopping system the whole band must be jammed. Most new products in wireless LAN technology are currently being developed with FHSS technology. Its only problem is its low bandwidth.

3.5.2 Medium Access Layer

With more and more companies and individuals requiring portable and mobile computing, the need for wireless local area networks continues to rise throughout the world. Because of this growth, IEEE formed a working group to develop a Medium Access Control (MAC) and Physical Layer (PHY) standard for wireless connectivity for stationary, portable, and mobile computers within a local area. This working group is IEEE 802.11.

Medium Access Control Protocol

- Most wired LANs products use Carrier Sense Multiple Access with Collision Detection (CSMA/CD) as the MAC protocol. Carrier Sense means that the station will listen before it transmits. If there is already someone transmitting, then the station waits and tries again later.

- If no one is transmitting then the station goes ahead and sends what it has. But what if two stations send at the same time ? The transmissions will collide and the information will be lost. This is where Collision Detection comes into play. The station will listen to ensure that its transmission made it to the destination without collisions. If a collision occurred then the stations wait and try again later.

- The time the station waits is determined by the back off algorithm.

- This technique works great for wired LANs but wireless topologies can create a problem for CSMA/CD.

- The problem is the hidden node problem.

Hidden Node problem

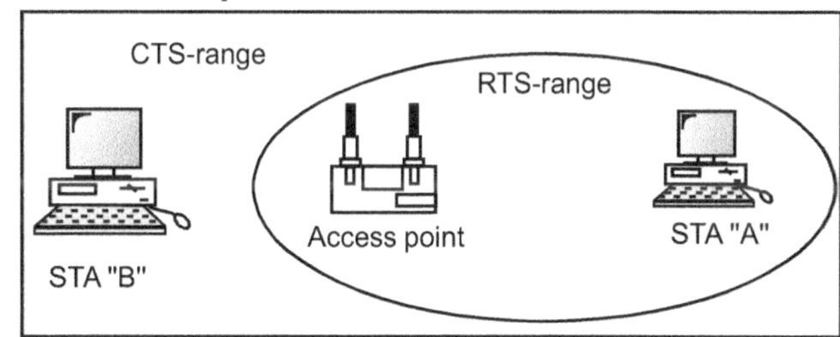

Fig. 3.4 : Hidden Node Problem

- Since not all the stations are within radio range of each other, transmission going on in one part of a cell may not be received elsewhere in the same cell. In this Hidden Node problem, STA-B cannot hear request of STA-A since RTS-Range of STA-A is limited. So if node STA-A is transmitting, STA-B will not know and may transmit as well.

- This will result in collisions. To combat this problem, a second carrier sense mechanism is available. Virtual Carrier Senseenables a station to reserve the medium for a specified period of time through the use of RTS/CTS frames. In the case described above, STA-A sends a RTS frame to the AP. The RTS will not be heard by STA-B. The RTS frame contains a duration and ID field, which specifies the period of time for which the medium is reserved for a subsequent transmission.

- The reservation information is stored in the Network Allocation Vector.

- (NAV : internal reminders to keep quiet for a certain period of time) of all stations detecting the RTS frame.Upon receipt of the RTS, the AP responds with a CTS frame, which also contains a duration and ID field specifying the period of time for which the medium is reserved. While STA-B did not detect the RTS, it will detect the CTS and update its NAV accordingly. Thus, collision is avoided even though some nodes are hidden from other stations. This solution is called Carrier Sense Multiple Access with Collision Avoidance or CSMA/CA. Now consider Fig. 3.5 in which station-2 is transmitting to station-1. If station-3 senses the medium, then since station-2 comes in range with station-3. So station-3 will sense the transmission going on. So it will falsely decide that it should not transmit to station-4.Note that transmission from 3 to 4 is OK, even when station-2 is transmitting. This problem is called as exposed station problem.

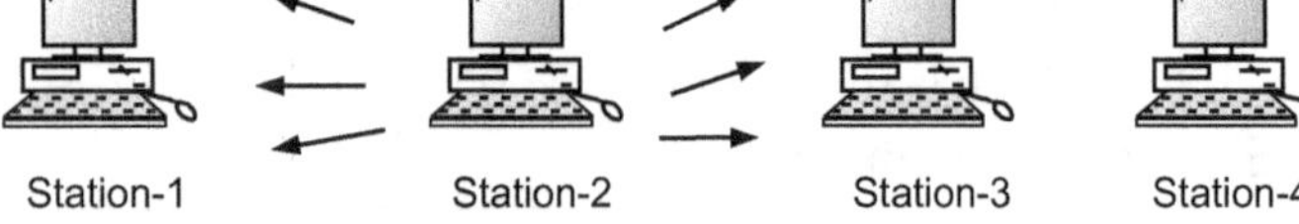

Fig. 3.5 : Exposed station problem

How CSMA/CA Work ?

- The station listens before it sends. If someone is already transmitting, wait for a random period and try again. If no one is transmitting then it sends a short message. This message is called the Ready To Send message (RTS). This message contains the destination address and the duration of the transmission.

- Other stations now know that they must wait that long before they can transmit. The destination (AP) then sends a short message, which is the Clear To Send message (CTS). This message tells the source that it can send without fear of collisions.

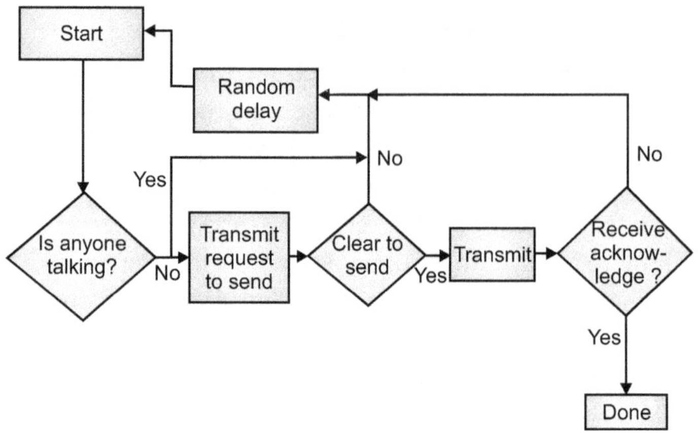

Fig. 3.6 : Working of CSMA/CA

- Each packet is acknowledged. If an acknowledgement is not received, the MAC layer retransmits the data. This entire sequence is called the 4-way handshake (RTS->CTS->Data->ACK).

3.6 BLUETOOTH

- Bluetooth is an emerging wireless communication technology that allows devices, within 10-100 meter proximity, to communicate with each other. The primary goal of this technology is to enable devices to communicate without physical cables. It is a worldwide specification for a small low-cost radio. It links mobile computers, mobile phones, other portable handheld devices, and provides Internet connectivity.

- It is developed, published and promoted by the Bluetooth Special Interest Group (SIG).Its key features are robustness, low complexity, low power and low cost. Bluetooth does not require direct line of sight and can also support multipoint communication in addition to point to point communication. The short range transceivers that are built into mobile gadgets to provide Bluetooth compatibility are designed to operate in the 2.45 GHz unlicensed radio band. It provides data rate up to 721 kbps as well as three 64 kbps voice channels. Through the use of frequency hopping, a Bluetooth transceiver can minimize the effect of interference from other signals by hopping to a new frequency after transmitting or receiving a packet.

- (Bluetooth hop frequency is 1600 hops/second). Each Bluetooth gadget has a unique 12 bit address. In order for Bluetooth gadget A to connect with Bluetooth gadget B, gadget A must know the (12-bit) address of gadget B. Bluetooth supports gadget authentication and communications encryption. Bluetooth uses GFSK (Gaussian frequency shift keying) modulation technique.

3.6.1 Architecture

Bluetooth defines two type of networks :

1. Piconet and

2. Scatternet

- We can assume piconet as a group of devices. A Bluetooth piconet consists of 1 master and 7 active slave device (all nodes must be within 10 meter range). All the slave devices are synchronized with the master device. There can be 255 parked nodes in the single piconet but at any time maximum 7 are communicating. A piconet is shown in Fig. 3.7.

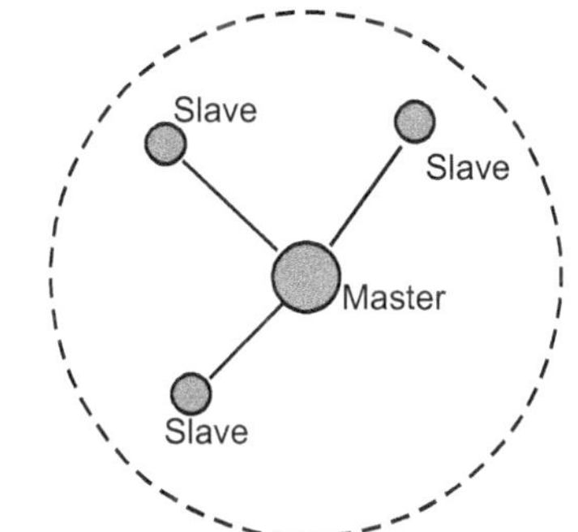

Fig. 3.7 : A simple Piconet

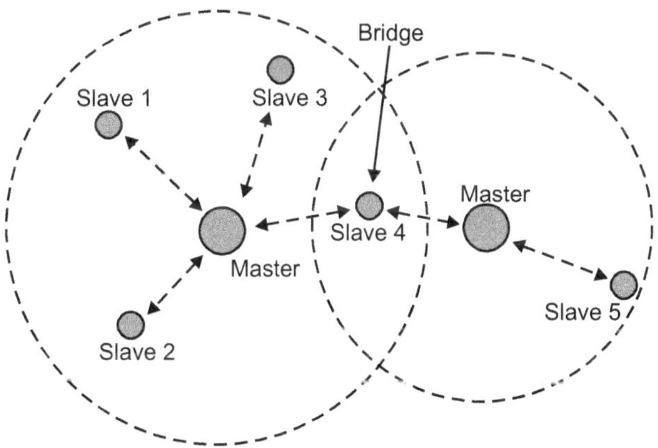

Fig. 3.8 : A simple scatternet

- Two piconets can be connected through a common Bluetooth device (a gateway or bridge) to form a scatternetas shown in Fig. 3.8. These interconnected piconets within the scatternet can enable devices which are not directly communicating with each other, or which are out of range of another device, to exchange data through several hops in the scatternet. Current implementations of Bluetooth depend primarily on simple point-to-point data links between Bluetooth devices within direct range of each other.

3.6.2 The Bluetooth Protocol Stack

The heart of the Bluetooth specification is the Bluetooth protocol stack. The protocol stack is shown in the figure given below :

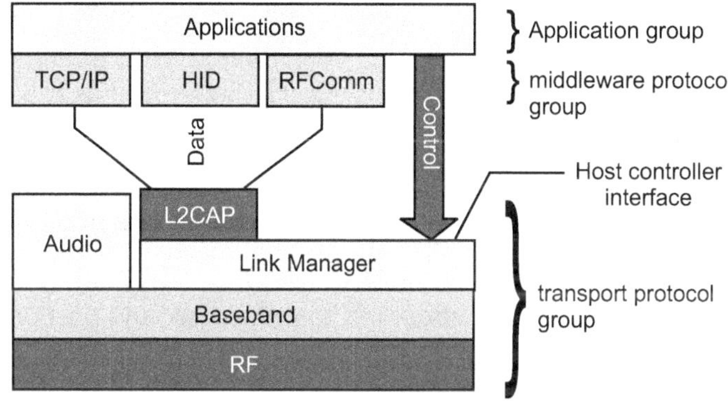

Fig. 3.9 : Bluetooth protocol stack

The Bluetooth specification divides the Bluetooth protocol stack into three logical groups.

1. Transport Protocol group,

2. Middleware Protocol group and

3. Application group

- The Transport group protocols allow Bluetooth devices to locate each other, and to manage physical and logical links with higher layer protocols and applications. The Radio, Baseband, Link Manager, Logical Link Control and Adaptation (L2CAP) layers and the Host Controller Interface (HCI) are included in the Transport Protocol group. These protocols support both asynchronous and synchronous transmission. All the protocols in the transport protocol group are required to support communications between Bluetooth devices.

- The Middleware Protocol group includes third-party and industry-standard protocols, as well as Bluetooth SIG developed protocols. These protocols allow existing and new applications to operate over Bluetooth links.

- Industry standard protocols include Point-to-Point Protocol (PPP), Internet Protocol (IP), Transmission Control Protocol (TCP), Wireless Application Protocols (WAP), and object exchange (OBEX) protocols.

Some of the layers in the transport protocol group are explained below :

Radio Layer : The specification of the Radio layer is primarily concerned with the design of the Bluetooth transceivers.

Baseband Layer : This layer defines how Bluetooth devices search for and connect to other devices.

The master and slave roles that a device may assume are defined here.

The master and slave communicate only in their pre-assigned time slots.

The devices use a time division duplexing (TDD), packet-based polling scheme for communication.

The Bluetooth specification doesn't establish a clear distinction between the responsibilities of the baseband and those of the link controller.

The best way to think about it is that the baseband portion of the layer is responsible for properly formatting data for transmission to and from the radio layer.

In addition, it handles the synchronization of links. The Baseband layer supports two types of links : Synchronous Connection- Oriented (SCO) and Asynchronous Connection-Less (ACL).

SCO links are characterized by a periodic, single-slot packet assignment, and are primarily used for voice transmissions that require fast, consistent data transfer. A device that has established a SCO link has, in essence, reserved certain time slots for its use. Its data packets are treated as priority packets, and will be serviced before any ACL packets.

A device with an ACL link can send variable length packets of 1, 3 or 5 time-slot lengths. But it has no time slots reserved for it.

Link Manager Layer : This layer implements the Link Manager Protocol (LMP). LMP manages bandwidth allocation for general data, bandwidth reservation for audio traffic, authentication using challenge response methods, and trust relationships between devices, encryption of data and control of power usage. Power usage control includes the negotiation of low power activity modes and the determination of transmission power levels.

L2CAP Layer : The Logical Link Control and Adaptation Protocol (L2CAP) layer provides the interface between the higher- layer protocols and the lower-layer transport protocols. L2CAP supports multiplexing of several higher layer protocols, such as RFComm and SDP. This allows multiple protocols and applications to share the air-interface. L2CAP is also responsible for packet segmentation and reassembly, and for maintaining the negotiated service level between devices.

HCI Layer : The Host Controller Interface (HCI) layer defines a standard interface for upper level applications to access the lower layers of the stack. This layer is not a required part of the specification. Its purpose is to enable interoperability among devices and the use of existing higher level protocols and applications.

3.6.3 Bluetooth Frame Structure

There are several frame formats, the most important of which is shown in following Fig. 3.10. It begins with an access code that usually identifies the master so that slaves within radio range of two masters can tell which traffic is for them. Then comes a 54-bit header containing typical MAC sublayer fields. Then comes the data field, of up to 2744 bits (for a five-slot transmission).

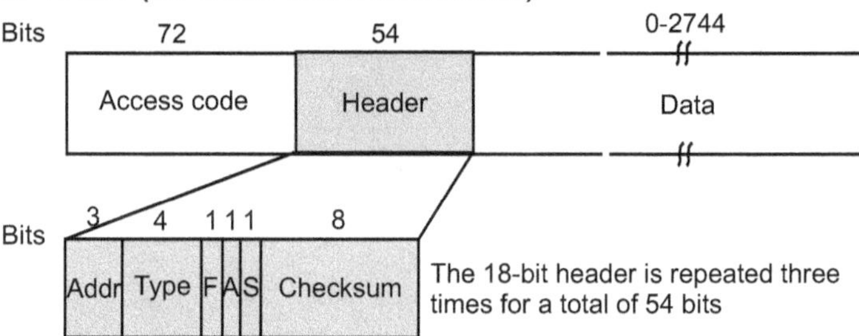

Fig. 3.10 : A Typical Bluetooth data frame

Header Field :

Address Field : The Address field identifies which of the eight active devices the frame is intended for.

Type : The Type field identifies the frame type (ACL, SCO, poll, or null).

F : The Flow bit is asserted by a slave when its buffer is full and cannot receive any more data.

A : The Acknowledgement bit is used to piggyback an ACK onto a frame.

S : The Sequence bit is used to number the frames to detect retransmissions. The protocol is stop-and-wait, so 1 bit is enough.

Checksum : Then comes the 8-bit header Checksum. The entire 18-bit header is repeated three times to form the 54-bit header shown in figure.

On the receiving side, a simple circuit examines all three copies of each bit. If all three are the same, the bit is accepted.

If not, the majority opinion wins.

3.6.4 Advantages of Bluetooth

It is an open specification that is publicly available and royalty free; Its short-range wireless capability allows peripheral devices to communicate over a single air-interface, replacing cables that use connectors with a multitude of shapes, sizes and numbers of pins; Bluetooth supports both voice and data, making it an ideal technology to enable many types of devices to communicate; Bluetooth uses an unregulated frequency band available anywhere in the world.

3.7 BLUETOOTH SPECIFICATION

3.7.1 Radio Specifications

This section describes the specifications of the Bluetooth link controller which carries out the baseband protocols and other low-level link routines. The Bluetooth radio specification is a short document that gives the details of radio transmission for Bluetooth devices. This specification defines three classes of transmitters based on output power.

- **Class 1:** Outputs 100mW for maximum range and 1mW minimum. In this class, power control is mandatory, ranging from 4 to 20dBm. This class provides the greatest distance.

- **Class 2:** Outputs 1.4mW at maximum and 0.25mW at minimum. Power control is optional.

- **Class 3:** Lowest power. Nominal output is 1.

Table 3.1 : Bluetooth Radio and Baseband Parameters

Topology	Up to 7 Simultaneous Links In a Logical Star
Modulation	GFSK
Peak data rate	1Mbps
RF bandwidth	220 kHz, 1Mhz
RF band	2.4 GHz, ISM band
RF carrier	3/79
Carrier spacing	1MHz
Transmit power	0.1 W
Piconet access	FH-TDD-TDMA
Frequency hop rate	1600 hops/s
Scatternet access	FH-CDMA

Bluetooth makes use of the 2.4-GHz band within the ISM (Industrial, Scientific, Medical) band. In most of the countries, the bandwidth is sufficient to define 79 1-MHz physical channels. Power control is used to control the devices so that they do not emit more RF power than necessary. The power control algorithm is implemented using link management protocol between a master and slaves in a piconet. Modulation for Bluetooth is Gaussian FSK, with a binary 1 is represented by a positive frequency deviation and a binary 0 is represented by a negative frequency deviation from the centre frequency.

Table 3.2: International Bluetooth Frequency Allocation

Area	Range	RF Channels
US, Most of Europe	2.4 to 2.4835 GHz	f= 2.402+n MHz, n=0 to 78
Japan	2.471 to 2.497 GHz	f= 2.473+n MHz, n=0 to 22
Spain	2.445 to 2.475 GHz	f= 2.449+n MHz, n=0 to 22
France	2.4465 to 2.4835 GHz	f= 2.454+n MHz, n=0 to 22

3.7.2 Baseband Specification

This is one of the most complex specifications of Bluetooth.

Frequency Hopping (FH)

1. It provides resistance to interference and multipath effects

2. It provides a form of multiple access among co-located devices in different piconets.

Channel Definition

- The channel is represented by a pseudo-random hopping sequence hopping through the 79 or 23 RF channels. The hopping sequence is unique for the piconet and is determined by the Bluetooth device address of the master; the phase in the hopping sequence is determined by the Bluetooth clock of the master. The channel is divided into time slots where each slot corresponds to an RF hop frequency. Consecutive hops correspond to different RF hop frequencies.

- The nominal hop rate is 1600 hops/s. All Bluetooth units participating in the piconet are time- and hop-synchronized to the channel.

Time Slots

- The channel is divided into time slots, each 625 µs in length. The time slots are numbered according to the Bluetooth clock of the piconet master. The slot numbering ranges from 0 to 227-1 and is cyclic with a cycle length of 227.

- In the time slots, master and slave can transmit packets. A TDD scheme is used where master and slave alternatively transmit, see Fig. 3.11. The master shall start its transmission in even numbered time slots only, and the slave shall start its transmission in odd numbered time slots only. The packet start shall be aligned with the slot start. Packets transmitted by the master or the slave may extend over up to five time slots. The RF hop frequency shall remain fixed for the duration of the packet.

- For a single packet, the RF hop frequency to be used is derived from the current Bluetooth clock value. For a multi-slot packet, the RF hop frequency to be used for the entire packet is derived from the Bluetooth clock value in the first slot of the packet. The RF hop frequency in the first slot after a multi-slot packet shall use the frequency as determined by the current Bluetooth clock value.

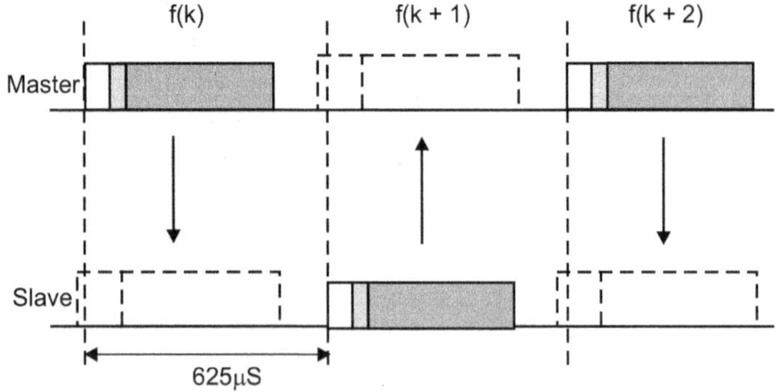

Fig. 3.11 : TDD and Timing

- Fig. 3.12 illustrates the hop definition on single and multi-slot packets. If a packet occupies more than one time slot, the hop frequency applied shall be the hop frequency as applied in the time slot where the packet transmission was started.

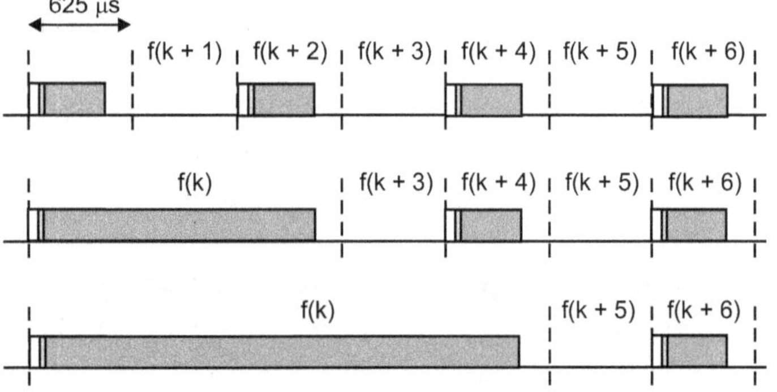

Fig. 3.12 : Multislot packets

3.7.3 Physical Links

Between master and slave(s), different types of links can be established. Two link types have been defined:

1. Synchronous Connection-Oriented (SCO) link
2. Asynchronous Connection-Less (ACL) link

The SCO link is a point-to-point link between a master and a single slave in the piconet. The master maintains the SCO link by using reserved slots at regular intervals. The ACL link is a point-to-multipoint link between the master and all the slaves participating on the piconet. In the slots not reserved for the SCO link(s), the master can establish an ACL link on a per-slot basis to any slave, including the slave(s) already engaged in an SCO link.

1. SCO LINK

- The SCO link is a symmetric, point-to-point link between the master and a specific slave. The SCO link reserves slots and can therefore be considered as a circuit-switched connection between the master and the slave. The SCO link typically supports time-bounded information like voice. The master can support up to three SCO links to the same slave or to different slaves. A slave can support up to three SCO links from the same master, or two SCO links if the links originate from different masters. SCO packets are never retransmitted.

- The master will send SCO packets at regular intervals, the so-called SCO interval T_{SCO} (counted in slots) to the slave in the reserved master-to-slave slots.

- The SCO slave is always allowed to respond with an SCO packet in the following slave-to-master slot unless a different slave was addressed in the previous master-to-slave slot. If the SCO slave fails to decode the slave address in the packet header, it is still allowed to return an SCO packet in the reserved SCO slot.

- The SCO link is established by the master sending an SCO setup message via the LM protocol. This message will contain timing parameters such as the SCO interval T_{SCO} and the offset D_{SCO} to specify the reserved slots.

- In order to prevent clock wrap-around problems, an initialization flag in the LMP setup message indicates whether initialization procedure 1 or 2 is being used.

- The slave shall apply the initialization method as indicated by the initialization flag. The master uses initialization 1 when the MSB of the current master clock

(CLK27) is 0; it uses initialization 2 when the MSB of the current master clock

(CLK27) is 1. The master-to-slave SCO slots reserved by the master and the slave shall be initialized on the slots for which the clock satisfies the following equation:

CLK27-1 mod T_{SCO} = D_{SCO} for initialization 1

(CLK27,CLK26-1) mod T_{SCO} = D_{SCO} for initialization 2

The slave-to-master SCO slots shall directly follow the reserved master-toslave SCO slots. After initialization, the clock value CLK(k+1) for the next master- to-slave SCO slot is found by adding the fixed interval TSCO to the clock value of the current master-to-slave SCO slot: CLK(k+1) = CLK(k) + T_{SCO}

2. ACL LINK

- In the slots not reserved for SCO links, the master can exchange packets with any slave on a per-slot basis. The ACL link provides a packet-switched connection between the master and all active slaves participating in the piconet.

- Both asynchronous and isochronous services are supported. Between a master and a slave only a single ACL link can exist. For most ACL packets, packet retransmission is applied to assure data integrity.

- A slave is permitted to return an ACL packet in the slave-to-master slot if and only if it has been addressed in the preceding master-to-slave slot. If the slave fails to decode the slave address in the packet header, it is not allowed to transmit. ACL packets not addressed to a specific slave are considered as broadcast packets and are read by every slave. If there is no data to be sent on the ACL link and no polling is required, no transmission shall take place.

3.7.4 Packets

General Format

- The bit ordering when defining packets and messages in the *Baseband Specification*, follows the *Little Endian format*, i.e., the following rules apply:
  - ➢ The *least significant bit* (LSB) corresponds to ;
  - ➢ The LSB is the first bit sent over the air;
  - ➢ In illustrations, the LSB is shown on the left side;

- The link controller interprets the first bit arriving from a higher software layer as ; i.e. this is the first bit to be sent over the air. Furthermore, data fields generated internally at baseband level, such as the packet header fields and payload header length, are transmitted with the LSB first. For instance, a 3-bit parameter $X-3$ is sent as over the air where 1 is sent first and 0 is sent last.

- The data on the piconet channel is conveyed in packets. The general packet format is shown in Fig. 3.13. Each packet consists of 3 entities: the access code, the header, and the payload. In the figure, the number of bits per entity is indicated.

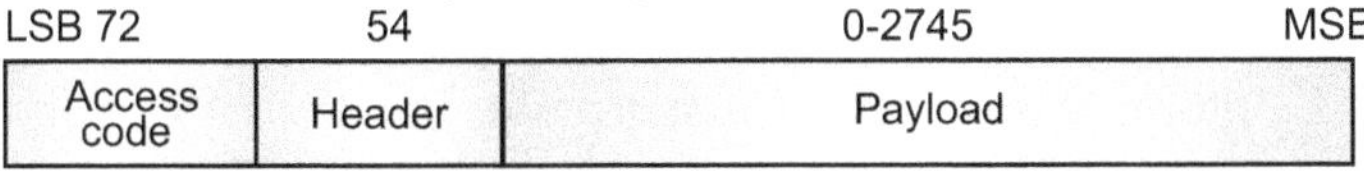

Fig. 3.13 : Standard packet Format

- The access code and header are of fixed size : 72 bits and 54 bits respectively.

- The payload can range from zero to a maximum of 2745 bits. Different packet types have been defined.

Packets may consist of the (shortened) access code only, of the access code + header, or of the access code + header + payload.

Access Code

- Each packet starts with an access code. If a packet header follows, the access code is 72 bits long, otherwise the access code is 68 bits long. This access code is used for synchronization, DC offset compensation and identification. The access code identifies all packets exchanged on the channel of the piconet: all packets sent in the same piconet are preceded by the same channel access code. In the receiver of the Bluetooth unit, a sliding correlator correlates against the access code and triggers when a threshold is exceeded.

- This trigger signal is used to determine the receive timing. The access code is also used in paging and inquiry procedures. In this case, the access code itself is used as a signalling message and neither a header nor a payload is present. The access code consists of a preamble, a sync word, and possibly a trailer, see Fig. 3.14.

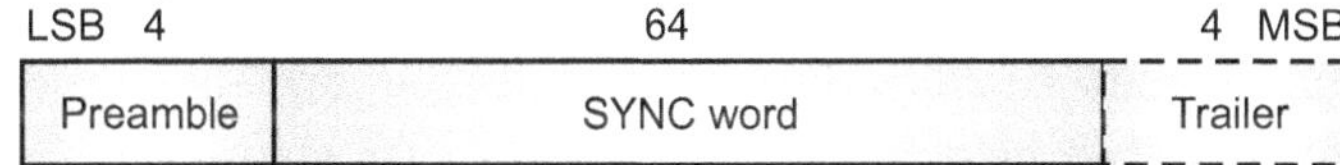

Fig. 3.14 : Access code format

Access Code Types

- There are three different types of access codes defined:
  1. Channel Access Code (CAC)
  2. Device Access Code (DAC)
  3. Inquiry Access Code (IAC)

- The CAC consists of a preamble, sync word, and trailerand its total length is 72 bits. When used as self-contained messages without a header, the DAC and IAC do not include the trailer bits and are of length 68 bits.

- The preamble is a fixed zero-one pattern of 4 symbols used to facilitate DC compensation. The sequence is either 1010 or 0101, depending whether the LSB of the following sync word is 1 or 0, respectively.

Sync Word

- The sync word is a 64-bit code word derived from a 24 bit address (LAP); for the CAC the master's LAP is used; for the GIAC and the DIAC, reserved, dedicated

- LAPs are used; for the DAC, the slave unit LAP is used. The construction guarantees large Hamming distance between sync words based on different LAPs. In addition, the good auto correlation properties of the sync word improve on the timing synchronization process.

Trailer

- The trailer is appended to the sync word as soon as the packet header follows the access code. This is typically the case with the CAC, but the trailer is also used in the DAC and IAC when these codes are used in FHS packets exchanged during page response and inquiry response procedures.

- The trailer is a fixed zero-one pattern of four symbols. The trailer together with the three MSBs of the syncword form a 7-bit pattern of alternating ones and zeroes which may be used for extended DC compensation. The trailer sequence is either 1010 or 0101 depending on whether the MSB of the sync word is 0 or 1, respectively. The choice of trailer is illustrated in Fig. 3.15.

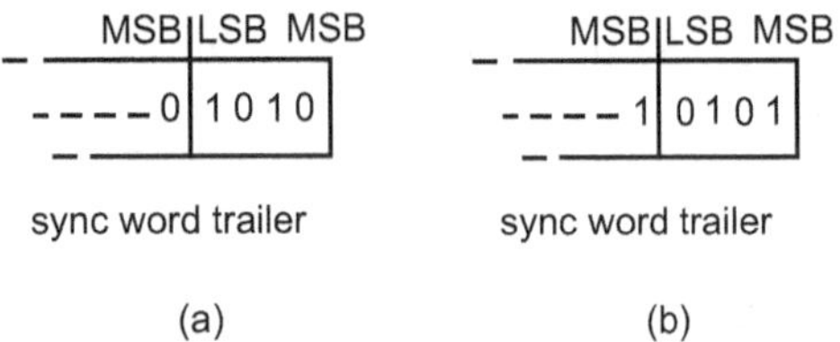

Fig. 3.15 : Trailer in CAC

3.7.5 Packet Header

The header contains link control (LC) information and consists of six fields:

1. **AM_ADDR:** 3- bit active member address
2. **TYPE:** 4-bit type code
3. **FLOW:** 1-bit flow control
4. **ARQN:** 1-bit acknowledge indication
5. **SEQN:** 1-bit sequence number
6. **HEC:** 8-bit header error check

The total header, including the HEC, consists of 18 bits, see Fig. 3.16 and is encoded with a rate 1/3 FEC resulting in a 54-bit header. Note that the AM_ADDR and TYPEfields are sent with their LSB first. The function of the different fields will beexplained next.

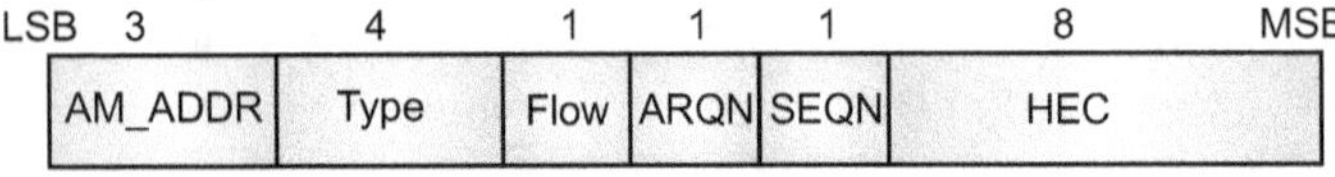

Fig. 3.16 : Header format

1. AM_ADDR

- The AM_ADDR represents a member address and is used to distinguish between the active members participating on the piconet. In a piconet, one or more slaves are connected to a single master. To identify each slave separately, each slave is assigned a temporary 3-bit address to be used when it is active. Packets exchanged between the master and the slave all carry the AM_ADDR of this slave; that is, the AM_ADDR of the slave is used in both master-to-slave packets and in the slave-to-master packets. The all-zero address is reserved for broadcasting packets from the master to the slaves.

- An exception is the FHS packet which may use the all-zero member address but is *not* a broadcast message. Slaves that are disconnected or parked give up their AM_ADDR. A new AM_ADDR has to be assigned when they re-enter the piconet.

2. Type

- Sixteen different types of packets can be distinguished. The 4-bit TYPE code specifies which packet type is used. Important to note is that the interpretation of the TYPE code depends on the physical link type associated with the packet.

- First, it shall be determined whether the packet is sent on an SCO link or an ACL link. Then it can be determined which type of SCO packet or ACL packet has been received. The TYPE code also reveals how many slots the current packet will occupy. This allows the non-addressed receivers to refrain from listening to the channel for the duration of the remaining slots.

3. Flow

- This bit is used for flow control of packets over the ACL link. When the RX buffer for the ACL link in the recipient is full and is not emptied, a STOP indication (FLOW=0) is returned to stop the transmission of data temporarily. Note, that the STOP signal only concerns ACL packets. Packets including only link control information (ID, POLL and NULL packets) or SCO packets can still be received. When the RX buffer is empty, a GO indication (FLOW=1) is returned.

- When no packet is received, or the received header is in error, a GO is assumed implicitly. In this case, the slave can receive a new packet with CRC although its RX buffer is still not emptied. The slave shall then return a NAK in response to this packet even if the packet passed the CRC check.

4. ARQN

- The 1-bit acknowledgment indication ARQN is used to inform the source of a successful transfer of payload data with CRC, and can be positive acknowledge ACK or negative acknowledge NAK. If the reception was successful, an ACK (ARQN=1) is returned, otherwise a NAK (ARQN=0) is returned. When no return message regarding acknowledge is received, a NAK is assumed implicitly.

- NAK is also the default return information. The ARQN is piggy-backed in the header of the return packet. The success of the reception is checked by means of a cyclic redundancy check (CRC) code. An unnumbered ARQ scheme which means that the ARQN relates to the latest received packet from the same source, is used.

5. SEQN

- The SEQN bit provides a sequential numbering scheme to order the data packet stream. For each new transmitted packet that contains data with CRC, the SEQN bit is inverted. This is required to filter out retransmissions at the destination; if a retransmission occurs due to a failing ACK, the destination receives the same packet twice. By comparing the SEQN of consecutive packets, correctly received retransmissions can be discarded.

6. HEC

- Each header has a header-error-check to check the header integrity. The HEC consists of an 8-bit word generated by the polynomial 647 (octal representation).

- Before generating the HEC, the HEC generator is initialized with an 8-bit value. For FHS packets sent in master page response state, the slave upper address part (UAP) is used. For FHS packets sent in inquiry response, the default check initialization is used. In all other cases, the UAP of the master device is used. After the initialization, a HEC is calculated for the 10 header bits. Before checking the HEC, the receiver must initialize the HEC check circuitry with the proper 8-bit UAP (or DCI). If the HEC does not check, the entire packet is disregarded.

Table 3.3 : Bluetooth Packet Types

Type Code	Physical Link	Name	No. of Slots	Description
0000	Common	NULL	1	Has no payload. Used to return link information to the source regarding the success of the previous transmission or the success of RX buffer. Not acknowledged
0001	Common	POLL	1	Has no payload. Used by master to poll a slave. Acknowledged.
0010	Common	FHS	1	Special control packet for revealing device address and the clock of the sender. Used in page master response inquiry response, and frequency hop synchronization. 2/3 FEC encoded.
0011	Common	DM1	1	Supports control messages and can also carry user data. 16 bit CRC. 2/3 FEC encoded
0101	SCO	HV1	1	Carries 10 information bytes; typically used for 64-kbps voice. 1/3 FEC encoded
0110	SCO	HV2	1	Carries 20 information bytes; typically used for 64-kbps voice. 2/3 FEC encoded
0111	SCO	HV3	1	Carries 30 information bytes; typically used for 64-kbps voice. No FEC encoded
1000	SCO	DV	1	Combined data (150 bits) and voice (50 bits) packet. Data field 2/3 FEC encoded.
0100	ACL	DH1	1	Carries 28 information bytes plus 16 bit CRC; typically used for high speed data. Not FEC encoded
1001	ACL	AUX1	1	Carries 30 information bytes plus 16 bit CRC; typically used for high speed data. Not FEC encoded
1010	ACL	DM3	3	Carries 123 information bytes plus 16 bit CRC; 2/3 FEC encoded
1011	ACL	DH3	3	Carries 185 information bytes plus 16 bit CRC; Not FEC encoded
1110	ACL	DM5	5	Carries 226 information bytes plus 16 bit CRC; Not FEC encoded
1111	ACL	DH5	5	Carries 341 information bytes plus 16 bit CRC; Not FEC encoded

3.7.6 Payload Format

- For some packet types, the baseband specification defines a format for the payload field. For voice payloads, no header is defined. For all of the ACL packets and for data portion of the SCO DV packet, a header is defined. For data payload, the payload format consists of three fields.

1. **Payload Header:** An 8 bit header is defined for single slot packets and a 16 bit header is defined for multislot packets.

2. **Payload Body:** Contains user information

3. **CRC:** A 16 bit CRC code is used on all data payloads except AUX1 packet.

3.7.7 Error Correction

There are three error correction schemes defined for Bluetooth:

1. 1/3 rate FEC

2. 2/3 rate FEC

3. ARQ scheme for the data

The purpose of the FEC scheme on the data payload is to reduce the number of retransmissions. However, in a reasonable error-free environment, FEC gives unnecessary overhead that reduces the throughput. Therefore, the packet definitions have been kept flexible to useFEC in the payload or not, resulting in the DM and DH packets for the ACL link and the HV packets for the SCO link. The packet header is always protected bya 1/3 rate FEC; it contains valuable link information and should be able to sustain more bit errors.Correction measures to mask errors in the voice decoder are not included in this section.

FEC CODE: RATE 1/3

- A simple 3-times repetition FEC code is used for the header. The repetition code is implemented by repeating the bit three times, see the illustration in Fig. 3.17. The 3-bit repetition code is used for the entire header, and also for the voice field in the HV1 packet.

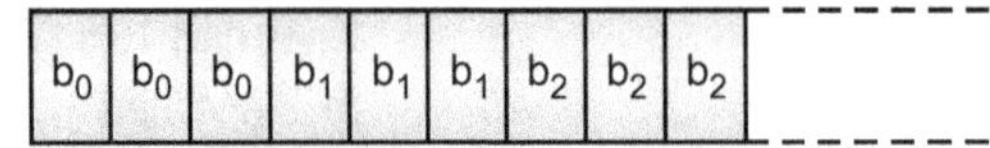

Fig. 3.17 : Bit repetition encoding scheme

FEC CODE: RATE 2/3

- The other FEC scheme is a (15,10) shortened Hamming code. The generator polynomial is . This corresponds to 65 in octal notation. The LFSR generating this code is depicted in 3.18. Initially all register elements are set to zero. The 10 information bits are sequentially fed into the LFSR with the switches S1 and S2 set in

position 1. Then, after the final input bit, the switches S1 and S2 are set in position 2, and the five parity bits are shifted out. The parity bits are appended to the information bits. Consequently, each block of 10 information bits is encoded into a 15 bit codeword. This code can correct all single errors and detect all double errors in each codeword. This 2/3 rate FEC is used in the DM packets, in the data field of the DV packet, in the FHS packet, and in the HV2 packet.

- Since the encoder operates with information segments of length 10, tail bits with value zero may have to be appended after the CRC bits.The total number of bits to encode, i.e., payload header, user data, CRC, and tail bits, must be a multiple of 10. Thus, the number of tail bits to append is the least possible that achieves this (i.e., in the interval 0...9). These tail bits are not included in the payload length indicator.The ARQ scheme is used with DM and DH packets and the data field of DV packets

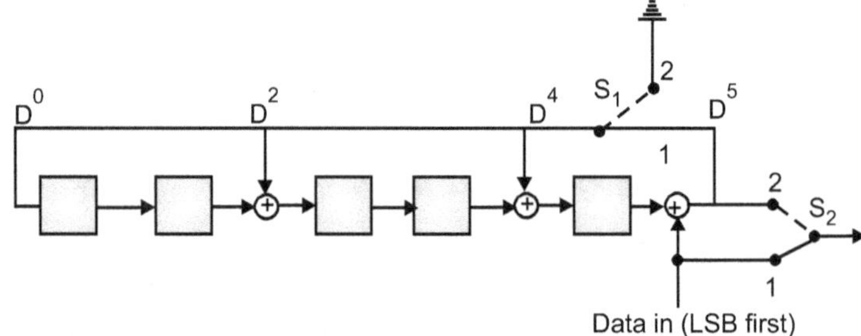

Fig. 3.18 : LFSR generating (15,10) shortened Hamming code

3.7.8 Error Detection

The destination detects errors and discards packets that are in error. Error detection is achieved with a CRC error detecting code supplemented with the FEC code.

Positive Acknowledgement :

- The destination returns a positive acknowledgement to successfully received, error free packets. Retransmission after timeout: The source retransmits a packet that has not been acknowledged after a predetermined amount of time. Negative acknowledgement and retransmission: The destination returns a negative acknowledgement to packets in which an error is detected. The source retransmits such packets.

Logical Channels

- Bluetooth defines five types of logical channels to carry out different type of payload traffic.

  ➤ **Link Control (LC):** It is used to manage the flow of packets over the link interface. The LC channel is mapped to the packet header. This channel carries low level link control information like ARQ, flow control and payload characterization. The LC channel I carried in every packet except in the ID packet, which has no packet header.

➤ **Link Manager (LM):** Transports link management information between participating stations. This logical channel supports LMP traffic and can be carried over either an SCO or ACL link.User asynchronous (UA): Carries asynchronous user data. This channel is normally carried over the ACL link but may be carried in a DV packet on the SCO link.User isochronous (UI): Carries user isochronous user data.

This channel is normally carried over the ACL link but may be carried in a DV packet on he SCO link. At the baseband level UI channels are treated in the same way as a UA channel.User synchronous (US): Carries synchronous user data. This channel is carried over the SCO link.

3.7.9 Channel Control

The operation of a piconet can b understood in terms of the states of operation during link establishment and maintenance.

Two main states are

1. **Standy:** The default state. This is a low power state in which only the native clock is running

2. **Connection :** The device is connected to a piconet as a master or a slave.

In addition to these two states there are seven substates that can be used to add slaves to a piconet. The substates are as follows:

1. **Page :** Device has issued a page. Used by the master to activate and connect to slave. Master sends page message by transmitting slave's device access code (DAC) in different hop channels.

2. **Page Scan :** Device is listening for a page with its on DAC.

3. **Master Response :** A device acting as a master receives a page response from a slave. The device can now enter the connection state or return to the page state to page for other slaves.

4. **Slave Response:** A device acting as a slave responds to a page from master. If connection setup succeeds, the device enters the connection state or it returns to the page scan state.

5. **Inquiry:** Device has issued an inquiry, to find the identity of the devices within range.

6. **Inquiry Scan :** Device is listening for an inquiry

7. **Inquiry Response :** A device that has issued an inquiry receives an inquiry response.

Inquiry Procedure

- The first step in establishing a piconet is for a potential master to identify devices in range that wish to participate in the piconet. A device begins an inquiry procedure for this purpose. Inquiry procedure begins when the potential master transmits an ID packet with an inquiry access code (IAC), which is code common to all Bluetooth devices. Out of 79 radio carrier, 32 are considered as wake up carriers. The master broadcasts the IAC over each of the 32 wake up carriers.

- This is done in Inquiry state. Meanwhile, devices in the Standby state periodically enter the Inquiry Scan state to search for IAC messages on the wake up carriers. When a device receives the inquiry, it enters the Inquiry response state and returns an FHS packet which contains its device address and timing information required by the master to initiate a connection. The master dos not respond to the FHS packet and may remain in the Inquiry state.

- Once a device has responded to an Inquiry, it moves to the page scan state to await a page from the master in order to establish a connection. If collision occurs in the Inquiry Response phase, no page will be received and device may return to the Inquiry scan state to attempt another inquiry and response.

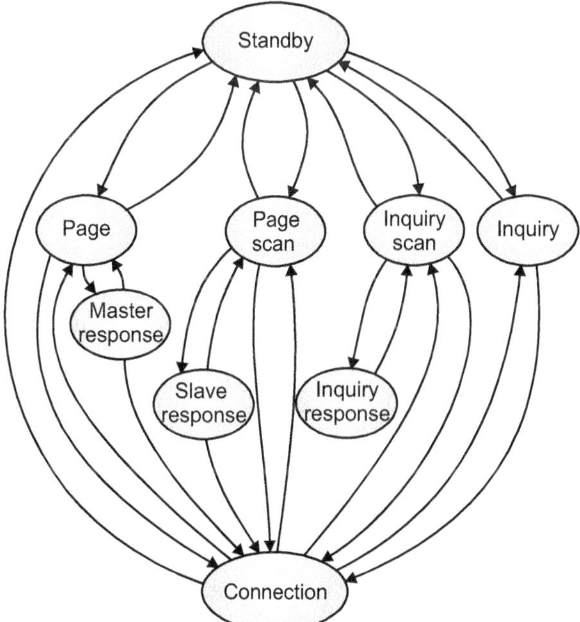

Fig. 3.19 : Bluetooth state transition diagram

Page Procedure

- Once the master has found devices within its range, it is able to establish connections to each device, setting up a piconet. Master uses the device's address (BD_ADDR) in order to page that device. The master pages by using an ID packet with a device access code (DAC) of the specific slave.

- The slave responds by returning the same DAC ID packet to the master. The master responds to this with its own FHS packet, containing its own device address and its real time Bluetooth clock value. Once again, the slave sends a response DAC ID packet to the master to confirm the receipt of master's FHS packet. At this point slave transits from Response state to the connection state. Master may continue to page until it has connected to all the desired slaves, then master enters the connection state.

ConnectionState

- In the CONNECTION state, the connection has been established and packets can be sent back and forth. In both units, the channel (master) access code and the master Bluetooth clock are used. The hopping scheme uses the *channelhopping sequence*. The master starts its transmission in even slots (CLK 1-0=00), the slave starts its transmission in odd slots (CLK 1-0=10)

- The CONNECTION state starts with a POLL packet sent by the master to verify the switch to the master's timing and channel frequency hopping. The slave can respond with any type of packet. If the slave does not receive the POLL packet or the master does not receive the response packet for *newconnectionTO* number of slots, both devices will return to page/page scan substates. he first information packets in the CONNECTION state contain control messages that characterize the link and give more details regarding the Bluetooth units. These messages are exchanged between the link managers of the units.

- For example, it defines the SCO links and the sniff parameters. Then the transfer of user information can start by alternately transmitting and receiving packets. The CONNECTION state is left through a detach or reset command. The detach command is used if the link has been disconnected in the normal way. All configuration data in the Bluetooth link controller is still valid. The reset command is a hard reset of all controller processes. After a reset, the controller has to be reconfigured. The Bluetooth units can be in several modes of operation during the CONNECTION state: active mode, sniff mode, hold mode, and park mode.

Active : The slave actively participates in the piconet by listening, transmitting and receiving packets. The master periodically transmits to he slaves to maintain synchronization.

Sniff : The slave does not listen on every receive slot but only on slots specified for its messages. The slave can operate in reduced power state for the rest of the time. In sniff mode the master designs specific number of slots for transmission to a specific slave.

Hold : The device in this mode does not support ACL packets and goes to reduced power status. The slave may still participate in SCO exchanges. During the period of inactivity slave is free to idle in reduced power state or participate in other piconet.

Park : When a slave does not need to participate on the piconet but still want to be the part of piconet, it can enter in park mode. This is a low power mode with very little activity.

3.8 ZIGBEE

- Zigbee is an IEEE 802.15.4-based specification for a suite of high-level communication protocols used to create personal area networks with small, low-power digital radios, such as for home automation, medical device data collection, and other low-power low-bandwidth needs, designed for small scale projects which need wireless connection. Hence, Zigbee is a low-power, low data rate, and close proximity (i.e., personal area) wireless ad hoc network.

- The technology defined by the Zigbee specification is intended to be simpler and less expensive than other wireless personal area networks (WPANs), such as Bluetooth or more general wireless networking such as Wi-Fi. Applications include wireless light switches, home energy monitors, traffic management systems, and other consumer and industrial equipment that requires short-range low-rate wireless data transfer.

- Its low power consumption limits transmission distances to 10–100 meters line-of-sight, depending on power output and environmental characteristics. Zigbee devices can transmit data over long distances by passing data through a mesh network of intermediate devices to reach more distant ones. Zigbee is typically used in low data rate applications that require long battery life and secure networking (Zigbee networks are secured by 128 bit symmetric encryption keys.) Zigbee has a defined rate of 250 kbit/s, best suited for intermittent data transmissions from a sensor or input device.

- Zigbee was conceived in 1998, standardized in 2003, and revised in 2006. The name refers to the waggle dance of honey bees after their return to the beehive.

- Zigbee is a low-cost, low-power, wireless mesh network standard targeted at battery-powered devices in wireless control and monitoring applications. Zigbee delivers low-latency communication. Zigbee chips are typically integrated with radios and with microcontrollers. Zigbee operates in the industrial, scientific and medical (ISM) radio bands: 2.4 GHz in most jurisdictions worldwide; though some devices also use 784 MHz in China, 868 MHz in Europe and 915 MHz in the USA and Australia, however even those regions and countries still use 2.4 GHz for most commercial Zigbee devices for home use. Data rates vary from 20 kb/s (868 MHz band) to 250 kb/s (2.4 GHz band).

- Zigbee builds on the physical layer and media access control defined in IEEE standard 802.15.4 for low-rate wireless personal area networks (WPANs). The specification includes four additional key components: network layer, application layer, Zigbee Device Objects (ZDOs) and manufacturer-defined application objects. ZDOs are responsible for some tasks, including keeping track of device roles, managing requests to join a network, as well as device discovery and security.

- The Zigbee network layer natively supports both star and tree networks, and generic mesh networking. Every network must have one coordinator device. Within star networks, the coordinator must be the central node. Both trees and meshes allow the use of Zigbee routers to extend communication at the network level. Another defining feature of Zigbee is facilities for carrying out secure communications, protecting establishment and transport of cryptographic keys, ciphering frames, and controlling device. It builds on the basic security framework defined in IEEE 802.15.4.

History

- Zigbee-style self-organizing ad-hoc digital radio networks were conceived in the 1990s. The IEEE 802.15.4-2003 Zigbee specification was ratified on December 14, 2004. The Zigbee Alliance announced availability of Specification 1.0 on June 13, 2005, known as the Zigbee 2004 Specification.

Cluster Library

- In September 2006, the Zigbee 2006 Specification was announced, obsoleting the 2004 stack The 2006 specification replaces the Message/Key Value Pair structure used in the 2004 stack with a cluster library. The library is a set of standardised commands, organised under groups known as clusters with names such as Smart Energy, Home Automation, Zigbee Light Link.

- In January 2017, Zigbee Alliance renamed the library to Dotdot and announced it as a new protocol to be represented by an emoticon (||:). They also announced it will now additionally run over other network types using Internet Protocoland will interconnect with other standards such as Thread. Since its unveiling, Dotdot has functioned as the default application layer for almost all Zigbee devices.

Zigbee PRO

- Zigbee PRO, also known as Zigbee 2007, was finalized in 2007. A Zigbee PRO device may join and operate on a legacy Zigbee network and vice versa. Due to differences in routing options, Zigbee PRO devices must become non-routing Zigbee end devices (ZEDs) on a legacy Zigbee network, and legacy Zigbee devices must become ZEDs on a Zigbee PRO network. It operates on not only the 2.4 GHz band, but also the sub-GHz band.

Use Cases

- Zigbee protocols are intended for embedded applications requiring low power consumption and tolerating low data rates. The resulting network will use very little power individual devices must have a battery life of at least two years to pass Zigbee certification.

Typical Application Areas Include:

- Home automation
- Wireless sensor networks
- Industrial control systems
- Embedded sensing
- Medical data collection
- Smoke and intruder warning
- Building automation
- Remote wireless microphone configuration

Zigbee is not for situations with high mobility among nodes. Hence, it is not suitable for tactical ad hoc radio networks in the battlefield, where high data rate and high mobility is present and needed.

3.8.1 Zigbee Alliance

- Established in 2002, the Zigbee Alliance is a group of companies that maintain and publish the Zigbee standard. The term Zigbee is a registered trademark of this group, not a single technical standard. The Alliance publishes application profiles that allow multiple OEM vendors to create interoperable products. The relationship between IEEE 802.15.4 and Zigbee is similar to that between IEEE 802.11 and the Wi-Fi Alliance.

- Over the years, the Alliance's membership has grown to over 500 companies, including the likes of Comcast, Ikea, Legrand, Samsung SmartThings, and Amazon. The Zigbee Alliance has three levels of membership: Adopter, Participant, and Promoter. The Adopter members are allowed access to completed Zigbee specifications and standards, and the Participant members have voting rights, play a role in Zigbee development, and have early access to specifications and standards for product development.

- The requirements for membership in the Zigbee Alliance cause problems for Free Software developers because the annual fee conflicts with the GNU General Public Licence. The requirements for developers to join the Zigbee Alliance also conflict with most other free software licenses. The Zigbee Alliance board has been asked to make their license compatible with GPL, but refused. Bluetooth has GPL licensed implementations.

1. Application Profiles

- The first Zigbee application profile, Home Automation, was announced November 2, 2007. The ZigBee Smart Energy 2.0 specifications define an Internet protocol to monitor, control, inform and automate the delivery and use of energy and water. It is an enhancement of the Zigbee Smart Energy version 1 specifications. It adds services for plug-in electric vehicle charging, installation, configuration and firmware download, prepay services, user information and messaging, load control, demand response and common information and application profile interfaces for wired and wireless networks. It is being developed by partners including:

  - ➤ HomeGrid Forum responsible for marketing and certifying ITU-T G.hn technology and products

  - ➤ HomePlug Powerline Alliance

  - ➤ International Society of Automotive Engineers SAE International

  - ➤ IPSO Alliance

  - ➤ SunSpec Alliance

  - ➤ Wi-Fi Alliance.

- Zigbee Smart Energy relies on Zigbee IP, a network layer that routes standard IPv6 traffic over IEEE 802.15.4 using 6LoWPAN header compression.

- In 2009, the RF4CE (Radio Frequency for Consumer Electronics) Consortium and Zigbee Alliance agreed to deliver jointly a standard for radio frequency remote controls. Zigbee RF4CE is designed for a broad range of consumer electronics products, such as TVs and set-top boxes. It promised many advantages over existing remote control solutions, including richer communication and increased reliability, enhanced features and flexibility, interoperability, and no line-of-sight barrier. The Zigbee RF4CE specification lifts off some networking weight and does not support all the mesh features, which is traded for smaller memory configurations for lower cost devices, such as remote control of consumer electronics.

- With the introduction of the second Zigbee RF4CE application profile in 2012 and increased momentum in MSO market, the Zigbee RF4CE team provides an overview on current status of the standard, applications, and future of the technology.

2. Radio Hardware

- The radio design used by Zigbee has few analog stages and uses digital circuits wherever possible.

- Though the radios themselves are inexpensive, the Zigbee Qualification Process involves a full validation of the requirements of the physical layer. All radios derived from the same validated semiconductor mask setwould enjoy the same RF characteristics. An uncertified physical layer that malfunctions could cripple the battery lifespan of other devices on a Zigbee network. Zigbee radios have very tight constraints on power and bandwidth. Thus, radios are tested with guidance given by Clause 6 of the 802.15.4-2006 Standard. Products that integrate the radio and microcontroller into a single module are available.

- This standard specifies operation in the unlicensed 2.4 to 2.4835 [30]GHz (worldwide), 902 to 928 MHz (Americas and Australia) and 868 to 868.6 MHz (Europe) ISM bands. Sixteen channels are allocated in the 2.4 GHzband, with each channel spaced 5 MHz apart, though using only 2 MHz of bandwidth. The radios use direct-sequence spread spectrum coding, which is managed by the digital stream into the modulator. Binary phase-shift keying (BPSK) is used in the 868 and 915 MHz bands, and offset quadrature phase-shift keying (OQPSK) that transmits two bits per symbol is used in the 2.4 GHz band.

- The raw, over-the-air data rate is 250 kbit/s per channel in the 2.4 GHz band, 40 kbit/s per channel in the 915 MHz band, and 20 kbit/s in the 868 MHz band. The actual data throughput will be less than the maximum specified bit rate due to the packet overhead and processing delays. For indoor applications at 2.4 GHz transmission distance may be 10–20 m, depending on the construction materials, the number of walls to be penetrated and the output power permitted in that geographical location. Outdoors with line-of-sight, range may be up to 1500 m depending on power output and environmental characteristics. The output power of the radios is generally 0-20 dBm (1-100 mW).

3.8.2 Device Types and Operating Modes

Zigbee Devices are of Three Kinds:

1. **Zigbee Coordinator (ZC):** The most capable device, the Coordinator forms the root of the network tree and might bridge to other networks. There is precisely one Zigbee Coordinator in each network since it is the device that started the network originally (the Zigbee LightLink specification also allows operation without a Zigbee Coordinator, making it more usable for off-the-shelf home products). It stores information about the network, including acting as the Trust Center & repository for security keys.

2. **Zigbee Router (ZR):** As well as running an application function, a Router can act as an intermediate router, passing on data from other devices.

3. **Zigbee End Device (ZED):** Contains just enough functionality to talk to the parent node (either the Coordinator or a Router); it cannot relay data from other devices. This relationship allows the node to be asleep a significant amount of the time thereby giving long battery life. A ZED requires the least amount of memory, and, therefore, can be less expensive to manufacture than a ZR or ZC.

- The current Zigbee protocols support beacon and non-beacon enabled networks. In non-beacon-enabled networks, an unslotted CSMA/CA channel access mechanism is used. In this type of network, Zigbee Routers typically have their receivers continuously active, requiring a more robust power supply. However, this allows for heterogeneous networks in which some devices receive continuously while others only transmit when an external stimulus is detected. The typical example of a heterogeneous network is a wireless light switch: The Zigbee node at the lamp may constantly receive, since it is connected to the mains supply, while a battery-powered light switch would remain asleep until the switch is thrown. The switch then wakes up, sends a command to the lamp, receives an acknowledgment, and returns to sleep. In such a network the lamp node will be at least a Zigbee Router, if not the Zigbee Coordinator; the switch node is typically a Zigbee End Device.

- In beacon-enabled networks, the special network nodes called Zigbee Routers transmit periodic beacons to confirm their presence to other network nodes. Nodes may sleep between beacons, thus lowering their duty cycle and extending their battery life. Beacon intervals depend on data rate; they may range from 15.36 milliseconds to 251.65824 seconds at 250 kbit/s, from 24 milliseconds to 393.216 seconds at 40 kbit/s and from 48 milliseconds to 786.432 seconds at 20 kbit/s. However, low duty cycle operation with long beacon intervals requires precise timing, which can conflict with the need for low product cost.

- In general, the Zigbee protocols minimize the time the radio is on, so as to reduce power use. In beaconing networks, nodes only need to be active while a beacon is being transmitted. In non-beacon-enabled networks, power consumption is decidedly asymmetrical: Some devices are always active while others spend most of their time sleeping.

- Except for the Smart Energy Profile 2.0, Zigbee devices are required to conform to the IEEE 802.15.4-2003 Low-Rate Wireless Personal Area Network (LR-WPAN) standard. The standard specifies the lower protocol layers—the physical layer (PHY), and the Media Access Control portion of the data link layer (DLL). The basic channel access mode is "carrier sense, multiple access/collision avoidance" (CSMA/CA). That is, the nodes talk in the same way that humans converse; they briefly check to see that no one is talking before he or she start, with three notable exceptions. Beacons are sent on a fixed timing schedule and do not use CSMA. Message acknowledgments also do not use CSMA. Finally, devices in beacon-enabled networks that have low latency real-time requirements may also use Guaranteed Time Slots (GTS), which by definition do not use CSMA.

3.8.3 Software

- The software is designed to be easy to develop on small, inexpensive microprocessors.

Network Layer

- The main functions of the network layer are to enable the correct use of the MAC sublayer and provide a suitable interface for use by the next upper layer, namely the application layer. Its capabilities and structure are those typically associated to such network layers, including routing. The Network Layer 's function is exactly as it sounds. It deals with network functions such as connecting, disconnecting, and setting up networks. It will add a network, allocate addresses, and add/remove certain devices. This layer makes use of star, mesh and tree topologies. It adds an interface to the application layer.

- On the one hand, the data entity creates and manages network layer data units from the payload of the application-layer and performs routing according to the current topology. On the other hand, there is the layer control, which is used to handle configuration of new devices and establish new networks: it can determine whether a neighboring device belongs to the network and discovers new neighbors and routers. The control can also detect the presence of a receiver, which allows direct communication and MAC synchronization.

- The routing protocol used by the network layer is AODV. In AODV, to find the destination device, AODV broadcasts out a route request to all of its neighbors. The neighbors then broadcast the request to their neighbors and onward until the destination is reached. Once the destination is reached, it sends its route reply via unicast transmission following the lowest cost path back to the source. Once the source receives the reply, it will update its routing table for the destination address of the next hop in the path and the path cost.

Application Layer

- The application layer is the highest-level layer defined by the specification and is the effective interface of the Zigbee system to its end users. It comprises the majority of components added by the Zigbee specification: both ZDO and its management procedures, together with application objects defined by the manufacturer, are considered part of this layer. This layer binds tables, sends messages between bound devices, manages group addresses, reassembles packets and also transports data. It is responsible for providing service to Zigbee device profiles.

Main Components

- The ZDO (Zigbee Device Object), a protocol in the Zigbee protocol stack, is responsible for overall device management, security keys, and policies. It is responsible for defining the role of a device as either coordinator or end device, as mentioned above, but also for the discovery of new (one-hop) devices on the network and the identification of their offered services. It may then go on to establish secure links with

external devices and reply to binding requests accordingly.

- The application support sublayer (APS) is the other main standard component of the layer, and as such it offers a well-defined interface and control services. It works as a bridge between the network layer and the other elements of the application layer: it keeps up-to-date binding tables in the form of a database, which can be used to find appropriate devices depending on the services that are needed and those the different devices offer. As the union between both specified layers, it also routes messages across the layers of the protocol stack.

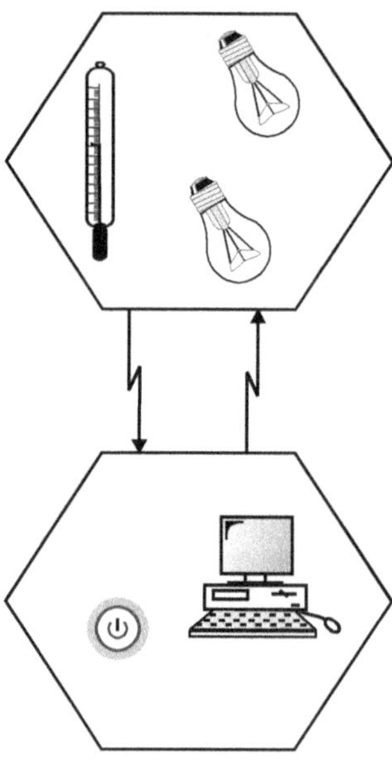

Fig. 3.20

Zigbee, High-Level Communication Model :

- An application may consist of communicating objects which cooperate to carry out the desired tasks. The focus of Zigbee is to distribute work among many different devices which reside within individual Zigbee nodes which in turn form a network (said work will typically be largely local to each device, for instance, the control of each household appliance).

- The collection of objects that form the network communicates using the facilities provided by APS, supervised by ZDO interfaces. The application layer data service follows a typical request-confirm/indication-response structure. Within a single device, up to 240 application objects can exist, numbered in the range 1-240. 0 is reserved for the ZDO data interface and 255 for broadcast; the 241-254 range is not currently in use but may be in the future.

Two Services are Available for Application Objects to use (in Zigbee 1.0):

1. The key-value pair service (KVP) is meant for configuration purposes. It enables description, request and modification of object attribute through a simple interface based on getting/set and event primitives, some allowing a request for a response. Configuration uses compressed XML (full XML can be used) to provide an adaptable and elegant solution.

2. The message service is designed to offer a general approach to information treatment, avoiding the necessity to adapt application protocols and potential overhead incurred on by KVP. It allows arbitrary payloads to be transmitted over APS frames.

- Addressing is also part of the application layer. A network node consists of an 802.15.4-conformant radio transceiver and one or more device descriptions (basically collections of attributes which can be polled or set, or which can be monitored through events). The transceiver is the base for addressing, and devices within a node are specified by an endpoint identifier in the range 1-240.

3.8.4 Communication and Device Discovery

- For applications to communicate, their comprising devices must use a common application protocol (types of messages, formats and so on); these sets of conventions are grouped in profiles. Furthermore, binding is decided upon by matching input and output cluster identifiers, unique within the context of a given profile and associated to an incoming or outgoing data flow in a device. Binding tables contain source and destination pairs.

- Depending on the available information, device discovery may follow different methods. When the network address is known, the IEEE address can be requested using unicast communication. When it is not, petitions are broadcast (the IEEE address being part of the response payload). End devices will simply respond with the requested address while a network coordinator or a router will also send the addresses of all the devices associated with it.

- This extended discovery protocol permits external devices to find out about devices in a network and the services that they offer, which endpoints can report when queried by the discovering device (which has previously obtained their addresses). Matching services can also be used.

- The use of cluster identifiers enforces the binding of complementary entities using the binding tables, which are maintained by Zigbee coordinators, as the table must always be available within a network and coordinators are most likely to have a permanent power supply. Backups, managed by higher-level layers, may be needed by some applications. Binding requires an established communication link; after it exists, whether to add a new node to the network is decided, according to the application and security policies.

- Communication can happen right after the association. Direct addressing uses both radio address and endpoint identifier, whereas indirect addressing uses every relevant field (address, endpoint, cluster, and attribute) and requires that they are sent to the network coordinator, which maintains associations and translates requests for communication. Indirect addressing is particularly useful to keep some devices very simple and minimize their need for storage. Besides these two methods, broadcast to all endpoints in a device is available, and group addressing is used to communicate with groups of endpoints belonging to a set of devices.

3.8.5 Security Services

- As one of its defining features, Zigbee provides facilities for carrying out secure communications, protecting establishment and transport of cryptographic keys, cyphering frames, and controlling devices. It builds on the basic security framework defined in IEEE 802.15.4. This part of the architecture relies on the correct management of symmetric keys and the correct implementation of methods and security policies.

Basic Security Model

- The basic mechanism to ensure confidentiality is the adequate protection of all keying material. Trust must be assumed in the initial installation of the keys, as well as in the processing of security information. For an implementation to globally work, its general conformance to specified behaviors is assumed.

- Keys are the cornerstone of the security architecture; as such their protection is of paramount importance, and keys are never supposed to be transported through an insecure channel. A momentary exception to this rule occurs during the initial phase of the addition to the network of a previously unconfigured device. The Zigbee network model must take particular care of security considerations, as ad hoc networks may be physically accessible to external devices. Also the state of the working environment cannot be predicted.

- Within the protocol stack, different network layers are not cryptographically separated, so access policies are needed, and conventional design assumed. The open trust model within a device allows for key sharing, which notably decreases potential cost. Nevertheless, the layer which creates a frame is responsible for its security. If malicious devices may exist, every network layer payload must be ciphered, so unauthorized traffic can be immediately cut off. The exception, again, is the transmission of the network key, which confers a unified security layer to the grid, to a new connecting device.

Security Architecture

- Zigbee uses 128-bit keys to implement its security mechanisms. A key can be associated either to a network, being usable by both Zigbee layers and the MAC sublayer, or to a link, acquired through pre-installation, agreement or transport. Establishment of link keys is based on a master key which controls link key correspondence. Ultimately, at least, the initial master key must be obtained through a secure medium (transport or pre-installation), as the security of the whole network depends on it. Link and master keys are only visible to the application layer. Different services use different one-way variations of the link key to avoid leaks and security risks.

- Key distribution is one of the most important security functions of the network. A secure network will designate one special device which other devices trust for the distribution of security keys: the trust center. Ideally, devices will have the center trust address and initial master key preloaded; if a momentary vulnerability is allowed, it will be sent as described above. Typical applications without special security needs will use a network key provided by the trust center (through the initially insecure channel) to communicate.

- Thus, the trust center maintains both the network key and provides point-to-point security. Devices will only accept communications originating from a key supplied by the trust center, except for the initial master key. The security architecture is distributed among the network layers as follows:

- The MAC sublayer is capable of single-hop reliable communications. As a rule, the security level it is to use is specified by the upper layers.

- The network layer manages routing, processing received messages and being capable of broadcasting requests. Outgoing frames will use the adequate link key according to the routing if it is available; otherwise, the network key will be used to protect the payload from external devices.

- The application layer offers key establishment and transport services to both ZDO and applications.

- The security levels infrastructure is based on CCM*, which adds encryption- and integrity-only features to CCM.

- According to a German computer magazine website, Zigbee Home Automation 1.2 is using fallback keys for encryption negotiation which are known and cannot be changed. This makes the encryption highly vulnerable.

3.9 CONNECTING DEVICES

- Devices that are used to connect other devices are called connecting devices. That is, connecting device is a device that connects.

There are 4 types of connecting devices. They are as follows,

1. Repeater:

A repeater is a device that receives signals and before it become too weak or corrupted the repeater regenerates the original bit pattern and sends the data. It is like a refreshing station for data while travelling. Some properties are as follows,

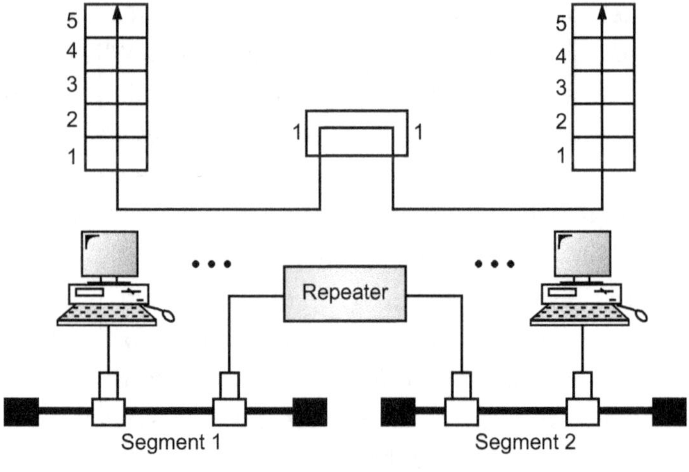

Fig. 3.21

- It operates only in the physical layer.
- It connects two segments of the same LAN.
- It forwards every bits it receives.
- It regenerate bits, does not amplify them.

2. Hub:

A hub is a multi-port repeater. That is a hub can connect more than two segments of a LAN where a repeater can connect only two segments. It also does not have any filtering capabilities like repeater.

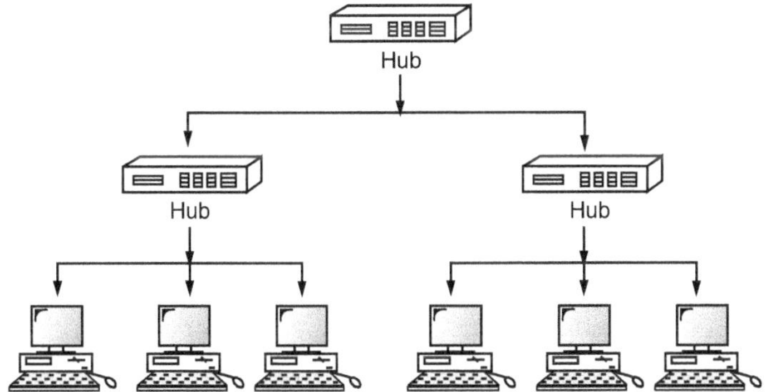

Fig. 3.22

3. Bridge:

- A bridge operates in both physical layer and data link layer. As a physical layer device, it can regenerate bits and as a data link layer device it can check the physical addresses contained in the frame. Bits can be carried to any distance by using properly spaced repeaters but the collision domain will not allow this and that's why brides are required.

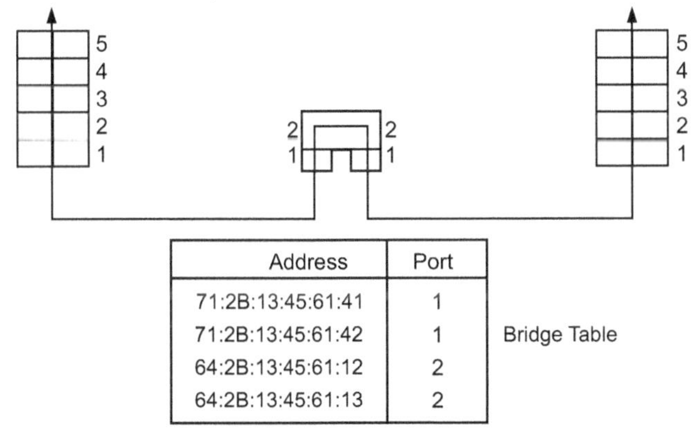

Address	Port	
71:2B:13:45:61:41	1	
71:2B:13:45:61:42	1	Bridge Table
64:2B:13:45:61:12	2	
64:2B:13:45:61:13	2	

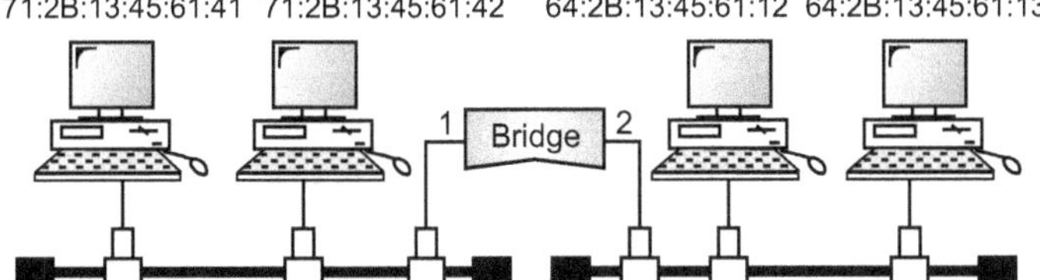

Fig. 3.23

- Bridge has filtering capability. It has table that maps addresses to port. A bridge can check the table to decide if a frame has to be forwarded or dropped. Bridges also work in a single LAN. There are some bridges called transparent bridge which can generate its address table automatically by learning the frame movement in the network.

4. Router:

- A router is three layer device; physical, data link and network layer. As a physical layer device it can regenerates bit, as data link layer device the router can check the physical addresses contained in a packet and as a network layer device it can check the network layer addresses (addresses in IP layer).

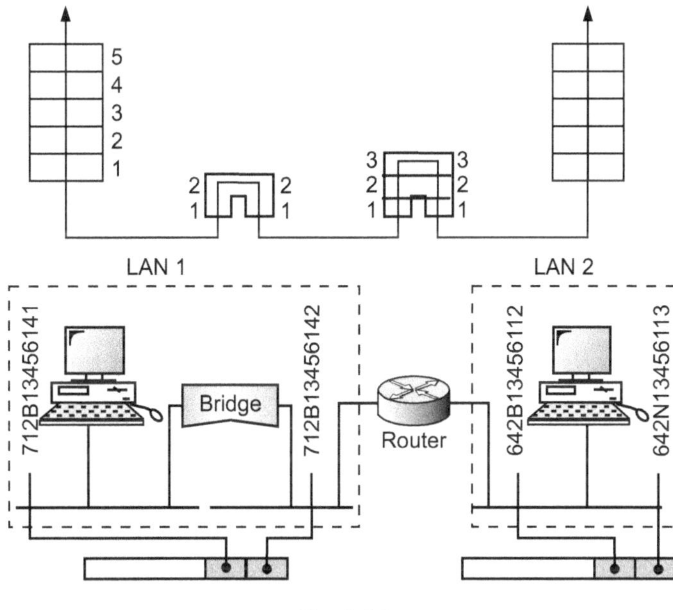

Fig. 3.24

- Routers are inter-networking devices as it can connects LAN-LAN, LAN-WAN, WAN-WAN.

- Routers can change the physical addresses of a packet. For example, considering a packet is getting sent from LAN 1 to LAN 2 as in the above given Fig. At this moment the source address is of the sender's address and the destination address is of a host of the LAN 2. Now when the packet reaches the router, the router changes the source address of the packet to its own address and then send the packet to the destination address.

3.10 VIRTUAL LANS

- A virtual LAN (VLAN) is any broadcast domain that is partitioned and isolated in a computer network at the data link layer (OSI layer 2). LAN is the abbreviation for local area network and in this context virtual refers to a physical object recreated and altered by additional logic.

- VLANs work by applying tags to network frames and handling these tags in networking systems – creating the appearance and functionality of network traffic that is physically on a single network but acts as if it is split between separate networks. In this way, VLANs can keep network applications separate despite being connected to the same physical network, and without requiring multiple sets of cabling and networking devices to be deployed.

- VLANs allow network administrators to group hosts together even if the hosts are not directly connected to the same network switch. Because VLAN membership can be configured through software, this can greatly simplify network design and deployment. Without VLANs, grouping hosts according to their resource needs necessitates the labor of relocating nodes or rewiring data links.

- VLANs allow networks and devices that must be kept separate to share the same physical cabling without interacting, improving simplicity, security, traffic management, or economy. For example, a VLAN could be used to separate traffic within a business due to users, and due to network administrators, or between types of traffic, so that users or low priority traffic cannot directly affect the rest of the network's functioning.

- Many Internet hosting services use VLANs to separate their customers' private zones from each other, allowing each customer's servers to be grouped together in a single network segment while being located anywhere in their data center. Some precautions are needed to prevent traffic "escaping" from a given VLAN, an exploit known as VLAN hopping.

- To subdivide a network into VLANs, one configures network equipment. Simpler equipment can partition only per physical port (if at all), in which case each VLAN is connected with a dedicated network cable. More sophisticated devices can mark frames through VLAN tagging, so that a single interconnect (trunk) may be used to transport data for multiple VLANs. Since VLANs share bandwidth, a VLAN trunk can use link aggregation, quality-of-service prioritization, or both to route data efficiently.

Uses

- VLANs address issues such as scalability, security, and network management. Network architects set up VLANs to provide network segmentation. Routers between VLANs filter broadcast traffic, enhance network security, perform address summarization, and mitigate network congestion.

- In a network utilizing broadcasts for service discovery, address assignment and resolution and other services, as the number of peers on a network grows, the frequency of broadcasts also increases. VLANs can help manage broadcast traffic by forming multiple broadcast domains. Breaking up a large network into smaller independent segments reduces the amount of broadcast traffic each network device and network segment has to bear. Switches may not bridge network traffic between VLANs, as doing so would violate the integrity of the VLAN broadcast domain.

- VLANs can also help create multiple layer 3 networks on a single physical infrastructure. VLANs are data link layer (OSI layer 2) constructs, analogous to Internet Protocol (IP) subnets, which are network layer (OSI layer 3) constructs. In an environment employing VLANs, a one-to-one relationship often exists between VLANs and IP subnets, although it is possible to have multiple subnets on one VLAN.

- Without VLAN capability, users are assigned to networks based on geography and are limited by physical topologies and distances. VLANs can logically group networks to decouple the users' network location from their physical location. By using VLANs, one can control traffic patterns and react quickly to employee or equipment relocations. VLANs provide the flexibility to adapt to changes in network requirements and allow for simplified administration.

- VLANs can be used to partition a local network into several distinctive segments, for instance:
  - ➤ Production
  - ➤ Voice over IP
  - ➤ Network management
  - ➤ Storage area network (SAN)
  - ➤ Guest Internet access
  - ➤ Demilitarized zone (DMZ)

- A common infrastructure shared across VLAN trunks can provide a measure of security with great flexibility for a comparatively low cost. Quality of service schemes can optimize traffic on trunk links for real-time (e.g. VoIP) or low-latency requirements (e.g. SAN). However, VLANs as a security solution should be implemented with great care as they can be defeated unless implemented carefully.

- In cloud computing VLANs, IP addresses, and MAC addresses in the cloud are resources that end users can manage. To help mitigate security issues, placing cloud-based virtual machines on VLANs may be preferable to placing them directly on the Internet.

- Network technologies with VLAN capabilities include:
  - ➤ Asynchronous Transfer Mode (ATM)
  - ➤ Fiber Distributed Data Interface (FDDI)
  - ➤ Ethernet
  - ➤ HiperSockets
  - ➤ InfiniBand

Configuration and design considerations

- Early network designers often segmented physical LANs with the aim of reducing the size of the Ethernet collision domain—thus improving performance. When Ethernet switches made this a non-issue (because each switch port is a collision domain), attention turned to reducing the size of the data link layer broadcast domain. VLANs were first employed to separate several broadcast domains across one physical medium. A VLAN can also serve to restrict access to network resources without regard to physical topology of the network.

- VLANs operate at the data link layer of the OSI model. Administrators often configure a VLAN to map directly to an IP network, or subnet, which gives the appearance of involving the network layer. Generally, VLANs within the same organization will be assigned different non-overlapping network address ranges. This is not a requirement of VLANs. There is no issue with separate VLANs using identical overlapping address ranges (e.g. two VLANs each use the private network 192.168.0.0/16). However, it is not possible to route data between two networks with overlapping addresses without delicate IP remapping, so if the goal of VLANs is segmentation of a larger overall organizational network, non-overlapping addresses must be used in each separate VLAN.

- A basic switch that is not configured for VLANs has VLAN functionality disabled or permanently enabled with a default VLAN that contains all ports on the device as members. The default VLAN typically uses VLAN identifier 1. Every device connected to one of its ports can send packets to any of the others. Separating ports by VLAN groups separates their traffic very much like connecting each group using a distinct switch for each group.

- Remote management of the switch requires that the administrative functions be associated with one or more of the configured VLANs.

- In the context of VLANs, the term trunk denotes a network link carrying multiple VLANs, which are identified by labels (or tags) inserted into their packets. Such trunks must run between tagged ports of VLAN-aware devices, so they are often switch-to-switch or switch-to-router links rather than links to hosts. (Note that the term 'trunk' is also used for what Cisco calls "channels" : Link Aggregation or Port Trunking). A router (Layer 3 device) serves as the backbone for network traffic going across different VLANs. It is only when the VLAN port group is to extend to another device that tagging is used. Since communications between ports on two different switches travel via the uplink ports of each switch involved, every VLAN containing such ports must also contain the uplink port of each switch involved, and traffic through these ports must be tagged.

- Switches typically have no built-in method to indicate VLAN to port associations to someone working in a wiring closet. It is necessary for a technician to either have administrative access to the device to view its configuration, or for VLAN port assignment charts or diagrams to be kept next to the switches in each wiring closet.

EXERCISE

1. Explain the concept of Wireless LANs with protocol stack of 802.11

2. Draw and explain Bluetooth architecture

3. Write short note on Bluetooth profiling

4. Draw and explain Zigbee architecture

5. Differentiate between Bluetooth and Zigbee

6. Explain the following connecting devices

   (a) Repeater

   (b) Hub

   (c) Bridge

   (d) Router

7. Draw suitable diagram

8. Explain the concept of virtual LANS with suitable application.

NETWORK LAYER

4.1 INTRODUCTION

- The network layer is concerned with getting packets from the source all the way to the destination. Getting to the destination may require making many hops at intermediates routers along the way. To achieve this goal, the network layer must know about the topology of the network and choose appropriate paths through it, even for large networks. This unit which is all about network layer firstly covers the switching techniques such as circuit switching, packet switching, message switching etc. Later on we have covered details IPv4 and IPv6 addressing scheme and subnetting.

- There are many protocols like distance vector, link state path vector which finds optimal path from source to destination. All these protocols are covered here. Protocols like RIP, OSPF and BGP in the context of routing in internet are also covered here. Later on, congestion control and QoS concepts are discussed. After that MPLS and mobile IP topics are covered in detail and finally routing protocols in MANET like AODV and DSR are covered at the end of the unit.

4.2 SWITCHING INTRODUCTION

- Switching is a technique by which various network devices are connected together on a network so that packets can be transmitted, received and processed at destination device.

- When multiple devices want to communicate with each other, then simple solutions are:
  - ➢ Mesh network formation.
  - ➢ Bus network formation.

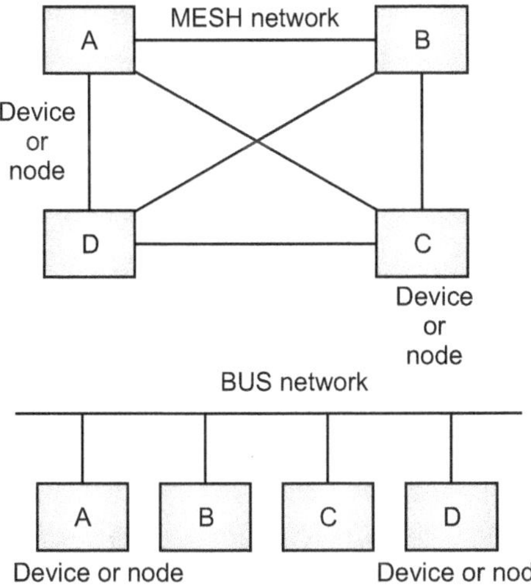

Fig. 4.1: Interdevice communication

- When above techniques are employed, it becomes impractical and wasteful when network size increases.

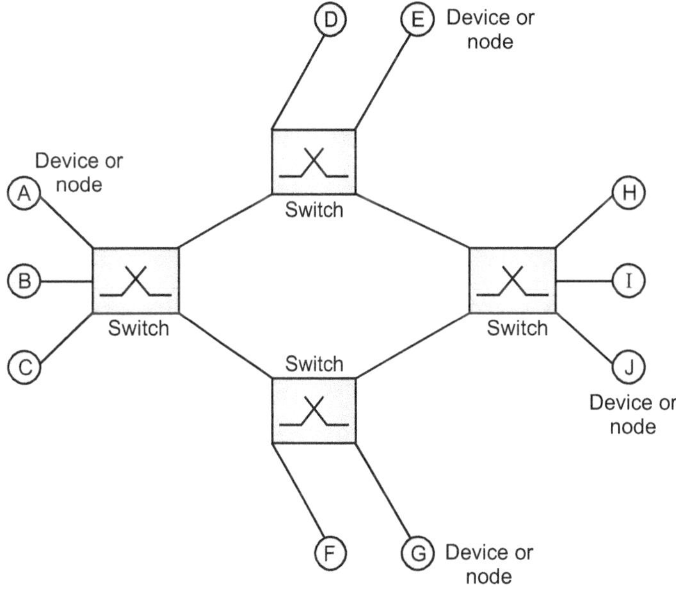

Fig. 4.2 : Typical switch based network

- Also network cost and maintenance becomes difficult for large network of devices.

- Then the solution to above problems is to use the switching system.

- Typical use of switch and switching system is as shown in Fig. 4.2.

- Thus, different devices/nodes can communicate with each other with the help of switches connected in the typical switch based network.

- The switches work on the principle of switching system.

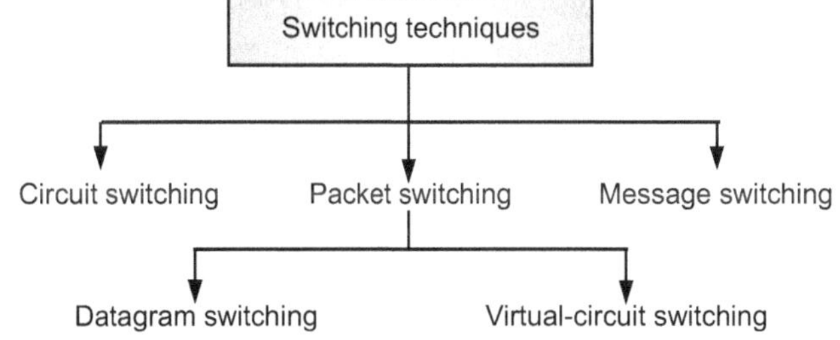

Fig. 4.3 : Switching techniques

4.3 CIRCUIT SWITCHING NETWORKS

- In circuit switching networks, nodes or devices are connected to each other by physical links via switches.

- Typical circuit switched network is as shown in Fig. 4.4.

- In this diagram, multiplexer (MUX) symbol is explicitly shown and it is implicitly included in switch fabric itself.

- The end system can be computer node or device like telephone set.

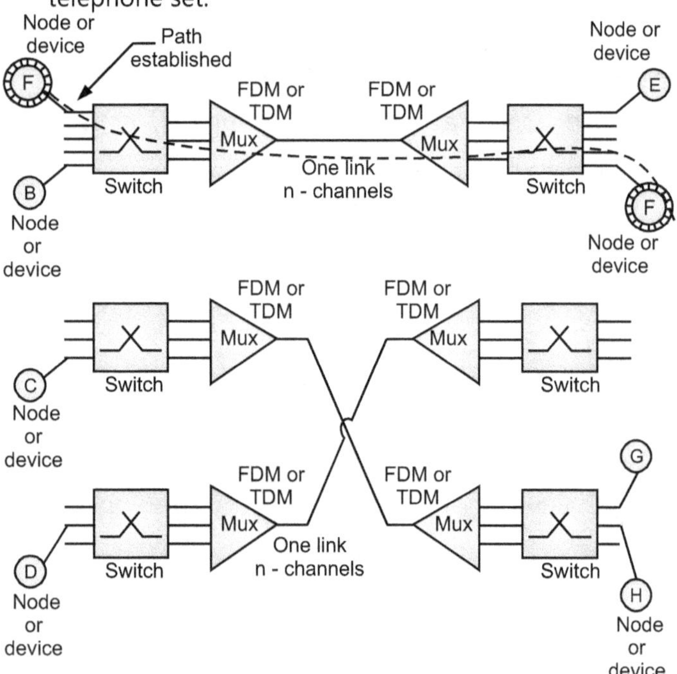

Fig. 4.4: Typical circuit switched network and communication between two end systems A and F

- In Fig. 4.4, the communication between two end systems A and F is highlighted.
- In circuit switched network, the communication between node 'A' and node 'F' is done through 3 phases as follows:
    1. Set-up phase (also known as connection establishment).
    2. Data transfer phase.
    3. Teardown phase (also known as connection release).

1. **In Set-Up Phase:** A dedicated circuit (i.e. combination of channels in links) needs to be established. For this node 'A' sends set-up request through switch fabric including multiplexer, to node 'F'. Then node 'F' receives this request and sends acknowledgement to node 'A' through same dedicated path. Thus, only after receiving this acknowledgement from node 'F', we can say that connection is established or set-up phase is completed. In this circuit switched network, end systems use addresses in TDM network whereas use telephone numbers in FDM network.

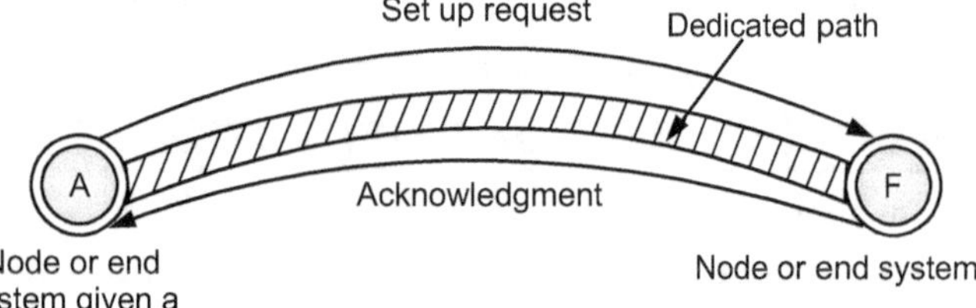

Computer Address – In TDM system

OR

Telephone Number – In FDM system

Fig. 4.5 : Typical set-up phase between end systems A and F

2. **In Data Transfer Phase:** End systems 'A' and 'F' can transfer data (or communication between two end systems is done).

3. **In Teardown Phase (or Connection Release Process):** Either 'A' system or 'F' system can stop the communication and release the common resources like dedicated link and switch etc.

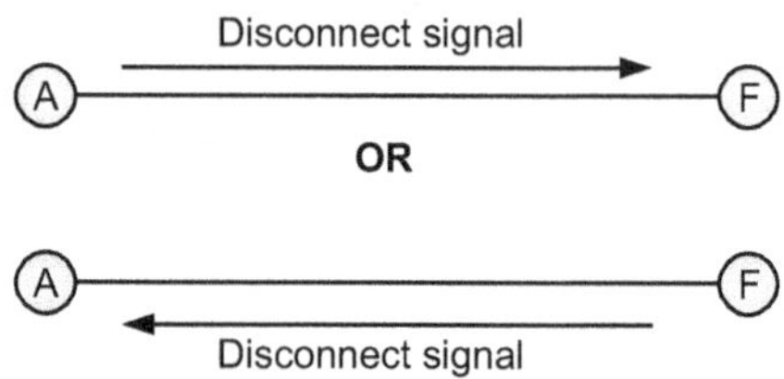

Fig. 4.6 : Teardown phase (or connection release process)

- The circuit switched network is less efficient as compared to others because network resources (switch and link) are allocated to 'A' and 'F' node and cannot be used by others or other connections are deprived.

- The total delay in communication between 'A' and 'F' end systems is given as :

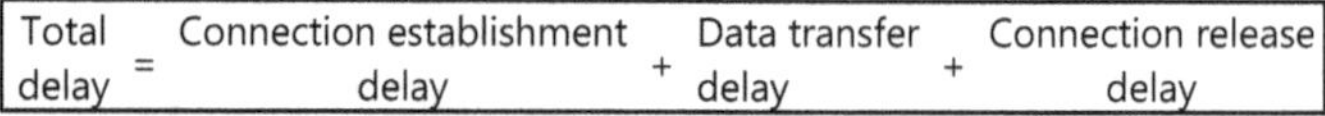

Total delay	=	Connection establishment delay	+	Data transfer delay	+	Connection release delay

- Data transfer delay is also given as :

Data transfer delay = Propagation time delay + Data transfer time delay

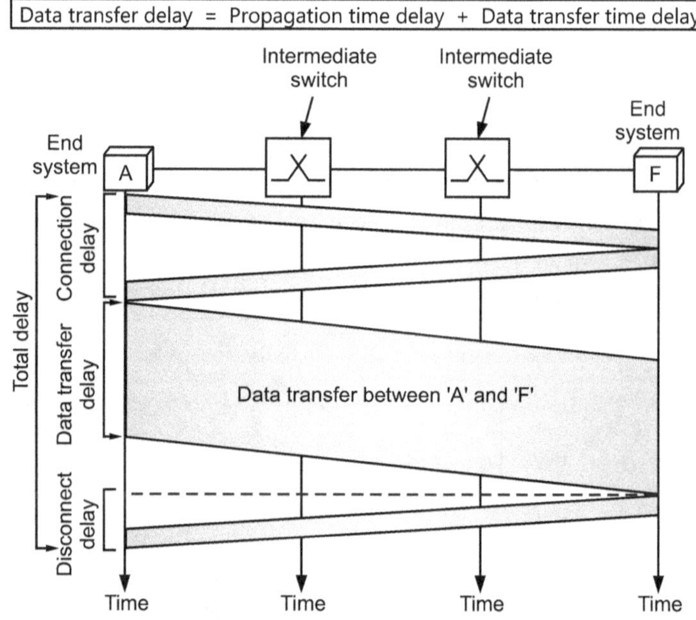

Fig. 4.7: Typical delays in circuit switched network

(Here between two end systems like 'A' and 'F')

- Thus, typical example or application of circuit switched network in telephone communication is as shown in Fig. 4.8.

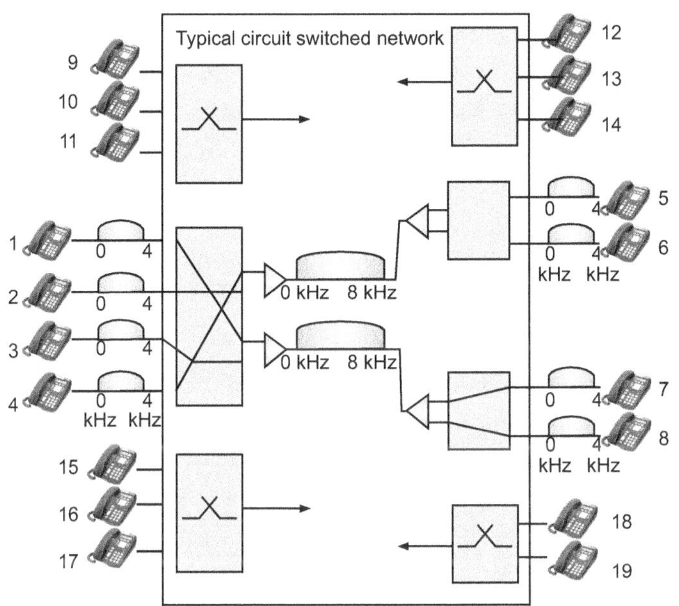

Fig. 4.8 : Typical circuit switched network example

- There is a common misunderstanding that circuit switching is used only for connecting voice circuits (analog or digital).

- The concept of a dedicated path persisting between two communicating parties or nodes can be extended to signal content other than voice.

- Its advantage is that it provides for non-stop transfer without requiring packets and without most of the overhead traffic usually needed, making maximal and optimal use of available bandwidth for that communication.

- The disadvantage of inflexibility tends to reserve it for specialized applications, particularly with the overwhelming proliferation of Internet-related technology.

- For call setup and control (and other administrative purposes), it is possible to use a separate dedicated signalling channel from the end node to the network. ISDN is one such service that uses a separate signalling channel while Plain Old Telephone Service (POTS) does not.

- The method of establishing the connection and monitoring its progress and termination through the network may also utilize a separate control channel as in the case of links between telephone exchanges which use SS7 packet-switched signalling protocol to communicate the call setup and control information and use TDM to transport the actual circuit data. Signalling System Number 7 (SS7) is a set of telephony signalling protocols which are used to set up most of the world's public switched telephone network

telephone calls. The main purpose is to set up and tear down telephone calls

- Early telephone exchange is a suitable example of circuit switching. The subscriber would ask the operator to connect to another subscriber, whether on the same exchange or via an inter-exchange link and another operator. In any case, the end result was a physical electrical connection between the two subscribers's telephones for the duration of the call. The copper wire used for the connection could not be used to carry other calls at the same time, even if the subscribers were in fact not talking and the line was silent.

- Thus, generally, resources are frequency intervals in a Frequency Division Multiplexing (FDM) scheme or more recently time slots in a Time Division Multiplexing (TDM) scheme. The set of resources allocated for a connection is called a circuit. A path is a sequence of links located between nodes called switches.

- The path taken by data between its source and destination is determined by the circuit on which it is flowing, and does not change during the lifetime of the connection. The circuit is terminated when the connection is closed.

- In circuit switching, resources remain allocated during the full length of a communication, after a circuit is established and until the circuit is terminated and then the allocated resources are freed.

- Resources remain allocated even if no data is flowing on a circuit, hereby wasting link capacity when a circuit does not carry as much traffic as the allocation permits.

- This is a major issue since frequencies (in FDM) or time slots (in TDM) are available in finite quantity on each link, and establishing a circuit consumes one of these frequencies or slots on each link of the circuit.

- As a result, establishing circuits for communication that carry less traffic than allocation permits can lead to resource exhaustion and network saturation, preventing further connections from being established.

- If no circuit can be established between a sender and a receiver because of a lack of resources, the connection is blocked.

- A second characteristic of circuit switching is the time cost involved when establishing a connection. In a communication network, circuit-switched or not, nodes need to lookup in a forwarding table to determine on which link to send incoming data, and to actually send data from the input link to the output link. Performing a lookup in a forwarding table and sending the data on an incoming link is called forwarding. Building the forwarding tables is called routing.

- In circuit switching, routing must be performed for each communication, at circuit establishment time. During circuit establishment, the set of switches and links on the path between the sender and the receiver is determined and messages are exchanged on all the links between the two end hosts of the communication in order to make the resource allocation and build the routing tables.

- In circuit switching, forwarding tables are hardwired or implemented using fast hardware, making data forwarding at each switch almost instantaneous. Therefore, circuit switching is well suited for long-lasting connections where the initial circuit establishment time cost is balanced by the low forwarding time cost.

- The circuit identifier (a range of frequencies in FDM or a time slot position in a TDM frame) is changed by each switch at forwarding time so that switches do not need to have a complete knowledge of all circuits established in the network but rather only local knowledge of available identifiers at a link. Using local identifiers instead of global identifiers for circuits also enables networks to handle a larger number of circuits.

- Traffic Engineering (TE) consists in optimizing resource utilization in a network by choosing appropriate paths followed by flow of data, according to static or dynamic constraints. A main goal of traffic engineering is to balance the load in the network, i.e., to avoid congestion on links on a network while other links are under-utilized.

- To achieve such goals, traffic engineering methods can vary from offline capacity planning algorithms to automatic, dynamic changes. Since circuit switching allocates a fixed path for each flow, circuits can be established according to traffic engineering algorithms.

- On the other hand, circuit switching networks are not reactive when a network topology change occurs. For instance, on a link failure, all circuits on a failed link are cut and communication is interrupted. Special mechanisms that handle such topological changes have to be devised. Traffic engineering can alleviate the consequences of a link failure by pre-planning failure recovery.

- A backup circuit can be established at the same time or after the primary circuit used for a communication is set up, and traffic can be rerouted from the failed circuit to the backup circuit if a link of the primary circuit fails. Circuit switching networks are intrinsically sensitive to link failures and rerouting must be performed by additional traffic engineering mechanisms.

Examples of Circuit Switched Networks

- Public Switched Telephone Network (PSTN).
- ISDN B-channel.
- Circuit Switched Data (CSD) and High-Speed Circuit-Switched Data (HSCSD) service in cellular systems such as GSM.
- Datakit [It supports file transfers, remote login, remote printing, and remote command execution. At the physical layer, it can operate over multiple media, from slow speed EIA-232 to 500 Mbit fiber optic links (called FIBERKIT)].
- X.21 (Used in the German DATEX-L and Scandinavian DATEX circuit switched data network).

4.4 DATAGRAM SWITCHING NETWORKS

- In packet switching networks, voice, video or data is converted into packet. Packet can be of fixed size or variable size, decided by network used and protocol used at both ends.

- In datagram switching, there is no resource allocation for a packet travelling from sender to receiver.

- This means there is no bandwidth reservation on links and no scheduled processing time for each packet.

- Thus, resources are allocated on demand and this allocation is done on a first come first serve basis.

- As a simple analogy consider two hotels (or restaurants). One which requires reservation and another that neither require reservation and nor accept them.

- For the hotel (or restaurant) which requires reservation, we have to go through the hassle of calling person of restaurant before we leave home and reach to restaurant.

- But when we arrive at restaurant we get table, can communicate with waiter and order for food.

- Thus, for the other restaurant which does not require reservation, we don't need to bother to reserve anything.

- In this restaurant when we arrive, we may have to wait for table, we may have to wait for communicating with waiter to order the food.

- Thus, restaurant with reservation and without reservation, this analogy is applicable for circuit switched network and datagram switched network respectively.

The typical packet flow in datagram packet switched network is as shown in Fig. 4.9.

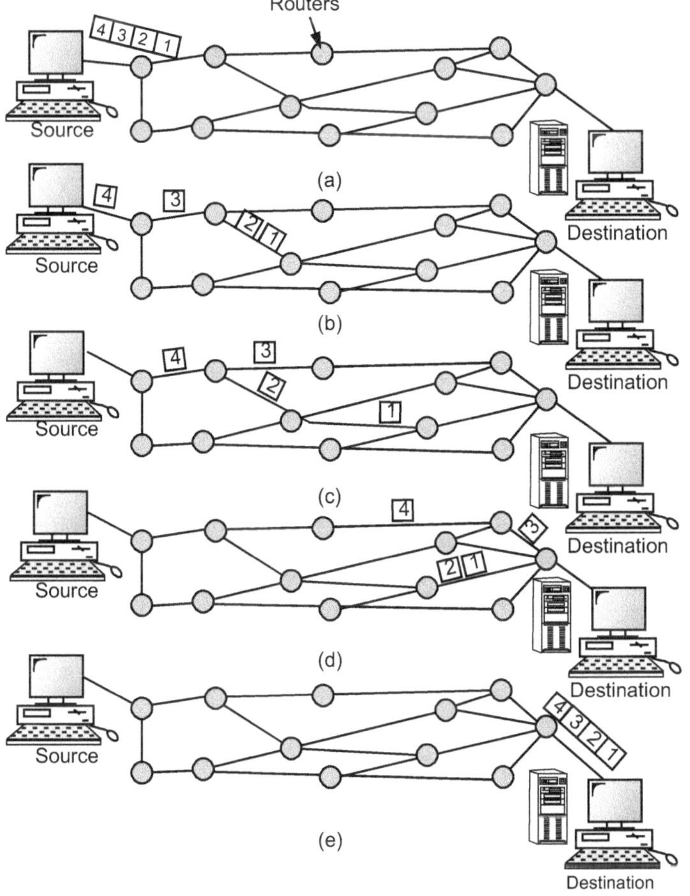

Fig. 4.9 : Datagram network packet transfer from source to destination

Thus, in datagram packet networks, following characteristics are important:

- Each packet is treated independently.

- Packet can take any practical route.

- Packets may arrive out of order at destination router.

- Packets may go missing in the datagram network journey.

- Receiver at another end is responsible to reorder the packets and recover the missing packets.

- Transport layer at both (sender and receiver) ends is responsible to reorder the packet sequence and recover the missing packets.

- We have already discussed the connection oriented and connectionless services in first chapter. Datagram based networks are also known as connectionless networks.

- There is no set-up phase or teardown phase present in datagram switching network. When data is ready, it is transferred with full source and destination address from sender to receiver. For this each intermediate router maintains the routing table as shown in Fig. 4.10.

- The routing tables are dynamic in nature and are updated periodically. (Specific time is set by router administrator).

Table 4.1 : Routing table maintained by Router Device

Destination Address	OutputPort
12347	1
34569	2
22130	3
⋮	⋮
⋮	⋮
⋮	⋮
75759	4

Fig. 4.10 : Router device uses routing table based on destination address

- The destination address is carried by the header among other information (or control information) of packet. This address remains same during the entire journey of the packet from source to destination or sender to receiver.

- Efficiency of datagram switch network is better than circuit switched network, because network resources

- are allocated only when there are packets to be transferred from source to destination.

- If source to destination packet transfer is finished or delayed then these resources can be used by other nodes or systems connected to this network.

- The delay between sender and receiver is given by,

$$\text{Total delay } (T_D) = \text{Transmission delay } (T_t) + \text{Waiting delay } (T_w)$$

- Typical datagram based packet switching network uses two routers in between sender and receiver as shown in Fig. 4.11.

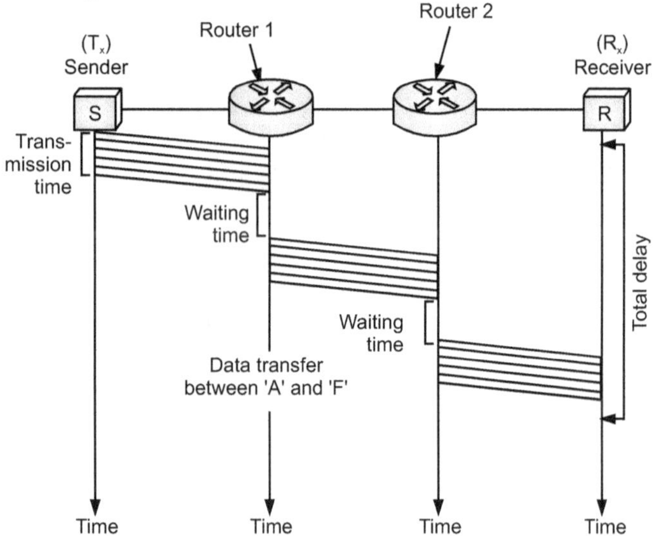

Fig. 4.11: Delay present in datagram network

- Thus, total delay in above typical datagram network is given as,

$$T_D = 3T_t + 3T_p + T_{w_1} + T_{w_2}$$

where,

$3T_t$ = Three transmission times available (hence $3T_t$)

$3T_p$ = Three propagation delays available (hence $3T_p$)

$\left.\begin{array}{c} T_{w_1} \\ T_{w_2} \end{array}\right\}$ = Two waiting time delays available (hence T_{w_1} and T_{w_2})

- The best example of datagram network is Internet or TCP/IP protocol based LAN or Internet communication.

- Thus, a routing table contains a mapping between the possible final destination of packets and the outgoing link on their path to the destination. Routing tables can be very large because they are indexed by possible destinations, making lookups and routing decisions computationally expensive, and the full forwarding process relatively slow compared to circuit switching.

- In datagram packet switching networks, each packet must carry the address of the destination host and use the destination address to make a forwarding decision. Consequently, routers do not need to modify the destination addresses of packets when forwarding packets.

- Since each packet is processed individually by a router, all packets sent by a host to another host are not guaranteed to use the same physical links. If the routing algorithm decides to change the routing tables of the network between the instants two packets are sent, then these packets will take different paths and can even arrive out of order.

- Second, on a network topology change such as a link failure, the routing protocol will automatically re-compute routing tables so as to take the new topology into account and avoid the failed link.

- As opposed to circuit switching, no additional traffic engineering algorithm is required to reroute traffic. Since routers make routing decisions locally for each packet, independent of the flow to which a packet belongs.

- Therefore, traffic engineering techniques, which heavily rely on controlling the route of traffic, are more difficult to implement with datagram packet switching than with circuit switching.

There are three primary types of datagram packet switches:

1. **Store and Forward:** Buffers data until the entire packet is received and checked for errors. This prevents corrupted packets from propagating throughout the network but increases switching delay.

2. **Fragment Free:** Filters out most error packets but doesn't necessarily prevent the propagation of errors throughout the network. It offers faster switching speeds and lower delay than store-and-forward mode.

3. **Cut Through:** Does not filter errors; it switches packets at the highest throughput, offering the least forwarding delay.

- A datagram network is a best effort network. Delivery is not guaranteed. Reliable delivery must be provided by the end systems (i.e. user's computers) using additional protocols.

- The most common datagram network is the Internet, which uses the IP network protocol. Applications which do not require more than a best effort service can be supported by direct use of packets in a datagram network, using the User Datagram Protocol (UDP) transport protocol.

- Applications like voice and video communications and notifying messages to alert a user that she/he has received new email are using UDP. Applications like e-mail, web browsing and file upload and download need reliable communications, such as guaranteed delivery, error control and sequence control.

- This reliability ensures that all the data is received in the correct order without errors. It is provided by a protocol such as the Transmission Control Protocol (TCP) or the File Transfer Protocol (FTP).

4.5 VIRTUAL CIRCUIT NETWORKS (VC NETWORKS)

- Virtual circuit network is another type of packet switched network.

- Virtual circuit network is a cross between datagram switching network and circuit switching network. It has characteristics of both the networks.

Table 4.2

Characteristics of Circuit Switched Network	Characteristics of Datagram Network
1. It has three phases like : (i) Set-up phase (connection establishment). (ii) Data transfer. (iii) Teardown phase (connection release).	1. Resource allocation can be on demand in VC networks.
2. Resources can be allocated during the set-up phase or connection establishment phase.	2. In datagram packet header, destination IP addresses are mentioned, whereas in VC packet header next switch VCI (Virtual Circuit Identifier) number is mentioned.
3. All packets follow the same path established during set-up phase or connection establishment phase.	

In today's technology

- Circuit switched network is implemented in physical layer.
- Datagram switched network is implemented in network layer.
- VC switched network is implemented in data link layer.

- In VC switched networks, two types of addressing used are as follows:

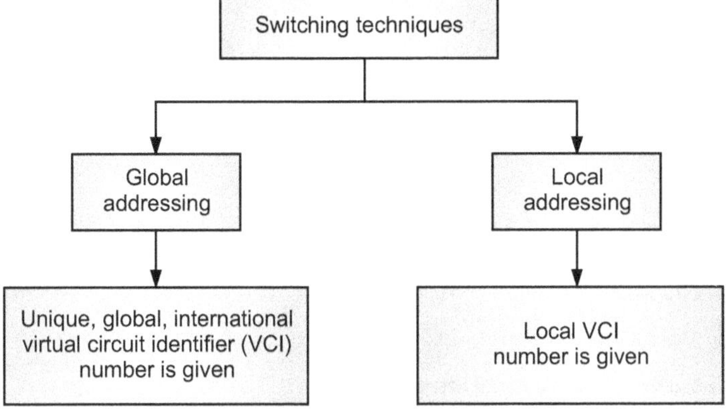

Fig. 4.12 : VC switched networks

4.5.1 Virtual Circuit Packet Network

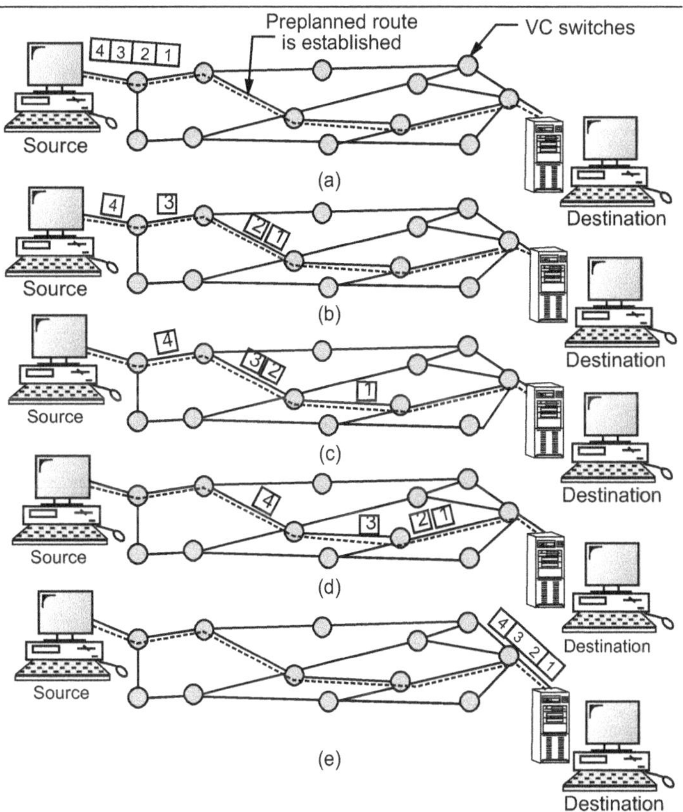

Fig. 4.13 (a) : Packet flow in typical virtual circuit packet switched network

- Typical VC switched network is as shown in Fig. 4.13 (b).

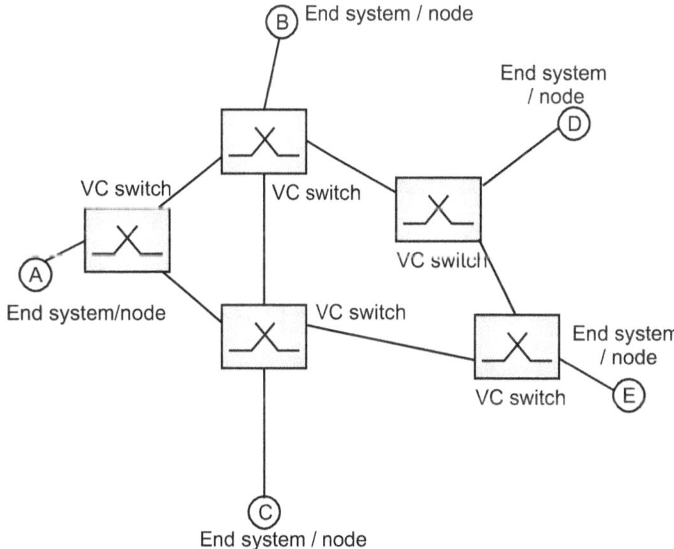

Fig. 4.13 (b) : Typical VC switched network

- **Virtual Circuit Identifier (VCI)** is used for data transfer from one end to another.

- VCI is used by frames between two VC switches. When frame arrives at one VC switch, it has a VCI, when it leaves it has a different VCI.

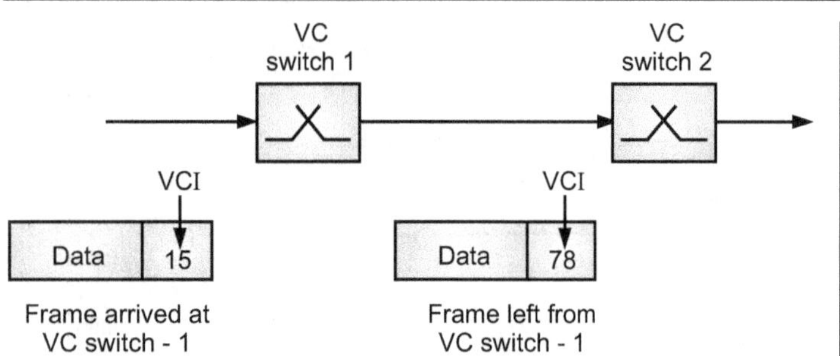

Fig. 4.14: VC switch decides vci number of frames

- The VC switched network uses three steps for communication:

1. Set-up phase (connection establishment phase).

2. Data transfer phase.

3. Teardown phase (connection release phase).

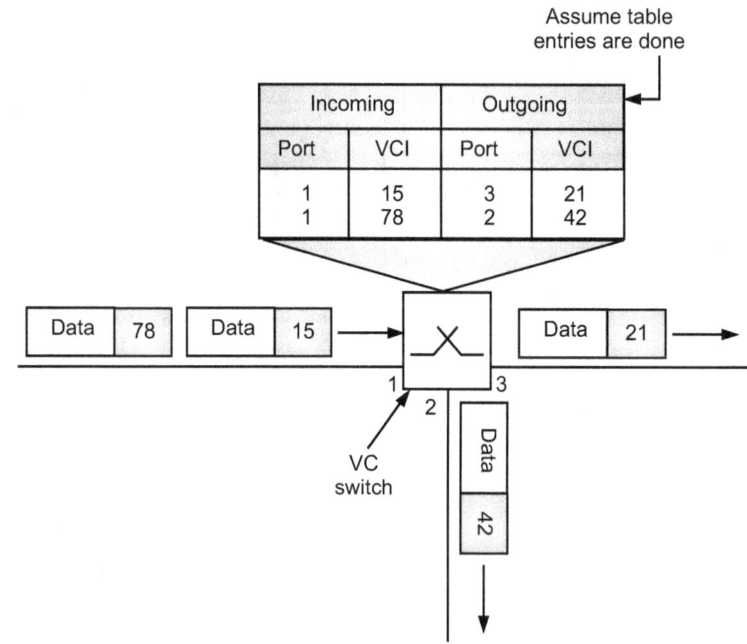

(a) Switching table maintained by typical VC switch

(Assume that entries are done initially)

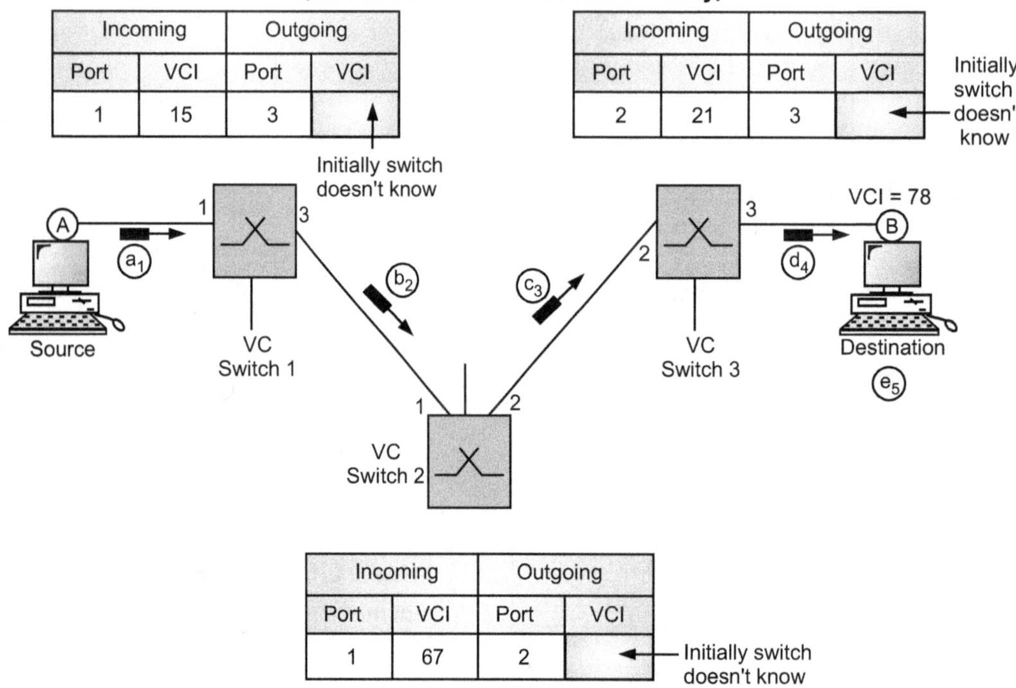

(b) Set-up request process from source 'A' to destination 'B'

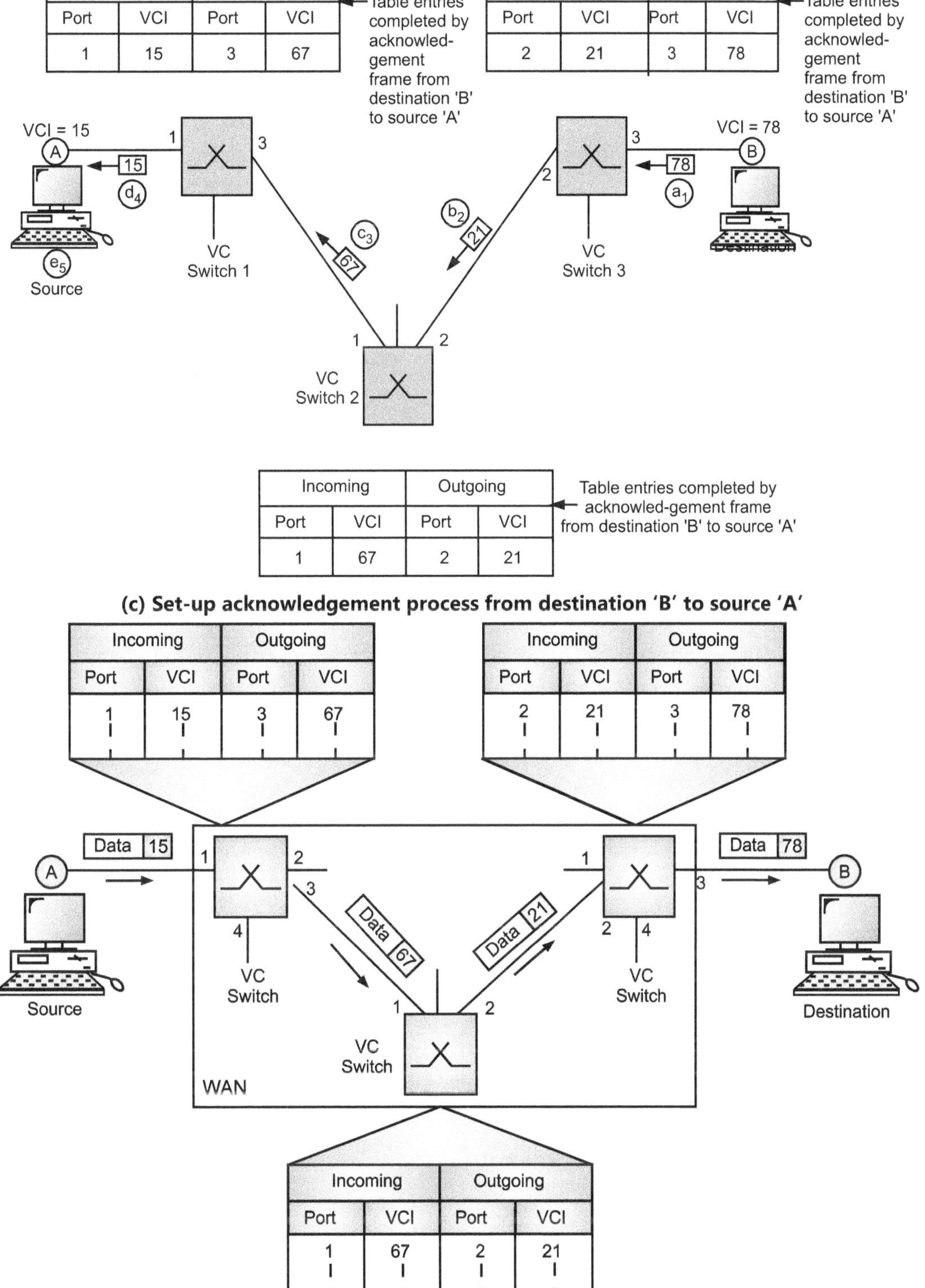

(c) Set-up acknowledgement process from destination 'B' to source 'A'

(d) Source to destination data transfer from A' to 'B' end system according to updated table entries

Fig. 4.15 : Communication between source 'A' and destination 'B' in VC switched network

Fig. 4.15 (a) says :

- When incoming frame 1 with VCI = 15 and frame 2 with VCI = 78 arrives to VC switch, depending upon switching table entries mentioned, frame 1-VCI is changed from 15 → 21 and sent on port 3. Also frame 2-VCI is changed from 78 → 42 and sent on port 2.

Fig. 4.15 (b) says :

- Set-up request process from source 'A' to destination 'B'.

- It clearly shows random incoming VCI number and undefined outgoing VCI numbers.

- Set-up request starts from source 'A' → VC switch 1 → VC switch 2 → VC switch 3 → destination 'B'.

Fig. 4.15 (c) says :

- Special acknowledgement frame is sent from destination 'B' to source 'A'.
- This special acknowledgement frame completes the table entries like

  VC switch 3 → Outgoing VCI = 78
      Incoming VCI = 21
  VC switch 2 → Outgoing VCI = 21
      Incoming VCI = 67
  VC switch 1 → Outgoing VCI = 67
      Incoming VCI = 15

- Thus, source 'A' gets the VCI = 15 as source frame VCI number and hence source 'A' sends | data | 15 | frame like to immediate VC switch 1 in data transfer phase.

Fig. 4.15 (d) says :

- Once table entries are confirmed the first frame generated from | data | 15 | source 'A' is with VCI = 15.
- Thus, data transfer takes place from source 'A' → VC switch 1 → VC switch 2 → VC switch 3 → destination 'B'.
- Hence, Fig. 4.15 shows communication between source 'A' and destination 'B' in VC switched network.

  Where, set-up phase request process is given by Fig. 4.15 (b).

  Set-up phase acknowledgement process is given by Fig. 4.15 (c).

  Data transfer between 'A' and 'B' is given by Fig. 4.15 (d).

- Finally, when data transfer is completed between 'A' and 'B' systems then the remaining process is teardown process or connection release process. In this process or in this phase source 'A' sends a special frame called a teardown request to destination 'B'. Destination 'B' also responds with teardown confirmation frame and sends to source 'A'. Thus, all corresponding switching table entries are deleted from VC switches.

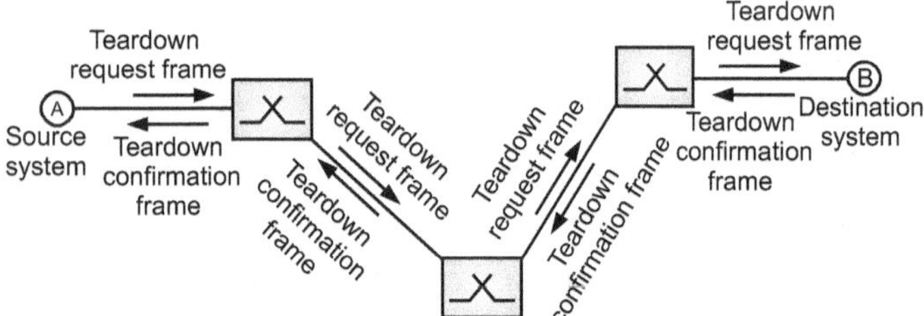

Fig. 4.16 : Typical teardown request and confirmation set-up between source 'A' and destination 'B' after data transfer is over

- Thus, the efficiency of the VC switched network is greater as compared to circuit switched network and datagram switched network.
- This happens because the main advantage of VC switched network is even if resource allocation is on demand; the source can check availability of the resources without actually reserving it.
- Thus, though path between source 'A' and destination 'B' is same, packets may arrive at the destination with different delays if resource allocation is done on demand as we discussed.
- The total delay in communication between source 'A' and destination 'B' is given by,

> Total delay = Transmission delay + Propagation delay
> + Set-up delay + Teardown delay

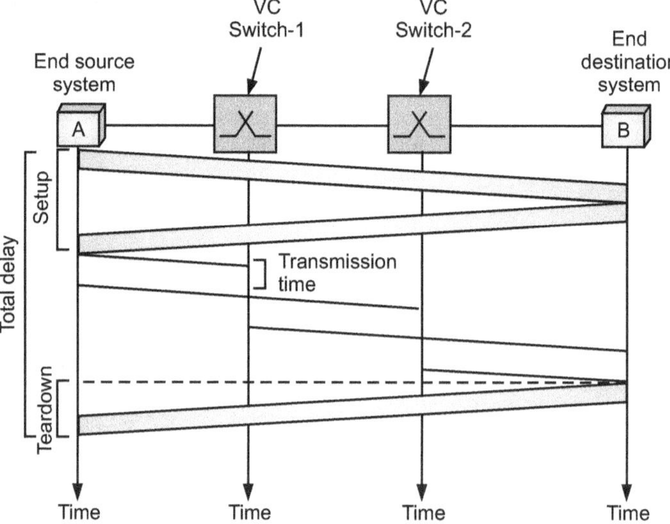

Fig. 4.17: Delay present in VC switched network

Thus, total delay in above typical VC switched network is given as,

$$T_D = 3T_t + 3T_p + \text{Set-up delay} + \text{Teardown delay}$$

where, $3T_t$ = Three transmission times available (hence $3T_t$)

           $3T_p$ = Three propagation delays available (hence $3T_p$)

Set-up delay consists of request + Acknowledged process delay.

Teardown delay consists of request + Acknowledged process delay.

Following are the typical VC switched networks:

- ➤ X.25 VC switched networks.
- ➤ Frame relay VC switched networks.
- ➤ ATM (Asynchronous Transfer Mode) networks.
- ➤ MPLS (Multiprotocol Label Switching) networks.

4.5.2 Concept of Virtual Circuit

- A *virtual circuit* is a logical connection created to ensure reliable communication between two network devices.

- A virtual circuit denotes the existence of a logical, bidirectional path from sender device to another receiver device across an VC switched network.

- Physically, the connection can pass through any number of intermediate nodes, such as VC switches.

- Multiple virtual circuits (logical connections) can be multiplexed onto a single physical circuit (a physical connection).

- Virtual circuits are demultiplexed at the remote end, and data is sent to the appropriate destinations. Fig. 4.18 illustrates four separate virtual circuits being multiplexed onto a single physical circuit.

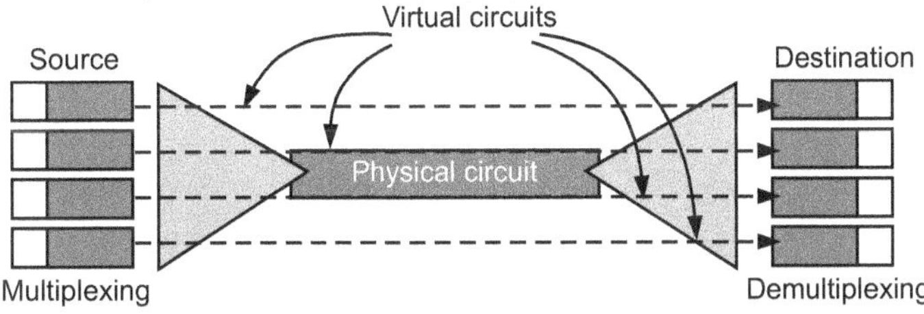

Fig. 4.18 : Virtual circuits and physical circuit between source and destination

- Thus, we have seen the trade-off between connection establishment and forwarding time costs that exist in circuit switching and datagram packet switching.

- In VC switching, routing is performed at circuit establishment time to keep fast packet forwarding.

- Other advantages of VC switching include the traffic engineering capability of circuit switching, and the resources usage efficiency of datagram packet switching.

- Nevertheless, a main issue of VC switched networks is the behavior on a topology change.

- As opposed to datagram packet switched networks which automatically re-compute routing tables on a topology change like a link failure, in VC switching all virtual circuits that pass through a failed link are interrupted.

- Hence, rerouting in VC switching relies on traffic engineering techniques.

4.5.3 Types of Virtual Circuits

1. **Switched Virtual Circuits :** (Physical connection is always present temporary logical connection is to be established every time for the data transfer) *Switched Virtual Circuits* (SVCs) are temporary connections used in situations requiring only sporadic data transfer between DTE devices across the Frame Relay network.

2. **Permanent Virtual Circuits:** (Physical connection is always present and permanent logical connection is already established so that any time data can be transferred) *Permanent Virtual Circuits* (PVCs) are permanently established connections that are used for frequent and consistent data transfers between DTE devices across the network. Communication across a PVC does not require the call setup and termination states that are used with SVCs. PVCs do not require that sessions be established and terminated. Therefore, DTEs can begin transferring data whenever necessary because the session is always active.

4.5.4 Typical Applications of VC Switched Networks

- **File Transfer**
  - ➤ Character-interactive traffic (e.g. text editing)
  - ➤ High Resolution graphics
- Access to Internet and Intranet
- Multimedia, Real Time Voice, Video, Fax
- LAN Peer-to-Peer, WAN Interconnection
- Multi-protocol networking applications
  - ➤ ATM, SNA, TCP/IP
- Private backbone networks.

4.6 COMPARISON OF DIFFERENT SWITCHING TECHNIQUES

4.6.1 Circuit Switching V/s Packet Switching

Table 4.3

Parameter	Circuit Switched	Packet Switched
Call setup requirement	Required	Not required
Dedicated physical path requirement	Yes, it is required	No, it is not required
Whether each packet follows the same path (route)	Yes, it follows the same path (route)	No, it does not follow the same path (route)
Bandwidth Available for Transmission	It is fixed	It is dynamic
Congestion can occur at	Setup time	On every packet
Bandwidth Wastage	Yes	No
Store and Forward transmission	Not available	Yes, it is available
Transparency in System	Yes, it is present	Not present
Charge applied	Per unit time	Per unit packet

4.6.2 Datagram (DG) V/s Virtual Circuit (VC) Switching

Table 4.4

Parameter	Datagram (DG) Network	Virtual Circuit (VC) Network
Requirement of Circuit Setup	Requirement of circuit setup is not needed	Requirement of circuit setup is needed.
Routing of Packet	Each packet is routed independently	Route is chosen when VC setup is over and all packet follows the same
Router failure	None	Less probability
If router fails	Packets are lost during the crash	All VCs that passed through the failed router are terminated
Achievement of Quality of service	It is difficult here	It is easy if enough resources can be allocated in advance for each VC
Congestion control	It is difficult over here	It is easy if enough resources can be allocated in advance for each VC
Examples of network	TCP/IP internet network	X.25, Frame Relay and ATM networks

4.7 INTERNET PROTOCOL (IP)   (Feb. 15, 16)

- Internet protocol is responsible for routing the data packets between the source machine and destination machine.

- It is simple, connectionless internetworking protocol.

- It does not give guarantee of reliable data transmission between source and destination.

- IP relies on protocols in other layers to establish the connection if connection oriented services is required, as well as to provide error detection and error recovery.

- Each IP datagram is handled independently and each one can follow a different route to the destination. So, there is always a possibility of receiving the packets out of order at the destination.

- IP relies on ICMP protocol to report errors in the processing of datagrams and provide additional administrative and status message.

4.7.1 IPv4 Header

- An IP datagram consists of a header part and a data part. The header has a 20 byte fixed part and a variable length optional part. The header format is shown in Fig. 4.19.

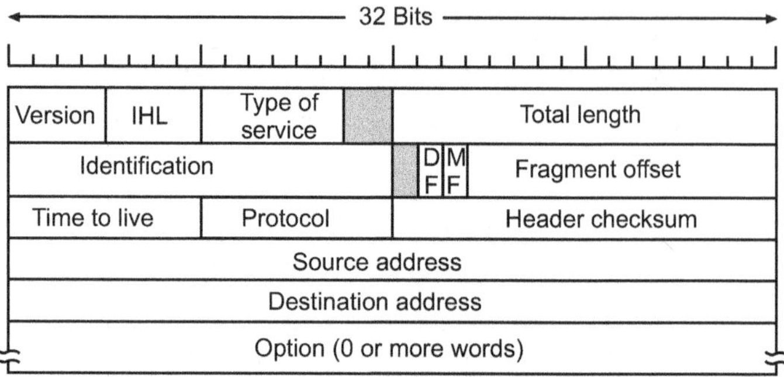

Fig. 4.19 : IP header

Various fields of IP header are as follows :

- **Version :** The Version field keeps track of which version of the protocol the datagram belongs to. Currently there are two versions of IP protocol, IPV4 and IPV6.

- **Header Length :** This is 4 bit field. It contains the length of the header expressed in 4 bytes. The size of header without including the options field is 20 bytes.

- **Type of Service :** This is 8 bit field contains a combination of 1-bit flags that can be used to request delay, throughput and reliability parameters.

- **Total Length :** This 16 bit field contains the total length of IP datagram. The total length includes the length of header as well as the data field. The maximum length is 65535 bytes.

- **Identifier:** It helps the destination host to determine, to which datagram the newly arrived fragment belongs to. All the fragments of the datagram contain the same identification number.

- **DF Flag (Don't Fragment) :** It is 1 bit field. DF stands for do not fragment. If destination is incapable of putting the fragments of the datagram, back together that time DF flag is set to 1. It instructs the router for not doing fragments of the datagram.

- **MF (More Fragments) :** This is 1 bit field. MF stands for more fragments. All fragments except the last one has this bit set. It is needed to know the destination that all fragments of the datagram are arrived.

- **Fragment Offset :** This is 13 bit field, shows the relative position of this fragment with respect to the whole datagram.

- It is the offset of the data in the original datagram measured in units of 8 bytes.

- Fig. 4.19 shows the datagram of 4000 bytes fragmented into 3 fragments.

- The bytes in the original datagram are numbered from 0 to 3999. In that the first fragment carries bytes from 0 to 1399. The offset for this datagram is 0/8 = 0.

- The second fragment carries bytes 1400 to 2799; the offset value for this fragment is 1400/8 = 175.

4.8 FRAGMENTATION   (May 16)

- A datagram can travel through different networks. Each router decapsulates the IP datagram from the frame it receives, processes it, and then encapsulates it in another frame.

- The format and size of the received frame depend on the protocol used by the physical network through which the frame has just traveled. The format and size of the sent frame depend on the protocol used by the physical network through which the frame is going to travel.

- For example, if a router connects a LAN to a WAN, it receives a frame in the LAN format and sends a frame in the WAN format.

4.8.1 Maximum Transfer Unit (MTU)

- Each data link layer protocol has its own frame format in most protocols. One of the fields defined in the format is the maximum size of the data field.

- In other words, when a datagram is encapsulated in a frame, the total size of the datagram must be less than this maximum size, which is defined by the restrictions imposed by the hardware and software used in the network

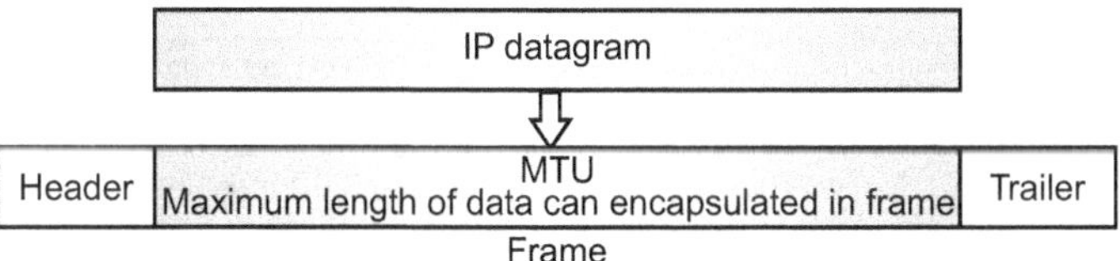

Fig. 4.20 : MTU

- The value of the MTU differs from one physical network protocol to another. For example, the value for the Ethernet LAN is 1500 bytes, for FDDI LAN is 4352 bytes, and for PPP is 296 bytes.

- In order to make the IP protocol independent of the physical network, the designers decided to make the maximum length of the IP datagram equal to 65,535 bytes.

- This makes transmission more efficient if we use a protocol with an MTU of this size. However, for other physical networks, we must divide the datagram to make it possible to pass through these networks. This is called fragmentation.

- The source usually does not fragment the IP packet. The transport layer will instead segment the data into a size that can be accommodated by IP and the data link layer in use.

- When a datagram is fragmented, each fragment has its own header with most of the fields repeated, but some changed. A fragmented datagram may itself be fragmented if it encounters a network with an even smaller MTU. In other words, a datagram can be fragmented several times before it reaches the final destination.

- A datagram can be fragmented by the source host or any router in the path. The reassembly of the datagram, however, is done only by the destination host because each fragment becomes an independent datagram.

- Whereas the fragmented datagram can travel through different routes, and we can never control or guarantee which route a fragmented datagram may take, all of the fragments belonging to the same datagram should finally arrive at the destination host.

- So, it is logical to do the reassembly at the final destination. An even stronger objection for reassembling packets during the transmission is the loss of efficiency it incurs.

- When a datagram is fragmented, required parts of the header must be copied by all fragments. The option field may or may not be copied as we will see in the next section.

- The host or router that fragments a datagram must change the values of three fields: flags, fragmentation offset, and total length. The rest of the fields must be copied.

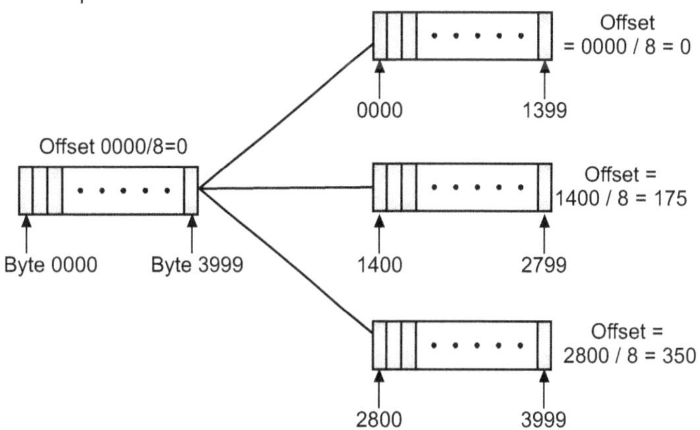

Fig. 4.21 : Fragmentation example

- Finally, the third fragment carries bytes 2800 to 3999. The offset value for this fragment is 2800/8 = 350.

- Remember that the value of the offset is measured in units of 8 bytes. This is done because the length of the offset field is 13 bits long and can not represent the sequence of bytes greater than 8191 (because $2^{13}=8192$).

- This forces hosts or routers that fragment datagrams, to choose the size of each fragment so that the first byte number is divisible by 8.

Time to Live

- This field is used as counter, which is used to limit packet lifetimes.

- If the maximum lifetime is 255, the counter is decremented on visiting each hop.

- When the counter becomes zero, the packet is discarded from the network.

Protocol

- This 8 bit field defines the higher level protocol that uses the service of the IP layer.

- An IP datagram can encapsulate the data from several higher level protocols such as TCP, UDP, ICMP and IGMP.

- This field specifies the final destination protocol to which the IP datagram should be delivered.

- The value of this field for different higher level protocols (in network as well as transport layer protocols) is shown in the following table 4.5 :

Table 4.5 : Protocols

Value	Protocol
1	ICMP
2	IGMP
6	TCP
17	UDP
89	OSPF

- **Header Checksum :** This field verifies the header only. This field is used to detect the error in the header.

- **Source Address :**

  - This 32 bit field defines the IP address of the source.

  - This field must remain unchanged during the time the IP datagram travels from the source host to the destination host.

- **Destination Address :**

  - This 32 bit field defines the IP address of the destination.

  - This field must remain unchanged during the time the IP datagram travels from the source host to the destination host.

- **Options :** Allows IP to support various options, such as security.

The header of the IP datagram is made of two parts :

  (i) Fixed part

  (ii) Variable part

- The fixed part is 20 bytes long and was discussed in the previous section. The variable part comprises the options, which can be a maximum of 40 bytes.

- Options, as the name implies, are not required for a datagram. They can be used for network testing and debugging. Although options are not a required part of the IP header, option processing is required of the IP software. This means that all implementations must be able to handle options if they are present in the header.

4.8.2 Format

Fig. 4.22 shows the format of an option. It is composed of a 1-byte type field, a 1-byte length field, and a variable-sized value field. The three fields are often referred to as type-length-value or TLV.

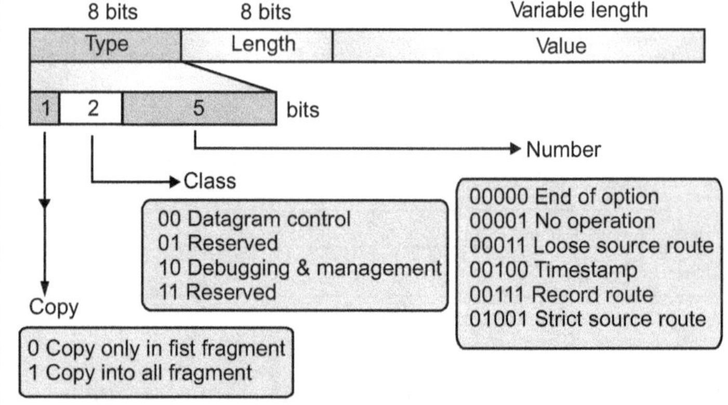

Fig. 4.22 : Option format

- **Type :** The type field is 8 bits long and contains three subfields: copy, class, and number.

- **Copy :** This 1-bit subfield controls the presence of the option in fragmentation. When its value is 0, it means that the option must be copied only to the first fragment. If its value is 1, it means the option must be copied to all fragments.

- **Class :** This 2-bit subfield defines the general purpose of the option. When its value is 00, it means that the option is used for datagram control. When its value is

10, it means that the option is used for debugging and management. The other two possible values (01 and 11) have not yet been defined.

- **Number :** This 5-bit subfield defines the type of option. Although 5 bits can define up to 32 different types, currently only 6 types are in use.
- **Length :** The Length Fielddefines the total length of the option including the type field and the length field itself. This field is not present in all of the option types.
- **Value :** The Value Field contains the data that specific options require. Like the length field, this field is also not present in all option types.

4.8.3 Option Types

As mentioned previously, only six options are currently being used. Two of these are 1-byte options, and they do not require the length or the data fields. Four of them are multiple-byte options; they require the length and the data fields

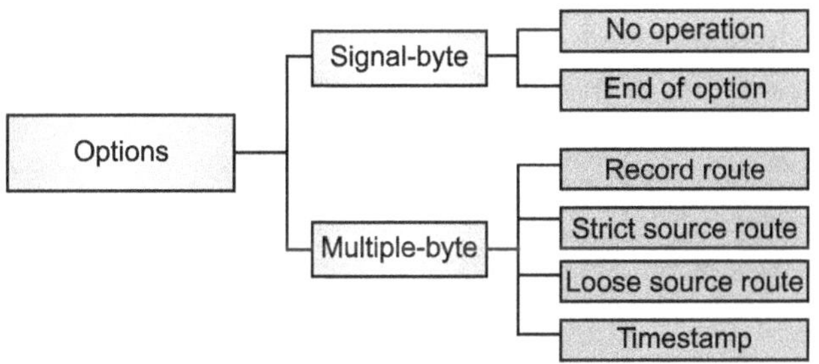

Fig. 4.23 : Option types

1. **No-Operation Option :** A no-operation optionis a 1-byte option used as a filler between options. For example, it can be used to align the next option on a 16-bit or 32-bit boundary

2. **End-of-Option Option :** An end-of-option optionis also a 1-byte option used for padding at the end of the option field. It, however, can only be used as the last option. Only one end-of-option can be used. After this option, the receiver looks for the payload data. This means that if more than 1 byte is needed to align the option field, some no-operation options must be used, followed by an end-of-option option

3. **Record-Route Option :** A record-route option is used to record the Internet routers that handle the datagram. It can list up to nine router IP addresses since the maximum size of the header is 60 bytes, which must include 20 bytes for the base header. This implies that only 40 bytes are left over for the option part. The source creates placeholder fields in the option to be filled by the visited routers.

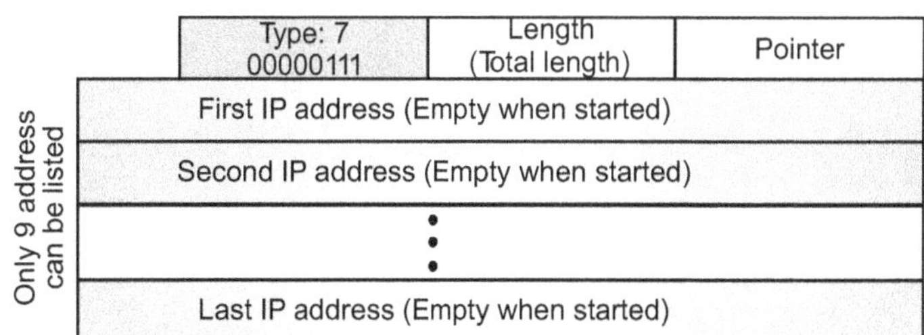

Fig. 4.24 : Record route option

- Both the code and length fields have been described above. The pointer field is an offset integer field containing the byte number of the first empty entry. In other words, it points to the first available entry.

- The source creates empty fields for the IP addresses in the data field of the option. When the datagram leaves the source, all of the fields are empty. The pointer field has a value of 4, pointing to the first empty field.

- When the datagram is traveling, each router that processes the datagram compares the value of the pointer with the value of the length. If the value of the pointer is greater than the value of the length, the option is full and no changes are made. However, if the value of the pointer is not greater than the value of the length, the router inserts its outgoing IP address in the next empty field (remember that a router has more than one IP address). In this case, the router adds the IP address of its interface from which the datagram is leaving. The router then increments the value of the pointer by 4.

4. **Strict-Source-Route Option :** A strict-source-route optionis used by the source to predetermine a route for the datagram as it travels through the Internet. Dictation of a route by the source can be useful for several purposes.

- The sender can choose a route with a specific type of service, such as minimum delay or maximum throughput. Alternatively, it may choose a route that is safer or more reliable for the sender's purpose. For example, a sender can choose a route so that its datagram does not travel through a competitor's network.

- If a datagram specifies a strict source route, all of the routers defined in the option must be visited by the datagram. A router must not be visited if its IP address is not listed in the datagram. If the datagram visits a router that is not on the list, the datagram is discarded and an error message is issued. If the datagram arrives at the destination and some of the entries were not visited, it will also be discarded and an error message issued.

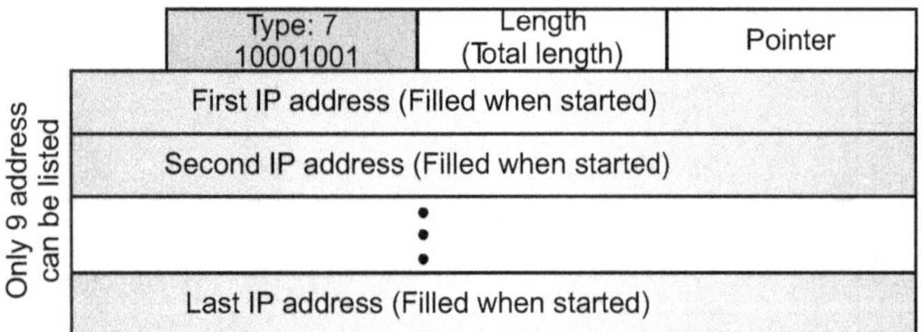

Fig. 4.25 : Strict source route option

- The format is similar to the record route option with the exception that all of the IP addresses are entered by the sender.

- When the datagram is traveling, each router that processes the datagram compares the value of the pointer with the value of the length. If the value of the pointer is greater than the value of the length, the datagram has visited all of the predefined routers.

- The datagram cannot travel anymore; it is discarded and an error message is created. If the value of the pointer is not greater than the value of the length, the router compares the destination IP address with its incoming IP address.

- If they are equal, it processes the datagram, swaps the IP address pointed by the pointer with the destination address, increments the pointer value by 4, and forwards the datagram. If they are not equal, it discards the datagram and issues an error message.

5. **Loose-Source-Route Option :** A loose-source-route optionis similar to the strict source route, but it is more relaxed. Each router in the list must be visited, but the datagram can visit other routers as well.

4.9 CHECKSUM

The error detection method used by most TCP/IP protocols is called the checksum. The checksum protects against the corruption that may occur during the transmission of a packet. It is redundant information added to the packet. The checksum is calculated at the sender and the value obtained is sent with the packet. The receiver repeats the same calculation on the whole packet including the checksum. If the result is satisfactory, the packet is accepted; otherwise, it is rejected.

4.9.1 Checksum Calculation at the Sender

- At the sender, the packet header is divided into n-bit sections (n is usually 16). These sections are added together using one's complement arithmetic resulting in a sum that is also n bits long. The sum is then complemented (all 0s changed to 1s and all 1s to 0s) to produce the checksum.

- To create the checksum the sender does the following :
  - The packet is divided into k sections, each of n bits.
  - All sections are added together using one's complement arithmetic.
  - The final result is complemented to make the checksum.

4.9.2 Checksum Calculation at the Receiver

- The receiver divides the received packet into n sections and adds all sections. It then complements the result. If the final result is 0, the packet is accepted; otherwise, it is rejected.

- When the receiver adds all of the sections and complements the result, it should get zero if there is no error in the data during transmission or processing. This is true because of the rules in one's complement arithmetic.

- Assume that we get a number called T when we add all the sections in the sender. When we complement the number in one's complement arithmetic, we get the negative of the number. This means that if the sum of all sections is T, the checksum is $-T$.

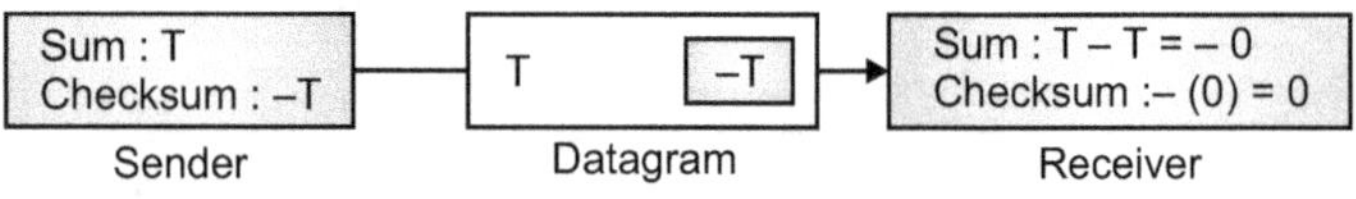

Fig. 4.26 : Checksum

- When the receiver receives the packet, it adds all the sections. It adds T and −T which, in one's complement, is −0 (minus zero). When the result is complemented, − 0 becomes 0. Thus, if the final result is 0, the packet is accepted; otherwise, it is rejected

4.10 IP ADDRESS　　　　　　(Feb. 16)

- Every host and router on the Internet has an IP address, which encodes its network number and host number.

- The combination is unique : in principle, no two machines on the Internet have the same IP address.

- All IP addresses are 32 bits long and are used in the Source address and Destination address fields of IP packets.

- 32 bits of IP address are divided into 4 parts, containing 8 bits in each part.

There are three methods of depicting IP address :

1. **Dotted Decimal :** IP address is formatted as in 11.15.10.23

2. **Binary :** IP address is formatted as in 00001100.10101000.00001101.00100101

3. **Hexadecimal :** IP address is formatted as in 8B.39.C2.43

- It is important to note that an IP address does not actually refer to a host. It really refers to a network interface, so if a host is on two networks (e.g. gateways can be on two networks as they may contain two network interface cards), it must have two IP addresses. However, in practice, most hosts are on one network and thus have one IP address.

- IP address is a structured or hierarchical address. It contains two main parts 1) Network address and 2) Host address (refer Fig. 4.27). The network addresses uniquely addresses each network.

- Every machine on same network shares that network address as a part of its IP address.

- In the IP address 137.57.30.57, 137.57 is network address and 30.57 is host address.

4.10.1 Classes of IP Address

- From several decades, IP addresses were divided into the five categories listed below.

- This allocation is called as classful addressing.

- IP addresses are classified into 5 types as given below :

  1. Class A
  2. Class B
  3. Class C
  4. Class D
  5. Class E

- The formats for all the classes of IP address are given in the following figure.

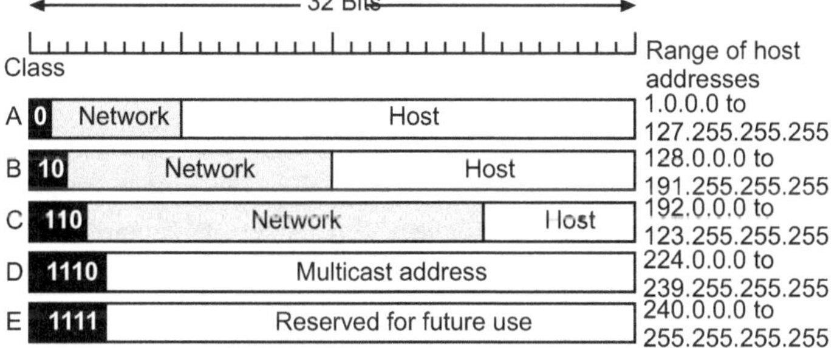

Fig. 4.27 : IP address formats

- Following table 4.6 shows the ranges of IP address belonging to the five classes.

Table 4.6

Class	Staring IP	Ending IP
Class A	1.0.0.0	127.255.255.255
Class B	128.0.0.0	191.255.255.255
Class C	192.0.0.0	223.255.255.255
Class D	224.0.0.0	239.255.255.255
Class E	240.0.0.0	255.255.255.255

Significance of Leading Bits of Network Address :

- See the Fig. 4.28 given below. When the router wants to make the routing process fast, router only observes the leading bits (i.e. 0, 10, 110, 1110 and 1111). By reading only these first bits it takes the decision of routing.

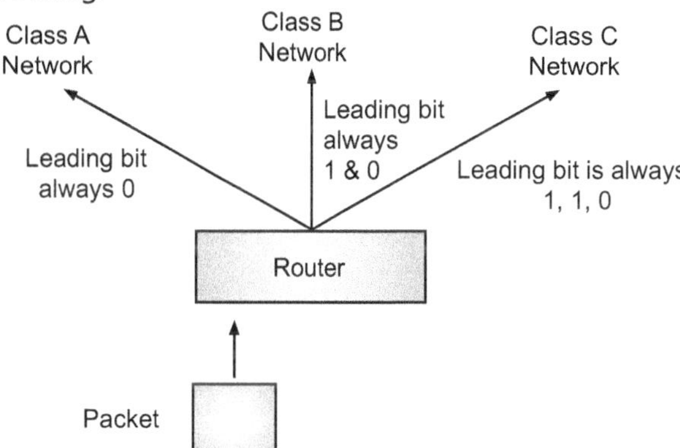

Fig. 4.28 : Leading bits of network address

Special purpose IP address and function of these addresses are listed below.

1. Network address with all zeros

   00000000. 00000000. 00000000. 00000000

   This address is used for the current network or host

2. Network address with all 1's

   11111111. 11111111. 11111111. 11111111

   Interpreted as broadcast address to all hosts on the network.

3. 127.XX.YY.ZZ : reserved for loop-back tests.

4. 0.0.0.0 : Used by the host at booting time, but it is not used afterwards.

4.10.2 Network Address

Following Fig. 4.29 shows the examples of Class A, B and Class C type of network addresses.

1. **Example of Class-A Network Address**

- As per previous discussion the range of Class-A network addresses is from 1.0.0.0 to 127.255.255.255.

- By referring Fig. 4.29 one can easily say that, Class-A IP address contains 8 bits of network id and remaining 24 bits gives the Host Id.

- Now observe the following figure, in which the example of Class-A network is given.

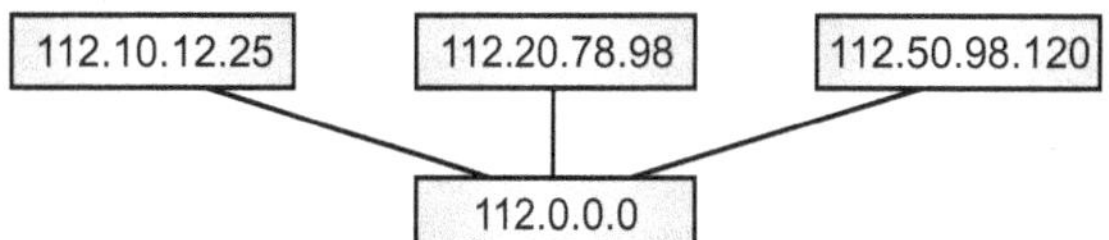

Fig. 4.29 : Class-a network address

Note : Here network address in this example is 112.0.0.0. The value of first octet (Network ID) remains same for all the hosts inside the network. That's why the host addresses are 112.10.12.25, 112.20.78.98 and 112.50.89.120.

2. Example of Class-B Network Address

As per previous discussion the range of Class-B network addresses is from 128.0.0.0 to 191.255.255.255.

By referring Fig. 4.30 one can easily say that, Class-B IP address contains 16 bits of network id and remaining 16 bits gives the Host Id.

Now observe the following Fig. 4.30, in which the example of Class-B network is given.

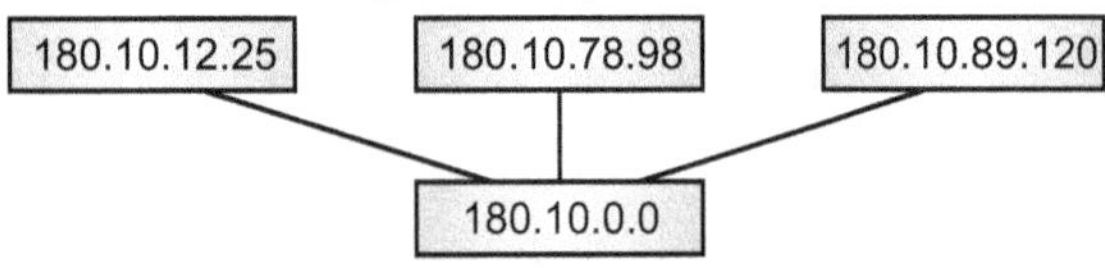

Fig. 4.30 : Class-B Network Address

Note : Here network address in this example is 180.10.0.0. The value of first 2 octets (Network ID) remains same for all the hosts inside the network. That's why the host addresses are 180.10.12.25, 180.10.78.98 and 180.10.89.120.

3. Example of Class-C Network Address

As per previous discussion the range of Class-C network addresses is from 192.0.0.0 to 223.255.255.255

By referring Fig. 4.31 can easily say that, Class-C IP address contains 24 bits of network id and remaining 8 bits gives the Host Id.

Now observe the following Fig. 4.31., in which the example of Class-C network is given.

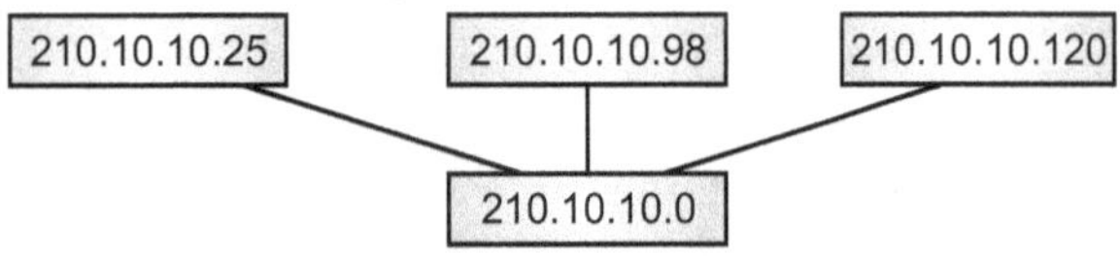

Fig. 4.31 : Class-C network address

Note : Here network address in this example is 210.10.10.0. The value of first 3 octets (Network ID) remains same for all the hosts inside the network. That's why the host addresses are 210.10.10.25, 210.10.10.98 and 210.10.10.120.

SOLVED EXAMPLES

Example 4.1 : *From the IP address 222.151.210.34, find out its class, Network ID and Host ID.*

Solution : The first octet value is 222. As it is 222, one can easily say that it is Class-C IP address.

In class C IP address, first 24 bits represents the Network ID, and last 8 bits represents the Host ID.

So accordingly, we can say that the network id is 222.151.210 and host id is 34.

4.11 GLOBALLY ROUTABLE (PUBLIC) AND PRIVATE NETWORK IP ADDRESSES

- There are two types of IP addresses
  1. Globally routable (included in the routing tables on the Internet), and
  2. Those which have been set aside for private networks.

- It is generally recommended that organizations use IP addresses from the blocks of private network addresses for hosts that require IP connectivity within their company network, but do not require external connections to the global Internet.

- The system with non-routable (private) IP addresses was introduced to help prevent a future shortage of IP addresses due to the explosive growth of the Internet.

- Because addresses belonging to these address blocks are not routed through the Internet routing system, the same numbers can be used at the same time by many different organizations.

- The three blocks of IP addresses which have been reserved for private networks are

10.0.0.0	10.255.255.255	(24-bit block/Class A)
172.16.0.0	172.31.255.255	(20-bit block/Class B)
192.168.0.0	192.168.255.255	(16-bit block/Class C)

- There are no official rules for when to use which of the three private network IP address blocks, but generally the one of the most suitable size is used. For obvious reasons there is no need to use 10.x.x.x if it is unthinkable that your LAN will ever grow to more than 254 hosts.

- Hosts with private network IP addresses cannot communicate directly with the Internet, because the Internet refuses to receive and transmit data with such origin or destination address.

- For a host with a private network IP address to be allowed to communicate with the Internet, it must have its data stream to the Internet handled by an intermediary host, which can act as an 'Internet representative' for the private host.

- The intermediary host must have ways to relay data between the global Internet and the host on the private network. Therefore, it must have a globally routable IP address that it uses when communicating with the Internet and a private network IP address that is uses for communication with the private host.

- There are a number of different types of intermediary hosts that fit this description. The most common types of intermediary hosts are proxy servers.

- Public IP address of the proxy server is the identity of the organizations private network on the internet. By using this IP, organization's private network communicates with the internet.

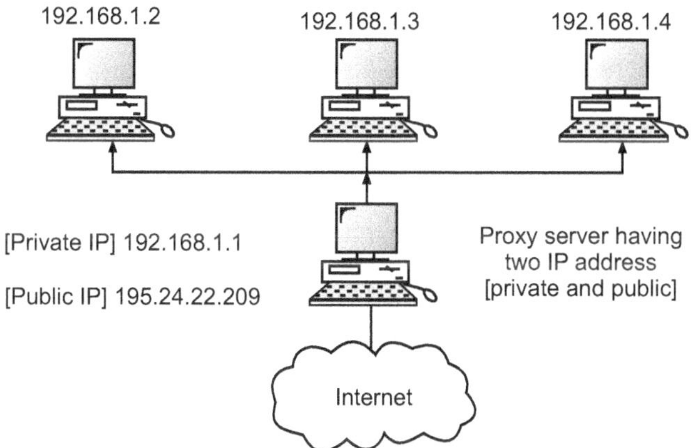

Fig. 4.32 : A proxy server translating private network IP addresses to globally routable IP addresses

- NAT (Network Address Translation) mechanism is used for conversion between the proxy server's private IP address to public IP address.

- Using private network IP addresses also gives a company a measure of security.

- Globally routable IP addresses are advertised in the routing tables on the Internet, making the system vulnerable to hackers.

- When private IP network addresses are used, however, the intermediary host (such as a proxy server) will work as a barrier against unwanted visits from the Internet.

- The current version of IP, IP version 4, defines a 32-bit address, which means that there are only 2^{32} (4,294,967,296) addresses available globally.

- Over the past few years, the number of available IP addresses on the Internet has started to run out, as the number of companies and people wishing to go on-line has exploded.

- As a consequence, a new generation of IP addresses (IPv6) is currently in the works. The current IP system will not become obsolete overnight; however, as the two systems will coexist for some time after the new version has been implemented.

4.12 IPv4 LIMITATIONS

Following are the major limitations of IPv4 addresses :

- **Shortage of IP Addresses :** IPv4 uses a 32 bit address, which generate 2^{32} (4 billion) possible addresses. Now a day's network is growing exponentially. Due to that IP addresses are getting tremendously consumed. So, the unique addresses are becoming inadequate.

- **Auto-Configuration and Mobility :** New technologies (mobile equipment, wireless network) are emerging and its use is quickly becoming common. There is no automatic way to automatically configuration this kind of equipment in the network.

- **Security :** The security option in IPv4 is optional, so it is not possible to keep all the data secure while it's routing through the network.

- **Support for Real Time Applications :** Services such as transmission of real time audio and video are becoming common nowadays. IPv4 does not provide ways for managing and reserving bandwidth for such real time transmissions.

4.13 IPv6 (INTERNET PROTOCOL VERSION 6)

(May 15)

Due to the limitations of IPv4, IPv6 comes into existence. This is next-generation internet protocol and had many advantages on the previously existing version of Internet protocol (IPv4). These are listed below :

- **Larger Address Space :** An IPv6 address is 128 bits long (compared to IPv4, IPv6 address is very long because IPv4 address was only of 32 bits).

  So total number of addresses generated using IPv6 is 2^{128}.

- **Better Header Format :** IPv6 uses a new header format in which options are separated from the base header and inserted when needed, between base header and upper layer data.

  This simplifies and speeds up the routing process because most of the options do not need to be checked by routers.

- **New Options :** IPv6 has new options to allow additional functionalities.

- **Allowance for Extension :** IPv6 is designed to allow the extension of the protocol if required by new technologies or applications.

- **Support for Resource Allocation :** In IPv6, the type of service field has been removed, but a mechanism (called flow label) has been added to enable the source to request special handling of packet.

This mechanism can be used to support traffic such as real time audio and video.

- **Support for More Security :** The encryption and authentication options in IPv6 provide confidentiality and integrity of the packet.

- **Plug and Play :** IPv6 includes plug and play in the standard specification. It therefore must be easier for novice user to connect their machines to network, it will be done automatically.

- **Clearer Specification :** IPv6 follows good practices of IPv4, and rejects its minor problems.

4.13.1 IPv6 Addresses

- An IPv6 address consists of 16 bytes (octets); it is 128 bits long. Refer following Fig. 4.33.

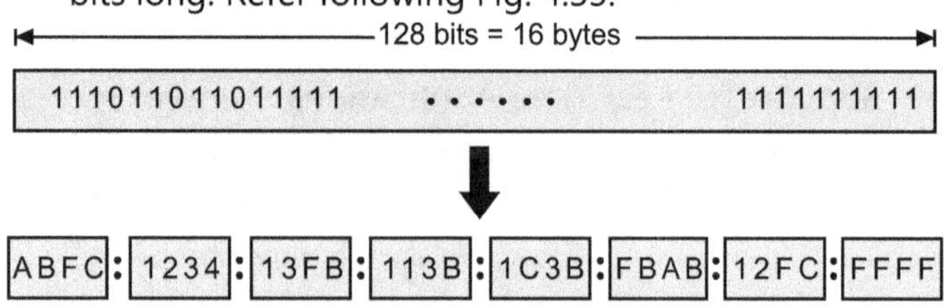

Fig. 4.33 : IPv6 address

- **Hexadecimal Colon Notation**
  - ➤ To make addresses more readable, IPv6 specifies hexadecimal colon notation.
  - ➤ In this notation, 128 bits are divided into eight sections, each of 2 bytes in length.
  - ➤ Two bytes in hexadecimal notation require four hexadecimal digits.
  - ➤ Therefore, the address consists of 32 hexadecimal digits, with every four digits separated by a colon.

- **Abbreviation**
  - ➤ Although, IP addresses in hexadecimal format are very long, many of the digits are zeros, in this case we can abbreviate the address.
  - ➤ The leading zeros of the section (four digits between two colons) can be omitted. Only leading zeros can be dropped, not the trailing zeros (Refer Fig. 4.34).

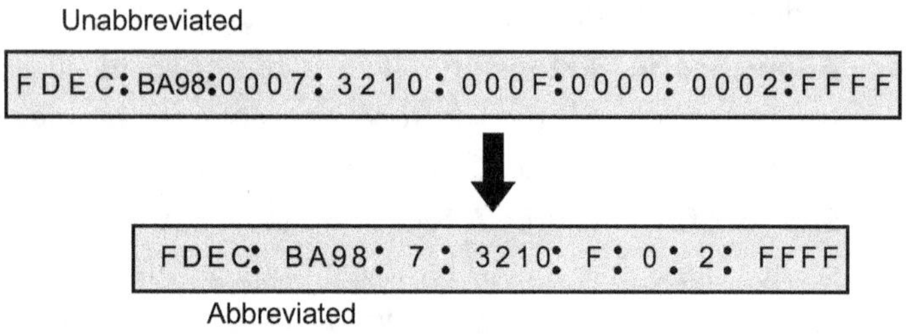

Unabbreviated

F D E C : BA98 : 0 0 0 7 : 3 2 1 0 : 0 0 0 F : 0 0 0 0 : 0 0 0 2 : F F F F

FDEC : BA98 : 7 : 3210 : F : 0 : 2 : FFFF

Abbreviated

Fig. 4.34 : Abbreviated address

- Using this form of abbreviation, 0074 can be written as 74, 000F as F, and 0000 as 0. Note that 3200 can not be abbreviated.

- Further abbreviation is possible, if there are consecutive sections consisting of zeros.

- We can remove the zeros altogether and replace them with double semicolon. Refer following figure. Note that this type of abbreviation is allowed only once per address. If there are two runs of zero sections, only one of them can be abbreviated.

Abbreciated

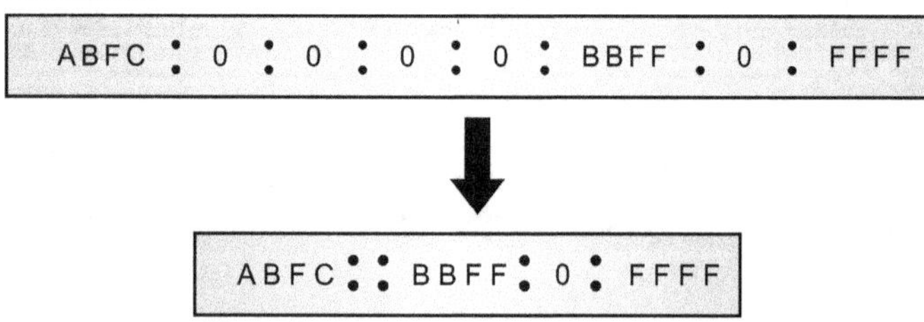

ABFC : 0 : 0 : 0 : 0 : BBFF : 0 : FFFF

ABFC :: BBFF : 0 : FFFF

More abbreviated

Fig. 4.35 : Abbreviated address with consecutive zeros

- Re-expansion of the abbreviated address is very simple: align the unabbreviated portions and insert zeros to get the original expanded address.

- **CIDR Notations :** IPv6 allows classless addressing and CIDR notation. For example, following Fig. 4.36 shows how we can define a prefix of 60 bits using CIDR.

ABFC :: BBFF : 0 : FFFF/60

Fig. 4.36 : CIDR address

- **Categories of Addresses :** IPv6 defines three types of addresses
  1. Unicast
  2. Anycast
  3. Multicast

These are explained below.

1. **Unicast :** A unicast address defines a single computer. The packet sent to unicast address must be delivered to that specific computer.

2. **Anycast :**
   - It defines a group of computers with addresses that have the same prefix. For example, all computers connected to the same physical network share the same prefix address.
   - A packet sent to an anycast address must be delivered to exactly one of the members of the group the closest or the most easily accessible.

3. **Multicast :** It defines a group of computers. The packet sent to a multicast address must be delivered to each member of the group.

4.13.2 IPv6 Header

- Following figure shows the base header of IPv6 having fixed length of 40 octets, consisting of following fields.

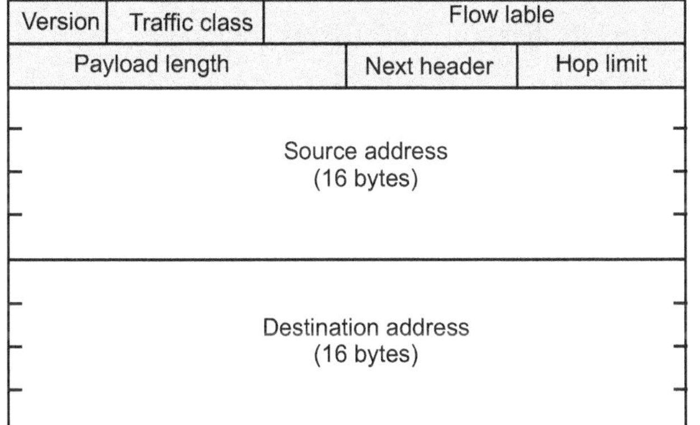

Fig. 4.37 : IPv6 header format

- **Version :** Size is 4 bits. It specifies the internet protocol version number. For IPv6 it is 6.

- **Priority (Traffic Class) :** Size is 4 bits. It defines the priority of the packet with respect to traffic congestion.

- **Flow Label :** The flow label is 3 byte field that is designed to provide special handling for a particular flow of data. It may be used by a host to label those packets for which it is requesting special handling by routers within a network.

- **Payload Length :** It is 16 it is field. It defines the total length of IP datagram including the base header.

- **Next Header :** It is a 8 bits field. It identifies the type of header immediately following IPv6 header.

- **Hop Limit :** This is 8 bits field; it serves the same purpose as the TTL field in IPv4.

- **Source Address :** Source address is 128 bits field. It identifies the original source of the datagram.

- **Destination Address :** Destination address is a 128 bits field. It identifies the destination of the datagram.

4.13.3 IPv4 V/s IPv6 Header

- The header length field is eliminated in IPv6 because the length of the header is fixed in this version.

- The service type field is eliminated in IPv6. The priority and flow label fields together take over the function of the service type field.

- The total length is eliminated in IPv6 and replaced by the payload length field.

- The identification, flag and offset fields are eliminated from the base header in IPv6. They are included in the fragmentation extension header.

- The TTL field is called Hop Limit in IPv6.

- The protocol field is replaced by the next header field.

- The header checksum is eliminated because the checksum is provided by upper layer protocols; it is therefore not needed at this level.

- The option fields in IPv4 are implemented as extension headers in IPv6.

4.13.4 Difference between IPv4 and IPv6

(Feb. 16, May 17)

Table 4.7

IPv4	IPv6
It is 32 bit source and destination addresses	It is 128 bit source and destination addresses.
There are maximum 2^{32} IP addresses.	There are maximum 2^{128} IP addresses.
IPv4 addresses are written by dotted decimal notation. e.g. 10.15.11.23	IPv6 addresses are written in hexadecimal colon notation. e.g. FADB:A2B2:A453:1212:AAB3: ADBD:BBCC:1234
Basic length of IPv4 header is 20 bytes (excluding option field)	Length of IPv6 header is 40 bytes.
IPv4 header has a checksum.	It has no header checksum.
Security is optional parameter.	It has been designed to satisfy the growing and expanded need for network security.
Headers include option field.	No option field is present in the basic header. All optional data is moved to extension header.
IPsec support is optional.	IPsec support is compulsory.
Manual or DHCP configuration is required.	Gets automatically configured, no need of manual and DHCP configuration.

4.14 ADDRESS MAPPING

4.14.1 ARP

- Most of the computer programs/applications use logical address (IP address) to send/receive messages, however the actual communication happens over the physical address (MAC address) i.e. from layer 2 of OSI model. So our mission is to get the destination MAC address which helps in communicating with other devices. This is where ARP comes into the picture, its functionality is to translate IP address to physical address.

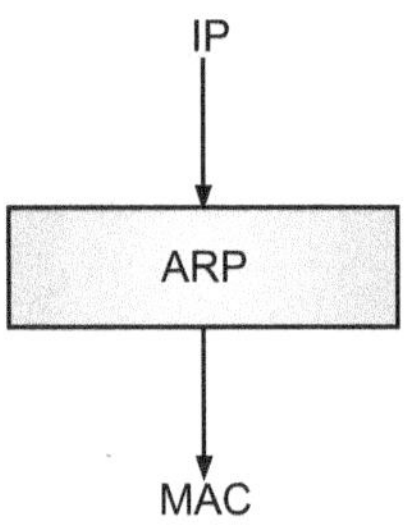

Fig. 4.38

- The acronym ARP stands for Address Resolution Protocol which is one of the most important protocols of the Network layer in the OSI model.

Note: ARP finds the hardware address, also known as Media Access Control (MAC) address, of a host from its known IP address.

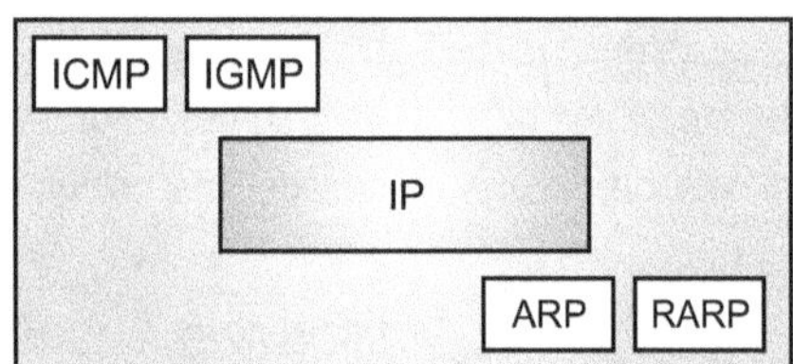

Fig. 4.39 : Network layer

Let's look at how ARP works.

- Imagine a device wants to communicate with the other over the internet. What ARP does? Is it broadcast a packet to all the devices of the source network. The devices of the network peel the header of the data link layer from the protocol data unit (PDU) called frame and transfers the packet to the network layer (layer 3 of OSI) where the network ID of the packet is validated with the destination IP's network ID of the packet and if it's equal then it responds to the source with the MAC address of the destination, else the packet reaches the gateway of the network and broadcasts packet to the devices it is connected with and validates their network ID

- The above process continues till the second last network device in the path to reach the destination where it gets validated and ARP, in turn, responds with the destination MAC address.

- The important terms associated with ARP are :

1. **ARP Cache:** After resolving MAC address, the ARP sends it to the source where it stores in a table for future reference. The subsequent communications can use the MAC address from the table

2. **ARP Cache Timeout:** It indicates the time for which the MAC address in the ARP cache can reside

3. **ARP Request:** This is nothing but broadcasting a packet over the network to validate whether we came across destination MAC address or not.

   ARP Request Packet Contains:

   (i)   The physical address of the sender.

   (ii) The IP address of the sender.

   (iii) The physical address of the receiver is 0s.

   (iv) The IP address of the receiver

   Note, that the ARP packet is encapsulated directly into data link frame.

4. **ARP Response/Reply:** It is the MAC address response that the source receives from the destination which aids in further communication of the data.

Cases when ARP is used:

CASE-1: The sender is a host and wants to send a packet to another host on the same network.

Use ARP to find another host's physical address

CASE-2: The sender is a host and wants to send a packet to another host on another network.

Sender looks at its routing table.

Find the IP address of the next hop (router) for this destination.

Use ARP to find the router's physical address

CASE-3: the sender is a router and received a datagram destined for a host on another network.

Router check its routing table.

Find the IP address of the next router.

Use ARP to find the next router's physical address.

CASE-4: The sender is a router that has received a datagram destined for a host in the same network.

Use ARP to find this host's physical address.

NOTE: An ARP request is a broadcast, and an ARP response is a Unicast.

Connect two PC, say A and B with cross cable. Now you can see the working of ARP by typing these commands:

1. A > arp -a

There will be no entry in table because they never communicated with each other.

2. A > ping 192.168.1.2

IP address of destination is 192.168.1.2

Reply comes from destination but one packet is lost because of ARP processing.

Now, entries of ARP table can be seen by typing the command.

This is how ARP table looks like:

4.14.2 RARP

- What happens if your own computer does not know its IP address, because it has no storage capacity, for example? In these cases, the Reverse Address Resolution Protocol (RARP) can help. The RARP is the counterpart to the ARP – the Address Resolution Protocol.

- The Reverse ARP is now considered obsolete, and outdated. Newer protocols such as the Bootstrap Protocol (BOOTP) and the Dynamic Host Configuration Protocol (DHCP) have replaced the RARP. However, it is useful to be familiar with the older technology as well. For instance, you can still find some applications which work with RARP today. It also helps to be familiar with the older technology in order to better understand the technology which was built on it.

- The RARP is a protocol which was published in 1984 and was included in the TCP/IP protocol stack. The RARP is **on the Network Access Layer** (i.e. the lowest layer of the TCP/IP protocol stack) and is thus a protocol used to send data between two points in a network. Each network participant has two unique addresses more or less: a logical address (the IP address) and a physical address (the MAC address). While the IP address is assigned by software, the MAC address is built into the hardware. You have already been assigned a Media Access Control address (MAC address) by the manufacturer of your network card.

- It is possible to not know your own IP address. This may happen if, for example, **the device could not save the IP address because there was insufficient memory available**. In such cases, the Reverse ARP is used. This protocol can use the known MAC address to retrieve its IP address. Therefore, its function is the complete opposite of the ARP. The ARP uses the known IP address to determine the MAC address of the hardware.

How does the RARP work?

- Who knows the IP address of a network participant if they do not know it themselves? A special **RARP server** does. This server, which responds to RARP requests, can also be a normal computer in the network. However, it must have stored all MAC addresses with their assigned IP addresses. If a network participant sends an RARP request to the network, only these special servers can respond to it.

- Since the requesting participant does not know their IP address, the data packet (i.e. the request) must be sent on the lowest layers of the network **as a broadcast**. This means that the packet is sent to all participants at the same time. However, only the RARP server will respond. If there are several of these servers, the requesting participant will only use the response that is first received. The request-response format has a similar structure to that of the ARP.

Table 4.8

Bits	Bytes	Field
0-7	1	Hardware Address Space
8-15	2	
16-23	3	Protocol Address Space
24-31	4	
32-39	5	Hardware Address Length
40-47	6	Protocol Address Length
48-55	7	Opcode
56-63	8	
64-71	9	Source Hardware Address
72-79	10	
80-87	11	
88-95	12	
96-103	13	
104-111	14	
112-119	15	Source Protocol Address
120-127	16	
128-135	17	
136-143	18	
144-151	19	Target Hardware Address
152-159	20	
160-167	21	
168-175	22	
176-183	23	
184-191	24	
192-199	25	Target Protocol Address
200-207	26	
208-215	27	
216-223	28	

In a standard IPv4 Ethernet network, the RARP messages are 28 bytes long.

The following information can be found in their respective fields:

- **Hardware Address Space**: These two bytes contain the type of hardware address.
- **Protocol Address Space**: This field, which is 2 bytes long, specifies the type of network protocol.
- **Hardware Address Length**: This is 8 bits and defines the length n of the hardware address.
- **Protocol Address Length**: This field defines the length m of the network address.
- **Opcode**: This field is two bytes long and defines the type of operation. An RARP request has the value 3 and the corresponding response the value 4.
- **Source Hardware Address**: This is where the MAC address of the sender is stored. The actual length of this field is n and is defined by the information under Hardware Address Length. A standard Ethernet network consists of 6 bytes.
- **Source Protocol Address**: This field would normally contain the IP address of the sender, but since the IP address is not known during a request, the field remains undefined. The response, however, will contain the IP address of the server. The length of this field is m and is dependent on the Protocol Address Length. Normally, though, the field is the same length as an IPv4 address (i.e. 4 bytes).
- **Target Hardware Address**: This field contains the target's MAC address. Since there is no specific target for an RARP request, this field also contains the sender's address. The server also includes the address of the requesting client in the response. The length of this field is also n and is specifically 6 bytes long for Ethernet networks.
- **Target Protocol Address**: This last field remains undefined during a request and contains in the response the information requested by the server: the participant's IP address. The length of this field is also m, which is usually defined as 4 bytes.

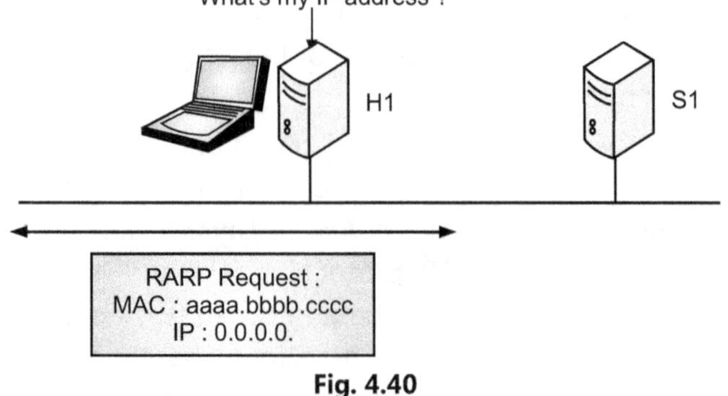

Fig. 4.40

There are important differences between the ARP and RARP. First and foremost, of course, the two protocols obviously differ in terms of their specifications. While the MAC address is known in an RARP request and is requesting the IP address, an ARP request is the exact opposite. The IP address is known, and the MAC address is being requested. The two protocols are also different in terms of the **content of their operation fields**: The ARP uses the value 1 for requests and 2 for responses. The RARP on the other hand uses 3 and 4. This means that a server can recognize whether it is an ARP or RARP from the operation code.

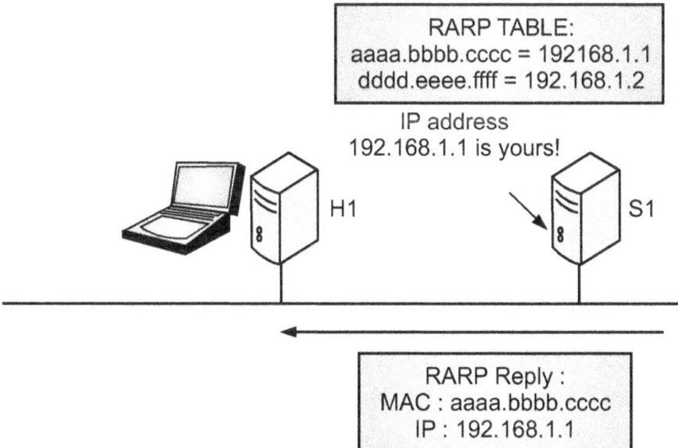

Fig. 4.41

Issues with the Reverse ARP

- The Reverse Address Resolution Protocol has some disadvantages which eventually led to it being replaced by newer ones. To be able to use the protocol successfully, the RARP server has to be located in the same physical network. The computer sends the RARP request on the lowest layer of the network. As a result, it is not possible for a router to forward the packet. In addition, the RARP cannot handle subnetting because no subnet masks are sent. If the network has been divided into multiple subnets, an RARP server must be available in each one.

- In addition, the network participant only receives their own IP address through the request. As previously mentioned, a subnet mask is not included and information about the gateway cannot be retrieved via Reverse ARP. Therefore, it is not possible to configure the computer in a modern network. These drawbacks led to the development of BOOTP and DHCP.

4.14.3 BOOTP

- The Bootstrap Protocol (BOOTP) is a computer networking protocol used in Internet Protocol networks to automatically assign an IP address to network devices from a configuration server. The BOOTP was originally defined in RFC 951.

- When a computer that is connected to a network is powered up and boots its operating system, the system software broadcasts BOOTP messages onto the network to request an IP address assignment. A BOOTP configuration server assigns an IP address based on the request from a pool of addresses configured by an administrator.

- BOOTP is implemented using the User Datagram Protocol (UDP) as transport protocol, port number 67 is used by the (DHCP) server to receive client requests and port number 68 is used by the client to receive (DHCP) server responses. BOOTP operates only on IPv4 networks.

- Historically, BOOTP has also been used for Unix-like diskless workstations to obtain the network location of their boot image, in addition to the IP address assignment. Enterprises used it to roll out a pre-configured client (e.g., Windows) installation to newly installed PCs.

- Originally requiring the use of a boot floppy disk to establish the initial network connection, manufacturers of network cards later embedded the protocol in the BIOS of the interface cards as well as system boards with on-board network adapters, thus allowing direct network booting.

- While some parts of BOOTP have been effectively superseded by the Dynamic Host Configuration Protocol (DHCP), which adds the feature of leases, parts of BOOTP are used to provide service to the DHCP protocol. DHCP servers also provide the legacy BOOTP functionality.

Operation

Case 1 : Client and server on same network

- The bootp server issues a passive open command on UDP port 67 and waits for client.

- The booted client issues active open command on port 68.This message is encapsulated in UDP user data-gram which in turn is encapsulated in IP data-gram. Client uses all 0's as source address and all 1's as destination address.

- The server now knows physical and IP address of the client. The server responds with either a broadcast or unicast UDP message with a source port of 67 and a destination port of 68.

Case 2 : Client and server on different networks

- Problem with the bootp request is that the request is broadcast. A broadcast IP data-gram cannot pass through any router. The router discards this packet.

- To solve this problem there is a need for an intermediary (relay).

- One of the host or router can be configured at application layer to operate as relay agent.

- The relay agent knows the uni-cast address of bootp server and listens for broadcast message on port 67.

- When it receives this broadcast packet, it encapsulates the message in uni-cast data-gram and sends request to bootp server.

- The packet carrying a uni-cast destination address is routed by any router and reaches the bootp server.

- The relay agent after receiving the reply, sends it to bootp client.

4.15 DYNAMIC HOST CONFIGURATION PROTOCOL (DHCP)   (May 15, 17, Nov. 16)

- BOOTP is not a dynamic configuration protocol.

- When client requests its IP address, BOOTP server consults a table that matches the physical address of the client with its IP address.

- This implies that the binding between the IP address and the physical address of the client already exists. The binding is predetermined.

- As BOOTP is static configuration protocol, it can not assign temporary IP address to the host; also it can not handle the situation when the host moves from one physical network to the other.

- To remove the limitations of BOOTP, DHCP protocol comes into existence, where DHCP provides static and dynamic address allocation that can be manual or automatic.

- DHCP is backward compatible with BOOTP, which means a host running the BOOTP client can request a static address from a DHCP server.

4.15.1 Dynamic Address Allocation

- DHCP database contains a pool of available IP addresses. This database makes DHCP dynamic.

- When a DHCP client requests a temporary IP address, DHCP server goes to the pool of available IP addresses and assigns an IP address for negotiable period of time.

- When DHCP client send a request to a DHCP server, the server first checks its static database.

- If an entry with the requested physical address exists in the static database, the permanent IP address of the client is returned.

- On the other hand, if the entry does not exist in the static database, the server selects an IP address from the available pool, assigns the IP address to the client, and adds the entry to the dynamic database.

- The dynamic aspect of DHCP is needed when the host moves from network to network or is connected and disconnected from the network.

- DHCP provides temporary IP address for a limited period of time.

- The address assigned from the pool is temporary addresses. The DHCP server issues a lease for a specific period of time.

- When the lease is expired, the client must either stop using the IP address or renew the lease.

- The server has a choice to agree or disagree with the renewal. If the server disagrees, the client stops using the address.

4.15.2 Packet Format

- As discussed previously, we know that DHCP is backward compatible with BOOTP.

- To make it backward compatible, the designers of DHCP have decided to use almost the same packet format as that of BOOTP protocol.

- They have only added one bit flag to the packet. However, for allowing different interaction with the server, extra options have been added to the option field. (Refer following Fig. 4.42).

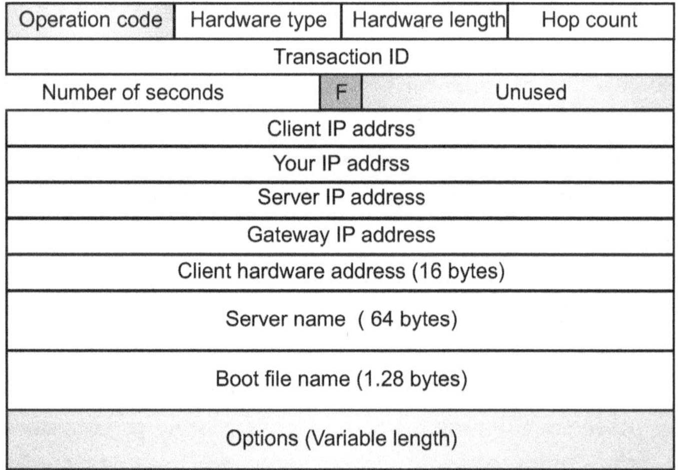

Fig. 4.42 : DHCP Packet

The fields are described below :

- **Operation Code :** This 8 bit field defines the type of BOOTP packet : 1) Request or 2) Reply.

- **Hardware Type :** This 8 bit field defines the type of physical network. For Ethernet, the value is 1.

- **Hardware Length :** This 8 bit field defines the length of physical address in bytes. For example, for Ethernet the value is 6.
- **Hop Count :** This 8 bit field defines the maximum number of hops the packet can travel.
- **Transaction ID :** This is a 4 byte field carrying an integer. This is set by the client and used to match reply with the request.
- **Number of Seconds :** This is a 16 bit field that indicates the number of seconds elapsed since the time the client started to boot.
- **Flag :** Server uses this field to specify the client that it is a forced broadcast reply. For unicast reply to the client, the destination IP address of the IP packet is the address assigned to the client. If the client does not know its IP address, it discards the packet. But if the IP datagram is broadcast, every host will receive and process the broadcast message.
- **Client IP Address :** This is a 4 byte field that contains the client IP address. If client does not have this information, this field has a value of 0.
- **Your IP Address :** This is a 4 byte field that contains the client IP address. It is filled by the server at the request of the client.
- **Server IP Address :** This is a 4 byte field that contains the server IP address. It is filled by the server in a reply message.
- **Gateway IP Address :** This is a 4 byte field that contains the router IP address. It is filled by the server in a reply message.
- **Client Hardware Address:** This is the physical address of the client.
- **Server Name :** This is an optional 64 byte field, filled by the server in reply packet.
- **Boot Filename:** This is an optional 128 byte field that can be filled by the server in a reply packet. It contains the full pathname of the boot file. The client can use this path to retrieve other booting information.
- **Options:** Several options have been added to the list of options. The option field in DHCP can be upto 312 bytes. Options for DHCP are shown in the following Table 4.9.

Table 4.9

Value	Value
1. DHCPDISCOVER	5. DHCPACK
2. DHCPOFFER	6. DHCPNACK
3. DHCPREQUEST	7. DHCPRELEASE
4. DHCPDECLINE	

4.15.3 DHCP Transition States

- Refer the DHCP transition diagram given in the following Fig. 4.43.
- At start, DHCP client is in the initializing state. The client broadcasts a DHCPDISCOVER message (it is a request message with DHCPDISCOVER option).
- After DHCPDISCOVER message, client goes into the selecting state, where server offers an IP address to the client by generating DHCPOFFER message.
- The server that sends DHCPOFFER locks the offered IP address so that it is not available for any other client within the lease duration.
- After accepting DHCPOFFER message from the DHCP server, client generates a DHCPREQUEST message for the DHCP server and goes to the requesting state.
- If the client does not receive DHCPOFFER message, it tries for four more times, each with time span of 2 seconds. Even though if the client did not get a reply, the client sleeps for 5 minutes before trying again.
- In the requesting state, client receives DHCPACK message when the server creates binding between client's physical address and its IP address. After receipt of DHCPACK, the client goes into the bound state.

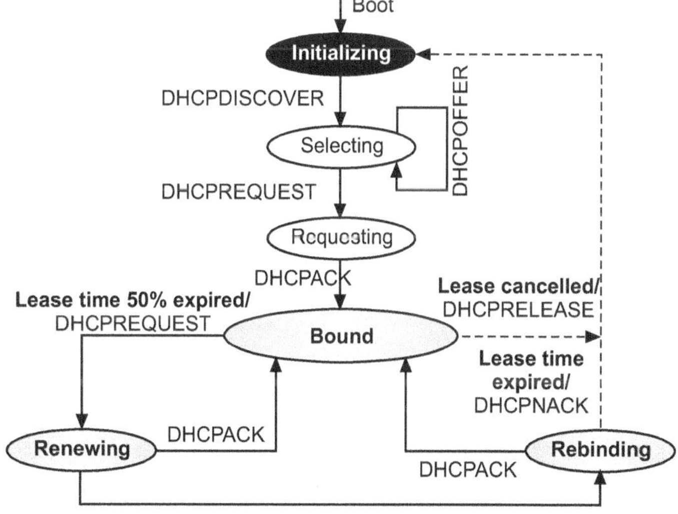

Fig. 4.43 : DHCP Transition Diagram

- In bound state, client can use the IP address until the lease expires. When 50% of the lease period is reached, the client sends another DHCPREQUEST to ask for renewal. It then goes to renewing state. In the bound state client can also cancel the lease and go to the initializing state.

- Client remains in the renewing state until it receives DHCPACK. If it does not receive DHCPACK and 87.5% of lease time expires, the client goes to rebinding state.

- In rebinding state, if client receives DHCPACK, it goes to the bound state and resets the timer; else it goes to initializing state when the lease gets expired.

- A large part of being a system administrator is collecting accurate information about your servers and infrastructure. There are a number of tools and options for gathering and processing this type of information. Many of them are built upon a technology called **SNMP**.

4.16 FORWARDING AND UNICAST ROUTING PROTOCOL

- An internet is a combination of networks connected by routers. When datagram goes from source to destination, it will probably pass through many routers until it reaches the router attached to the destination network.

- A router receives a packet from a network and passes it to another network. It is usually attached to several networks. When it receives a packet, to which network should it pass the packet ? The decision is based on optimization, i.e. which of the pathway is the optimum pathway.

- Routing is the process of finding a path from source machine to the destination machine in the network.

- The routing algorithmis the part of network layer software which is responsible for deciding on which output line an incoming packet should be transmitted.

- If the subnet is datagram subnet, then this decision must be made every time for every incoming packet as the best route might be different than that of the last time.

- If the subnet is virtual circuit subnet, routing decisions are made only when a new virtual circuit is being setup. After that data packets just follow the previously established route.

- The latter case is sometimes called as "Session Routing" because the route remains same for entire user session.

Regardless of whether routes are chosen independently for each packet or only when new connections are established. There are certain properties that are desirable in a routing algorithm:

- Correctness and simplicity hardly requires any comment, but the need of robustness may be less obvious at first.

- Once a major network comes in existence it may be expected to run continuously for many years without system wise failures. During that period, there will be hardware and software failures.

- The routing algorithm should be able to cope up with the changes in the topology and traffic without requiring all jobs in all hosts to be broadest and the network to be rebooted every time when some router crashes.

4.16.1 Optimality Principle

- General statement about optimal routes without regard to topology or traffic: "If router J is on the optimal path from router I to router K, then the optimal path from J to K also falls along the same route"

- The general statement about optimal routes without regard to network topology or traffic. This statement is known as the optimality principle.

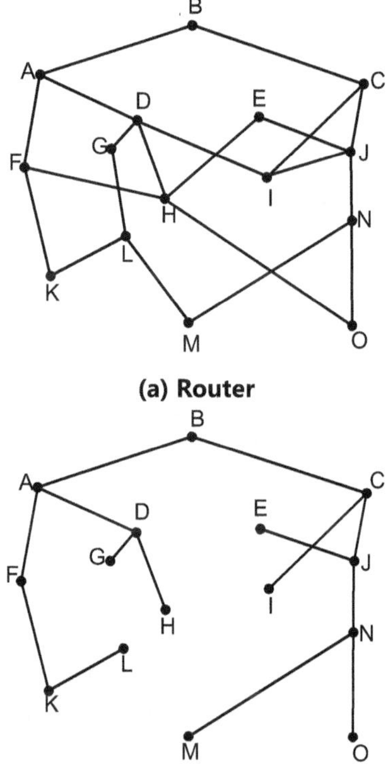

(a) Router

(b) Sink tree for router B

Fig. 4.44

- It states that if router J is on the optimal path from router I to router K, then the optimal path from J to K also falls along the same route. To see this, call the part of the route from I to Jr1 and the rest of the route r2.

- If a route better than r2 existed from J to K, it could be concatenated with r1 to improve the route from I to K, contradicting our statement that r1r2 is optimal. As a direct consequence of the optimality principle, we can see that the set of optimal routes from all sources to a given destination form a tree rooted at the destination.

- Such a tree is called a sink tree and is illustrated in Fig. 4.44, where the distance metric is the number of hops. Note that a sink tree is not necessarily unique; other trees with the same path lengths may exist. The goal of all routing algorithms is to discover and use the sink trees for all routers.

- Since a sink tree is indeed a tree, it does not contain any loops, so each packet will be delivered within a finite and bounded number of hops. In practice, life is not quite this easy.

- Links and routers can go down and come back up during operation, so different routers may have different ideas about the current topology. Also, we have quietly finessed the issue of whether each router has to individually acquire the information on which to base its sink tree computation or whether this information is collected by some other means.

4.16.2 Routing Algorithms Classification

- Routing algorithms are grouped into two major classes.

    1.   Adaptive Algorithms

    2.   Non-adaptive Algorithms

1.   Adaptive Algorithms

- Adaptive algorithms, in contrast, change their routing decisions to reflect the changes in the technology, and usually the traffic as well.

- Adaptive algorithms differ in where they get information (e.g. locally, from adjacent routers, or from all routers), when they change the routes (e.g. every ΔT seconds, when the load changes or when the topology changes), and what metric is used for optimization (e.g. distance, number of hopes, or estimated transit time).

- Adaptive algorithms are categorized into following types :

    (a)   Distance vector routing

    (b)   Link state routing

    (c)   Broadcast routing

    (d)   Multicast routing

    (e)   Hierarchical routing

2.   Non-Adaptive Algorithms:

- These algorithms do not base their routing decisions on measurement and estimates of the current traffic and topology.

- Instead, the choice of route to use to get from I to J (for all I and J) is computed in advance.

- This procedure is sometimes called as static routing.

Following are the non-adaptive algorithms :

    (a)   Shortest path routing

    (b)   Flooding

    (c)   Flow-based routing

4.17 DYNAMIC ROUTING ALGORITHMS (ADAPTIVE)

4.17.1 Distance Vector Routing   (Feb. 15, Nov. 15)

- In distance vector routing, the least cost route between any two nodes is the route with minimum distance.

- In this algorithm, each node (router) maintains a vector table of minimum distance to every node.

- The table at each node also guides the packets to the desired node by showing the next stop in the route (next-hop routing).

- Fig. 4.45 shows a system of five nodes with their corresponding tables. The table for node A shows how we can reach any node from this node. For example, our least cost to reach node E is 6 and the corresponding route passes through C.

Initialization :

- In Fig. 4.45, each node knows how to reach any other node and the cost.

- At the beginning, however, this is not the case. Each node can know only this distance between itself and the immediate neighbors, those directly connected to it.

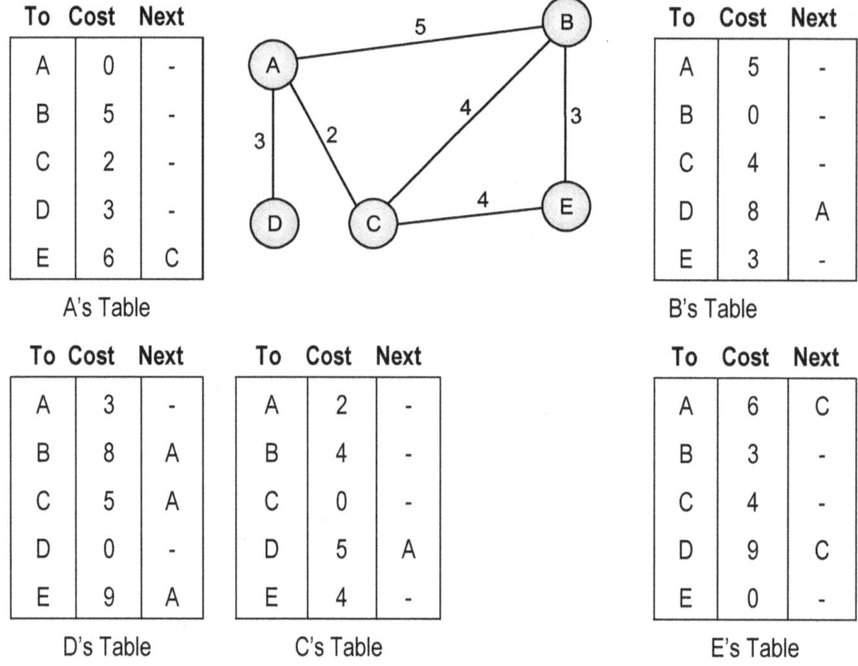

Fig. 4.45 : Distance vector routing table

So, for the moment, we assume that each node can send a message to the immediate neighbors and find the distance between itself and these neighbors.

Following Fig. 4.46 shows the initial tables for each node. The distance for any entry that is not a neighbor is marked as infinite (unreachable).

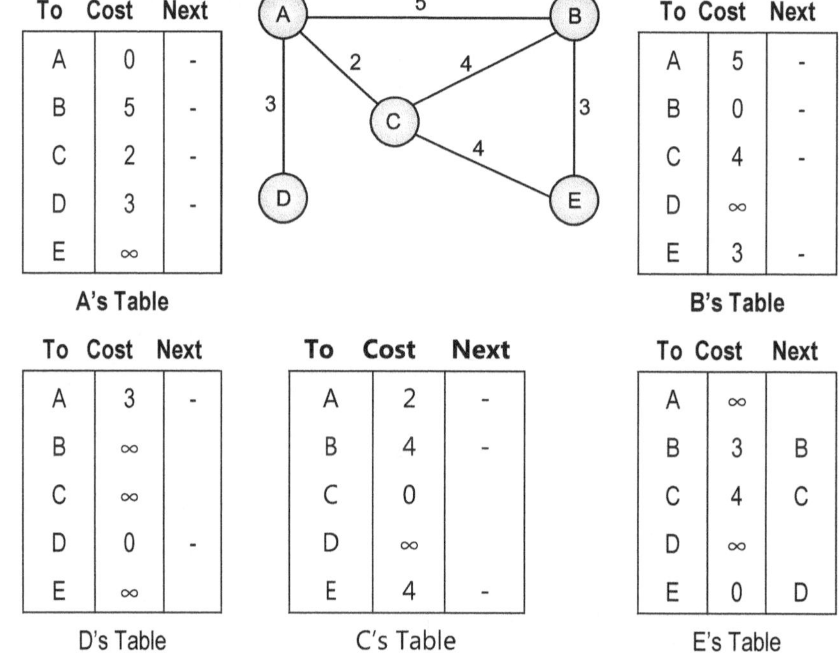

Fig. 4.46 : Initialization of tables in distance vector routing

Sharing

- The whole idea of distance vector routing is sharing of information between the neighbors.

- Although node A does not know about node E, node C does. So if node C shares routing table with A, node A can also know how to reach node E.

- On the other hand, node C does not know how to reach node D, but node A does. If node A shares its routing table with node C, node C also knows how to reach node D. In other words, node A and C as immediate neighbors, can improve their routing tables if they help each other.

- Still there is a problem. How much of the table must be shared with each neighbor? The best solution for each node is to send its entire table to the neighbor and let the neighbor decide what part to use and what to discard. However, the third column of the table is not useful to the neighbor.

- When the neighbor receives the table, this column needs to be replaced with the sender's name. A node therefore can send only first two columns of its table to any neighbor. In other words, sharing here means only sharing of first two columns.

Updating

- When a node receives a two column table from a neighbor, it needs to update its routing table. Updating takes three steps :

- The receiving node needs to add the cost between itself and the sending node to each value in the second column. The logic is clear. If node C claims that its distance to destination is x miles, and the distance between A and C is y miles, then the distance between A and that destination I via C, is x + y miles.

- The receiving node needs to add the name of the sending node to each row as the third column if the receiving node uses the information from any row. The sending node is the next node in the route.

- The receiving node needs to compare each row of its old table with the corresponding row of the modified version of the received table.

- If the next node entry is different, the receiving node chooses the row with the smaller cost. If there is a tie, the old one is kept.

- If the next node entry is the same, the receiving node chooses the new row. For example, if node C has previously advertised a route to node X with distance 3. Suppose that now there is no path between C and X; node C now advertises this route with the distance of infinity. Node a must not ignore this value even though its old entry is smaller. The old route does not exist any more. The new route has the distance of infinity.

- Following Fig. 4.47 shows how node A updates its routing table after receiving the partial table from node C.

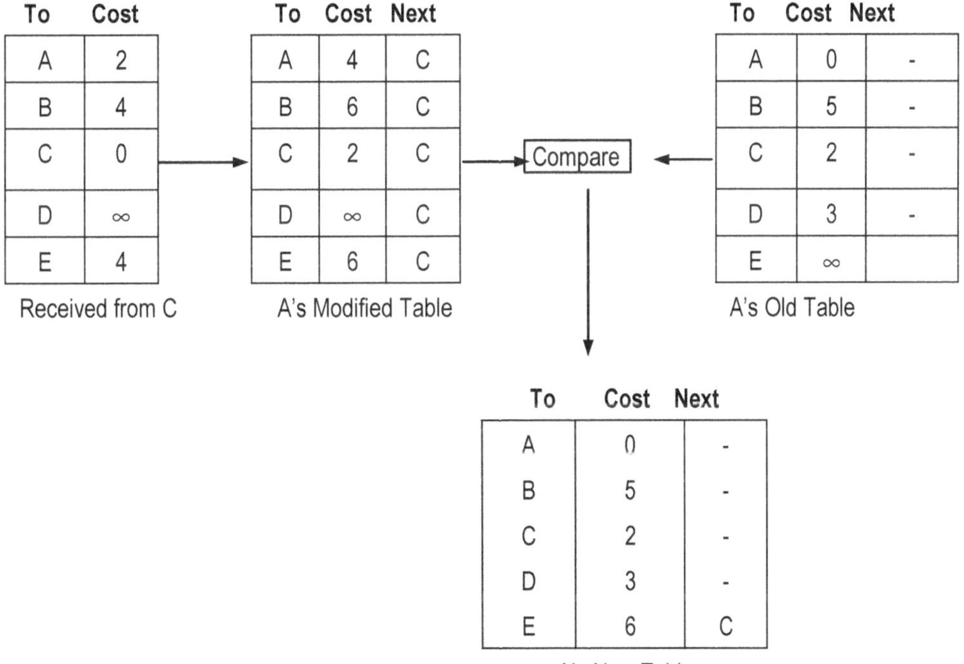

Fig. 4.47 : Updating in distance vector routing

- There are several points that we need to emphasize here.
  - ➤ As we know from mathematics, when we add any number to infinity, the result is still infinity.
  - ➤ The modified table shows how to reach A from A via C. If A needs to reach itself via C, it needs to go to C and then come back, a distance of 4.
  - ➤ The only benefit from this updating of node A is the last entry, how to reach E (distance of infinity); now it knows that cost is 6 via C.

- Each node can update its table using the tables received from other nodes. In short period of time, if there is no change in the network itself, such as a failure in a link, all nodes reach a stable condition in which the content of the table remains the same.

- When to share vector tables?
- The table is sent both periodically and when there is a change in the table. Nodes send its routing table, normally every 30 seconds in a periodic update.

Limitations

1. Count to Infinity Problem

- The core of the count-to-infinity problem is that if A tells B that it has a path somewhere, there is no way for B to know if the path has B as a part of it.
- To see the problem clearly, imagine a subnet connected like A-B-C-D-E-F, and let the metric between the routers be "number of jumps".
- Now suppose that A goes down (out of order). In the vector-update-process B notices that its once very short route of 1 to A is down - B does not receive the vector update from A.
- The problem is, B also gets an update from C, and C is still not aware of the fact that A is down - so it tells B that A is only two jumps from it, which is false.
- This slowly propagates through the network until it reaches infinity.

Explanation of Count to Infinity Problem

- Consider that we are having four routers L-M-N-O-P as shown in the Fig. 4.48 (a) given below :
- Initially consider that router L is down. So, from the point of all routers, it is at a distance of ∞.
- When L comes into network, other routers get this information through the vector exchange table.
- At the first vector exchange, router M gets the information that router L is one hop away from it and is at left hand side.
- Accordingly, router L updates its table. Refer Fig. 4.48 (b).

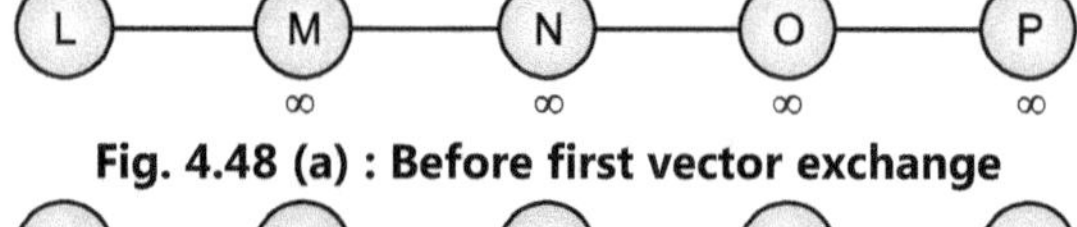

Fig. 4.48 (a) : Before first vector exchange

Fig. 4.48 (b) : After First vector exchange

- At the second vector exchange, router N comes to know about router L that, router L is two hops away from it. Based on this, router N updates its routing table. Thus, finally all routers will get the information about router L's presence on the network. Refer Fig. 4.48 (c).

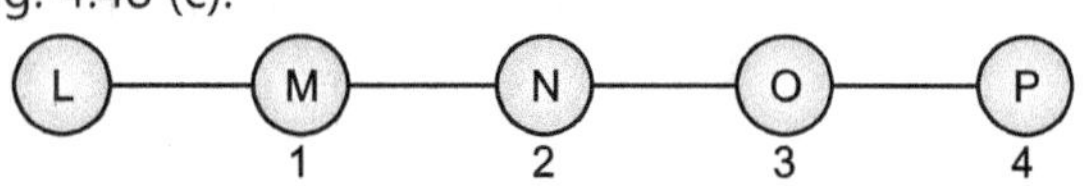

Fig. 4.48 (c) : Finally

- Now, if router L goes down suddenly due to any hardware problem.
- Now after that, the next vector exchange will tell router M that there is no direct path from L to M, but there is a path to router L from router M through router N (because still router N is not aware that router L is down, so it is maintaining its old routing table).
- As router L is 2 hops away from router N, so now by this vector exchange, router L will be 3 hops away from router M (this is absolutely wrong). This is shown in Fig. 4.48 (d).

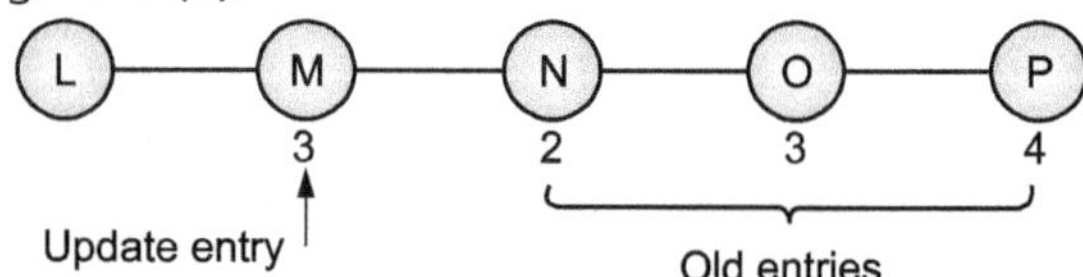

Fig. 4.48 (d) : After first exchange when router L is down

- On second exchange, router N comes to know that, both its neighbour M and N claim to have a path of length 3 to router L. So, it picks one of them at random and makes its new distance to L as 4. This is shown in Fig. 4.48 (e).

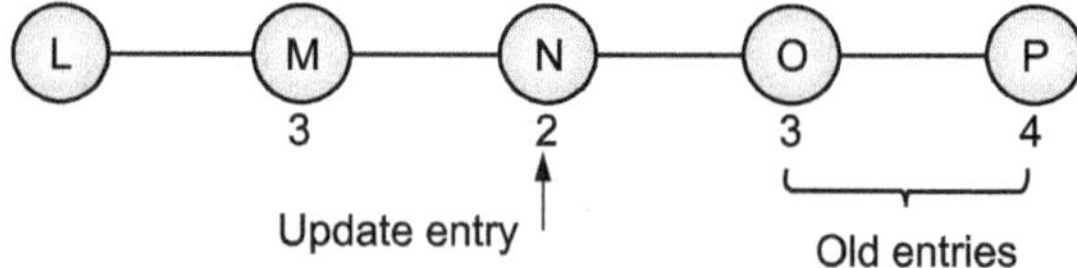

Fig. 4.48 (e) : After second exchange when Router L is down

- In the same way, other routers keep updating their tables after every exchange.
- Finally, after many exchanges, we will get ∞ distance of router L in the routing tables of router M, N, O and P.
- The conclusion of the entire process is that bad news propagates slowly. And this is a count-to-infinity problem.
- One can overcome this problem by using split horizon algorithm.

4.17.2 LinkState Routing   (May 17)

- Distance vector routing was used in the APRANET until 1979, when it was replaced by link state routing. Two primary problems caused its failure.
- Since the delay metric was queue length, it did not take line bandwidth into account when choosing routes.
- Initially, all the lines were 56 kbps, so line bandwidth was not an issue, but after some line had been upgraded to 230 kbps and others to 1.54 Mbps, not taking bandwidth into account was major problem. Of course, it would have been possible to change the delay metric to factor in line bandwidth.

- The algorithm often took too long to converge even with tricks like split horizon.
- Due to that, it was replaced by an entirely new algorithm which is now called as link state routing.
- The idea behind link state routing is simple and can be stated as five parts. Each router must:
- Discover its neighbors and learn their network addresses.
- Measure the delay or cost to each of its neighbor.
- Construct a packet telling all it has just learned.
- Send this packet to all other routers.
- Compute the shortest path to every other router.
- In effect the complete topology and all delay are experimentally measured and distributed to very router. Then to find the shortest path to every other router Dijkstra's algorithm is used by considering the five steps described below :

1. **Learning about the Neighbors:** When a router is booted, its first task is to learn who are its neighbors. It accomplishes this properly by sending a special HELLO packet on each point to point line. The router on the other end is expected to send back a reply telling who it is. These names must be globally unique.

2. **Measuring Line Cost :** The link state routing algorithm requires each router to know for at least has a reasonable estimate, of the delay to each of its neighbors. The most direct way to determine this delay is to send a special ECHO packet over the line that the other side is required to send back immediate. By measuring the round trip time and dividing it by two, the sending router can get a reasonable estimate of the delay. For even better results the test can be conducted several times and the average is used.

3. **BuildingLinkState Packets :** Once the information needed for the exchange has been collected the next step for each router is to build packet containing all the data. The packet starts with the identity of the sender, followed by a sequence number and age, and list of neighbors. For each neighbor, the delay to that neighbor is given. An example subnet is given in Fig. 4.49 (a) with delays shown in the lines. The corresponding link state packets for all six routes are shown in Fig. 4.49 (b).

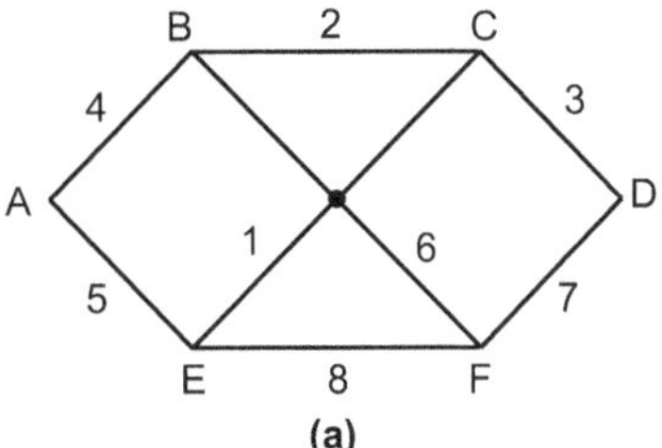

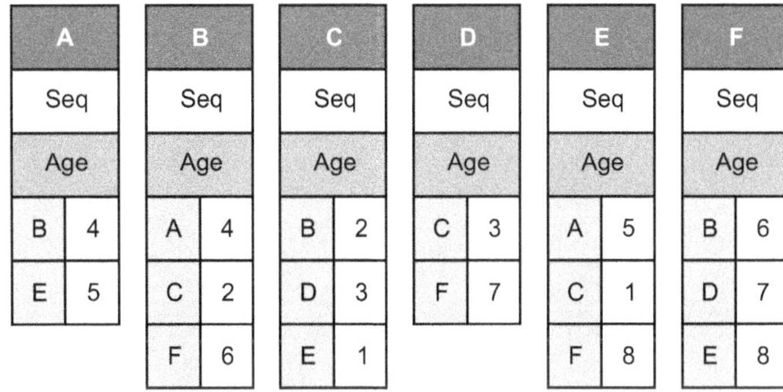

A		B		C		D		E		F	
Seq		Seq		Seq		Seq		Seq		Seq	
Age		Age		Age		Age		Age		Age	
B	4	A	4	B	2	C	3	A	5	B	6
E	5	C	2	D	3	F	7	C	1	D	7
		F	6	E	1			F	8	E	8

Fig. 4.49 : (a) A Subnet, (b) The link state packets for the subnet

- Building the link state packet is easy. The hard part is determining when to build them. One possibility is build them periodically at regular intervals. Another possibility is when some significant event occurs, such as a line or neighbor going down or coming back up again, or changing its properties appreciably.

4. **Distributing the LinkState Packets :** The trickiest part of the algorithm is easy. The hard part is determining when to build them. Once possibility is to build them periodically, i.e. at regular intervals. Another possibility is when some significant event occurs, such as a line or neighbor going down or coming back up again, or changing its properties appreciably.

5. **Computing the New Routes :** Once a router has accumulated a full set of link state packets, it can construct the entire subnet graph because every link is represented. Every link is infact represented twice, once for each direction. The two values can be average or used separately. Now Dijkstra's algorithm can be run locally to construct the shortest path to all possible directions. The results of this algorithm can be installed in routing tables and normal operation is resumed.

4.17.3 Path Vector

- The Fig. 4.50 illustrates a path vector algorithm. The goal is to find good paths to destination D. Each node advertises the path it prefers to get to D. For instance, when B gets two advertisements (CD and ED), it selects the one it prefers (CD) and advertises the path BCD to its neighbor.

- The selection of preferred path follows a set of rules (e.g., avoid E because it is expensive or unreliable; prefer a shorter path, and so on.) Since the paths are advertised, the scheme avoids loops.

- Note that nodes can implement incompatible rules that lead to instability of the algorithm, as the example below shows.

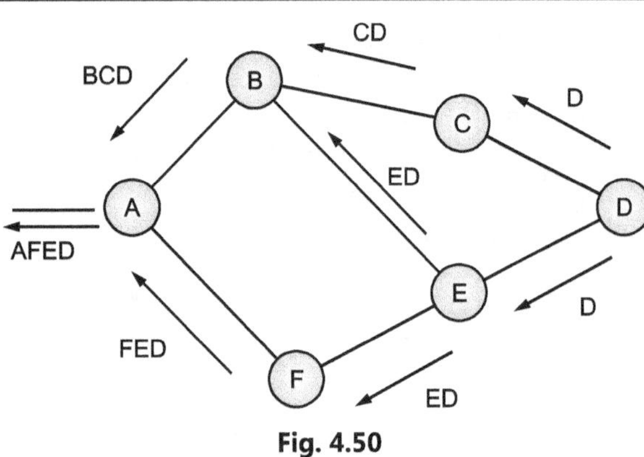

Fig. 4.50

- In this example, all the nodes want to connect to the destination D. Every node prefers a two-link counter-clockwise path to a direct path and does not want a 3-link path nor a path going clockwise. Thus, A prefers ACD over AD. At each step, a node advertises its selected path.

- According to these rules, whenever a node advertises that it selected a two-link path, its clockwise neighbour must select the direct path. Also, whenever a node advertises that it selected the direct path, its clockwise neighbor must select the two-link path. Hence, we expect the following pattern:

  A2 > B1 > C2 > A1 > B2 > C1 > A2 , etc

- Where A2 means that A selects a 2-link path and A1 that it selects a direct path, and similarly for the other nodes.

- Oscillations are bad because they can result in packets taking very different paths or even looping in the network. One can prevent such oscillations by agreeing on an ordering of sets of nodes.

- A path vector protocol does not rely on the cost of reaching a given destination to determine whether each path available is loop free or not. Instead, path vector protocols rely on analysis of the path to reach the destination to learn if it is loop free or not.

- A path vector protocol guarantees loop free paths through the network by recording each hop the routing advertisement traverses through the network. In this case, router A advertises reachability to the 10.1.1.0/24 network to router B.

- When router B receives this information, it adds itself to the path, and advertises it to router C. Router C adds itself to the path, and advertises to router D that the 10.1.1.0/24 network is reachable in this direction.

- Router D receives the route advertisement and adds itself to the path as well. However, when router D attempts to advertise that it can reach 10.1.1.0/24 to

router A, router A will reject the advertisement, since the associated path vector contained in the advertisement indicates that router A is already in the path. When router D attempts to advertise reachability for 10.1.1.0/24 to router B, router B also rejects it, since router B is also already in the path.

- Any time a router receives an advertisement in which it is already part of the path, the advertisement is rejected, since accepting the path would effectively result in a routing information loop.

4.18 ROUTING IN INTERNET

4.18.1 Routing Information Protocol

- This is an intradomain (interior) routing protocol used inside an Autonomous System (AS).

- It is very simple protocol based on Distance Vector Routing mechanism.

- Typical format of routing table 4.10 is shown in the table given below :

Table 4.10

Destination Network Address field	Hop Count field	Next Router field

- First column consists of destination networks address.

- The Hop Count Field the number of Hops that are required to reach the destination using shortest path algorithms.

- Next router field column contains the address of the next router to which the packet is to be delivered.

Example of a Domain Using RIP

- Fig. 4.51 shows an autonomous system with seven networks and four routers.

- The table of each router is also shown. Let's look at the routing table of R1.

- The table has seven entries to show how to reach each network in the autonomous system.

- Router 1 is directly connected to networks 130.10.0.0 and 130.11.0.0, which means that there are no Next router entries for these two networks.

- To send a packet to one of the three networks at the far left, router R1 needs to deliver the packet to router R2.

- The next router entry for these three networks is the interface of router R2 with IP address 130.10.0.1.

- To send a packet to the two networks at the far right, router R1 needs to send the packet to the interface of router R4 with IP address 130.11.0.1.

- Similarly, one can give explanations of other routing tables given in the following Fig. 4.51

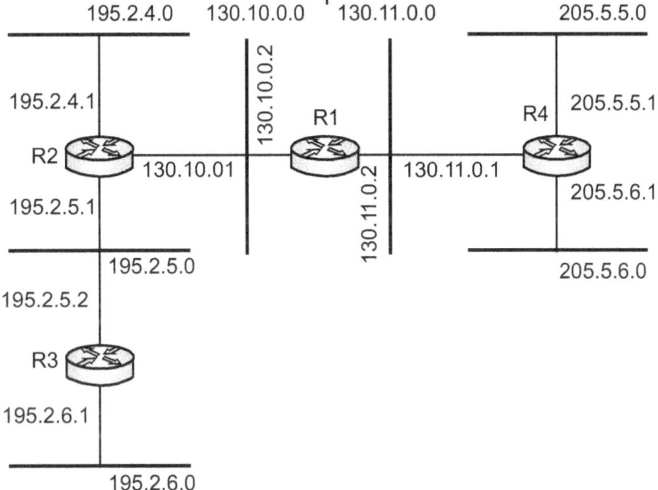

Table R1			Table R2			Table R3			Table R4		
Dest.	**Hop**	**Next**	**Dest.**	**Hop**	**Next**	**Dest.**	**Hop**	**Next**	**Dest.**	**Hop**	**Next**
130.10.0.0	1	—	130.10.0.0	1	—	130.10.0.0	2	195.2.5.1	130.10.0.0	2	130.11.0.2
130.11.0.0	1	—	130.11.0.0	2	130.10.0.2	130.11.0.0	3	195.2.5.1	130.11.0.0	1	—
195.2.4.0	2	130.10.0.1	195.2.4.0	1	—	195.2.4.0	2	195.2.5.1	195.2.4.0	3	130.11.0.2
195.2.5.0	2	130.10.0.1	195.2.5.0	1	—	195.2.5.0	1	—	195.2.5.0	3	130.11.0.2
195.2.6.0	3	130.10.0.1	195.2.6.0	2	195.2.5.2	195.2.6.0	1	—	195.2.6.0	4	130.11.0.2
205.5.5.0	2	130.11.0.1	205.5.5.0	3	130.10.0.2	205.5.5.0	4	195.2.5.1	205.5.5.0	1	—
205.5.6.0	2	130.11.0.1	205.5.6.0	3	130.10.0.2	205.5.6.0	4	195.2.5.1	205.5.6.0	1	—

Fig. 4.51 : Example of Domain using RIP

1. Initializing Routing Table

- When router is added to a network, it initializes its routing table.

- Such a table consists of only the directly attached networks and the hop counts. The next hop field which identifies the next router is empty.

- When RIP messages are received, each routing table is updated using RIP updating algorithm.

2. RIP Advertisement

- The routing table is updated very frequently, when the host machine gets a RIP response message from the neighbouring routers.

- In RIP, routing updates are exchanged between neighbours approximately every 30 seconds using Routing Response Message.

- These routing response messages are also known as RIP advertisements.

3. RIP Message Format (RIPv1)

- The format of RIP version 1 message is shown in the following Fig. 4.52.

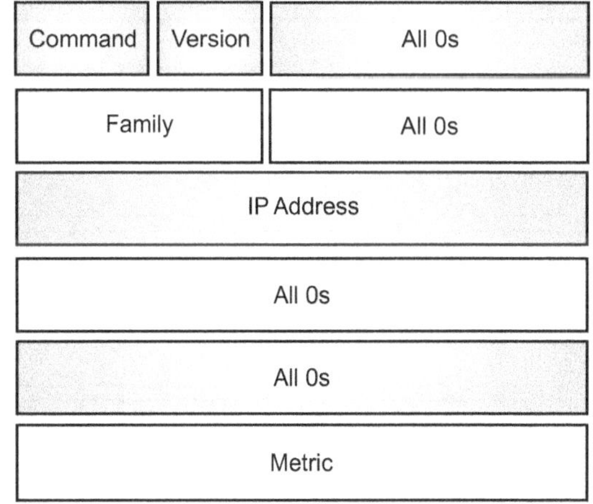

Fig. 4.52 : RIP message format (RIPv1)

Following are the fields of RIP message :

- **Command :** This is 8 bit field. It specifies the type of the message : (1) Request and (2) Response.

- **Version :** This 8 bit field defines a version. There are two versions of RIP message.
- **Family :** This 16 bit field defines the family of the protocol used. For TCP/IP protocol the value is 2.
- **IP Address :** The address field defines the IP address of the specific entry.
- **Metric :** This 32-bit field defines the hop count from the advertising router to the destination network. This value is between 1 to 15 for valid route and 16 for unreachable route.

4. Requests and Responses Messages

RIP has two types of messages

(i) Request Message

A request message is sent by a router that has just come up or by a router that has some timeout entries.

A request can ask about specific entries or all entries. See following Fig. 4.53.

(ii) Response Message

Response can be either solicited or unsolicited. A solicited response is sent only in answer to a request.

An unsolicited response, on the other hand, is sent periodically, every 30 seconds or when there is a change in the routing table.

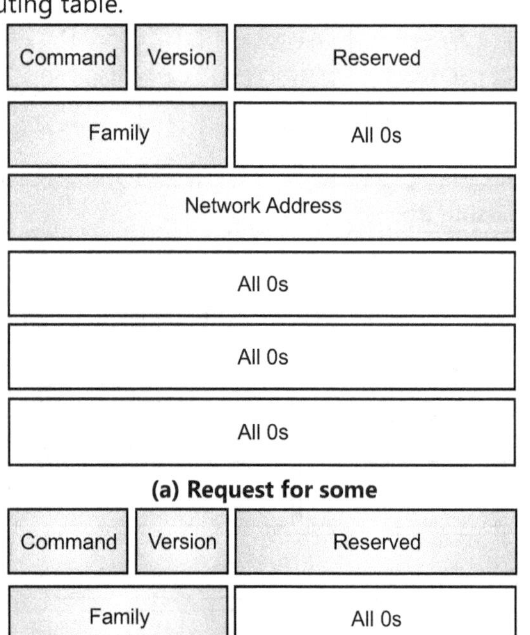

(a) Request for some

(b) Request for all

Fig. 4.53 : Request messages

5. Timers in RIP

RIP uses three timers:

(i) Periodic Timer

It controls advertising and regular update messages. Although the protocol specifies that this timer must be set to 30 seconds, the working model uses a random number between 25 and 35 seconds.

It is countdown timer; when 0 is reached, the update message is sent, and the timer is randomly set once again.

(ii) Expiration Timer

It governs the validity of the route.

When router receives update information for a route, expiration timer is set to 180 sec for that particular route.

Every time a new update for a particular route is received, the timer is reset.

In normal situations, it occurs after every 30 seconds.

If there is a problem on an internet and no update is received within the allotted 180 seconds, the route is considered as expired which means the destination is unreachable.

Every route has its own expiration timer.

(iii) Garbage Collection Timer:

This timer is used to remove the routing information from the table when the route becomes invalid. When this timer becomes zero, the route is washed out from the routing table.

6. RIP Version 2

RIPv2 was designed to overcome some of the shortcomings of RIPv1. The designers of version2 have only replaced those fields in version 1 that were filled with 0s by some new fields.

Message Format for RIPv2

- The message format is shown in the following Fig. 4.54.

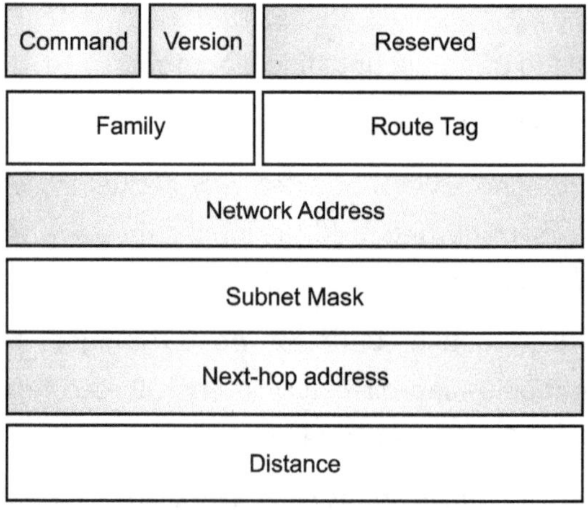

Fig. 4.54 : RIPv2

- The new fields that are not present in RIPv1 are :
- **Route Tag :** This field carries information such as autonomous system number.
- **Subnet Mask :** It is 4 byte field that carries the subnet mask. This means that RIP2 supports classless addressing and CIDR.
- **Next-Hop Address :** This field shows the address of the next hop.

Table 4.10 : Difference between RIPV1 and RIPV2

RIPv1	RIPv2
Uses distance vector routing algorithm.	Uses distance vector routing algorithm.
Maximum hop count is 15. If it exceeds 15, it means that destination is unreachable.	Maximum hop count is 15. If it exceeds 15, it means that destination is unreachable.
Uses Classful addressing mechanism.	Uses Classless addressing mechanism.
No support for Variable length subnet masks (VLSM).	Supports VLSM.
No support for discontiguous networks.	Support for discontiguous networks.

4.18.2 Open Shortest Path First (OSPF)   (Nov. 15)

- It is an intradomain/interior routing protocol based on link state routing. Its domain is an autonomous system.
- For handling routing efficiently, OSPF divides an autonomous system into 'Areas'.

What is Area ?

- An area is a collection of networks, hosts and routers within a specific autonomous system.
- An autonomous system can be divided into many different areas.

   All the networks inside an area must be connected.

- Normally two types of routers are used in the autonomous system

1. Area Border Routers

- These are the special routers used at the border of the area.
- These routers summarize the information about its area and send it to other areas.

2. Backbone Routers inside Backbone Area

- Among the areas inside the autonomous system, there is a special area called as backbone area.
- All of the areas inside an autonomous system must be connected to the backbone. In short, backbone works as primary area and the other areas work as secondary areas.

- The routers inside the backbone area are called as backbone routers.
- Backbone router can also be an area border router.
- In case, if some problem occurs and the connectivity between the backbone and the area is broken, virtual link between the routers must be created by the administrator to allow continuity of functions of the backbone as the primary area.
- Each area has area identification. The area identification of backbone is zero. Following Fig. 4.55 shows an autonomous system and its areas.

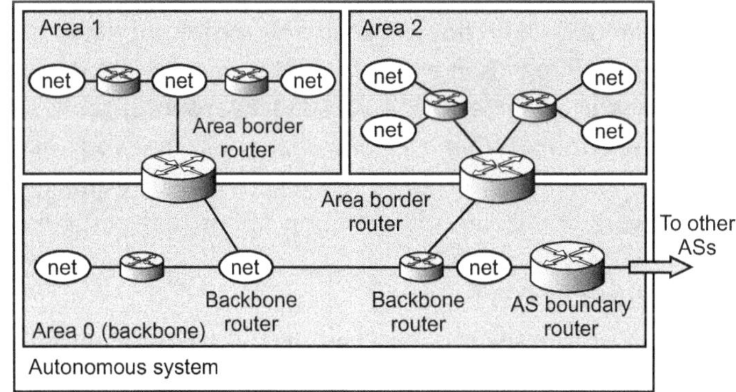

Fig. 4.55 : Areas in autonomous system

1. OSPF V/s RIP

- In earlier days, the most popular routing protocol was RIP. But it was only good when the network is small.
- It has some limitations which could problem in large networks. Comparison between RIP V/s OSPF is given below.

RIP

- RIP has limited HOP counts. It is 15. If a RIP Network spans more than 15 HOPS, it is considered as unreachable. In short network diameter is 15 hops.
- RIP doesn't support for variable length subnet masks (VLSM).
- Periodic update of routing table consumes lots of bandwidth especially on WAN clouds.
- RIP converges slower than OSPF does.
- RIP Network is a FLAT network. Here no concept of Autonomous systems, Areas, Boundaries and Summarization.
- RIP does not support multipath routing.
- It doesn't support authentication.

OSPF

- No limitations on the HOP count. Possibly network diameter is 65535 hops.
- Can use VLSM.

- Converges quickly.
- Can divide into Areas. This will help us to use summarization.
- Allows Authentication.
- It uses Dijkstra's algorithm (SPF algorithm).
- Reducing the usage of BW, by sending triggered updates to announce the Network changes.
- Sending periodic updates on long intervals.
- OSPF supports multipath routing.
- Rather than exchanging node (and network) reachability information, OSPF routers exchange link state information.
- Unlike RIP, OSPF doesn't send any routing updates on periodic intervals. It will only send triggered updates. It means every time it doesn't send full routing table to its neighbors. Whenever any changes in network, like new router added or a router removed from the network, it will send information about that particular network to its neighbor.

2. Features of OSPF

- **Type of Service :** Depending on the requirement and the nature of the application, it can provide different services to different routes. e.g. high throughput can be selected for one class of service, while minimum delivery delay is more critical for some other applications.
- **Load Balancing :** When multiple routes are available to a particular destination, traffic can be distributed on the routes so as to balance the traffic load. No path will carry a very high traffic and no path will carry very low traffic. Traffic will be distributed evenly.
- **Subdivision of Autonomous System :** Autonomous systems are further divided into smaller areas, which is always better from network administration point of view.
- **Security :** Routing information exchanges between the routers are authenticated. Malicious transmissions from foreign routers are discarded.
- **Special Features to Support LAN Environment :** Although the relationships between routers are maintained on a logical link basis, link state transmissions are minimized by the architecture.
- **OSPF is an Open Specification :** It means, anyone can implement this standard without paying royalties.
- **OSPF Area :** An area is a collection of networks, hosts and routers within a specific autonomous system. The topology of area is hidden from rest of the autonomous system. This technique reduces the storage of routing traffic information on a specific router.

3. Metric

- OSPF protocol allows the administrator to assign a cost, called the metric, to each route.
- A metric can be based on type of service (minimum delay, maximum throughput, and so on).
- As a matter of fact, a router can have multiple routing tables, each based on different type of service.

4. Types of Links

- In OSPF terminology, a connection is called as link. There are four types of links/connections available in OSPF.

  (i) Point to point (ii) Transient

  (iii) Stub   (iv) Virtual

(i) Point to Point Link

A point to point link connects two routers without any other host or router in between.

Graphically, routers are represented by nodes and the link is represented by bidirectional edge connecting the nodes (refer following Fig. 4.56).

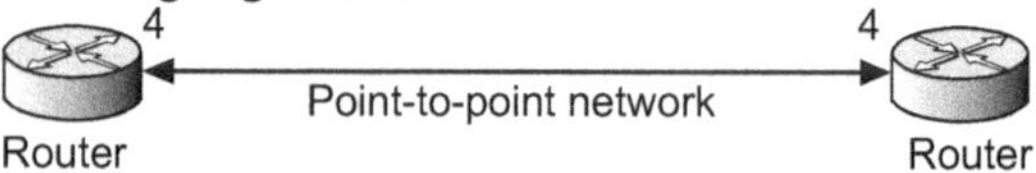

Fig. 4.56 : Point to point link

(ii) Transient Link

Transient link is a network with several routers attached to it.

The data can enter and exit through any of the router.

All LANs and some WANs with two or more routers are of this type (refer following Fig. 4.57 (a).

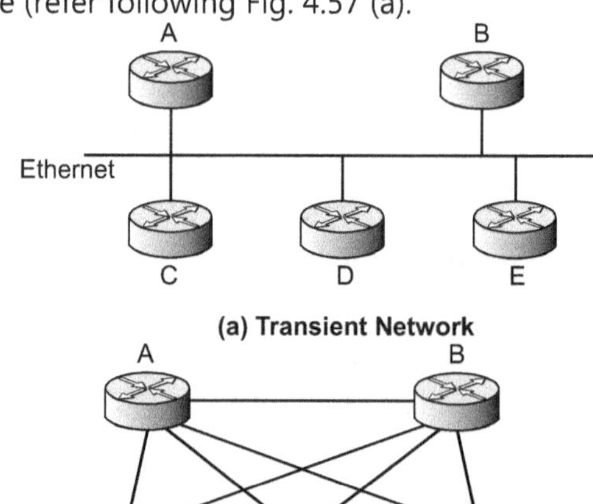

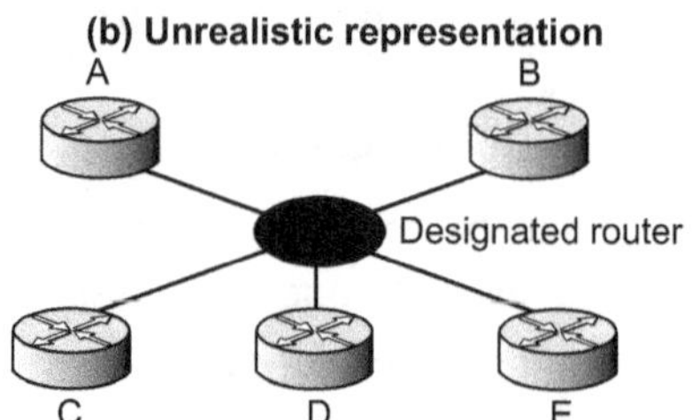

Fig. 4.57 : Transient link

(iii) Stub Link :

A stub link is a network that is connected to only one router.

Data packets enter and exit the network through this single router.

This is a special case of transient network (refer Fig. 4.58 given below) :

(a) Stubnetwork   (b) Representation

Fig. 4.58 : Stub link

(iv) Virtual Link :

When the link between two routers is broken, the administration may create a virtual link between them using a longer path that probably goes through several routers.

5. OSPF Packets

- OSPF uses five different types of packets as shown in the following Fig. 4.59 given below :

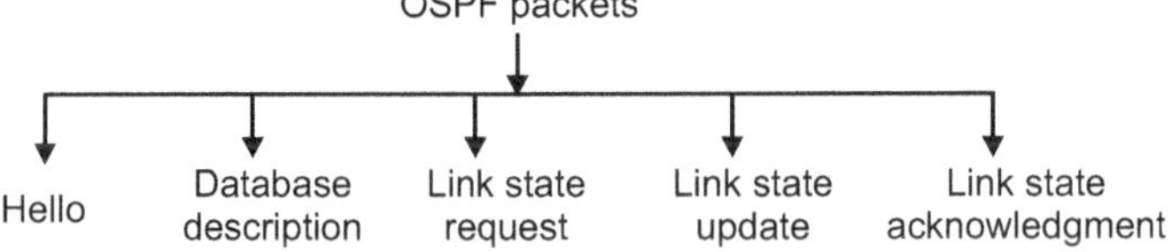

Fig. 4.59 : Types of OSPF packets

- The most important type of OSPF packet is Link State Update that itself has five different kinds which are as follows :
  (i)   Router Link
  (ii) Network Link
  (iii) Summary Link to Network
  (iv) Summary Link to Autonomous System boundary router
  (v)   External Link.

6. Common Header of OSPF Packet

All OSPF packets have the same common header as shown in the following Fig. 4.60. Let's take some more information about it.

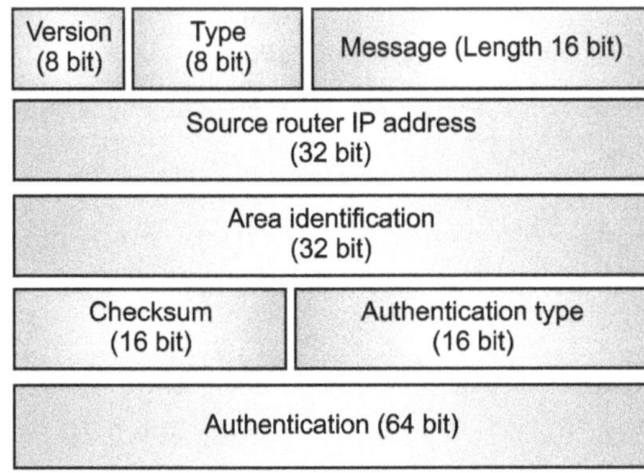

Fig. 4.60 : OSPF header

Fields of OSPF Header are as follows :

- **Version :** It defines the version of OSPF protocol. Currently it is version 2.

- **Type :** It defines the type of the packet. We have five types as defined earlier.

- **Message Length :** It defines the total length of the message including header.

- **Source Router IP address :** It defines an IP address of the router that sends the packet.

- **Area Identification :** It defines the area within which the routing takes place.

- **Checksum :** This field is used for error detection on the entire packet excluding the authentication type and authentication data field.

- **Authentication Type :** It defines the authentication protocol used in this area. At this time, two types of authentication are defined : 0 for none and 1 for password.

- **Authentication :** This 64-bit field is the actual value of the authentication data. If authentication type is 0, this field is filled with 0s. If authentication type is 1, this field carries an eight-character password.

LinkState Update Packet :

- It is the most important type of OSPF packet.

- It is the heart of OSPF operation.

- It is used by the router to advertise the states of its links.

- Each link state update packet may contain several different Link State Advertisements (LSAs) which are explained further :

7. LinkState Advertisements (LSAs)

As stated above there are five different types of Link State Advertisements (LSA) which are as follows :

  (i)   Router Link LSA
  (ii) Network Link LSA
  (iii) Summary Link to Network LSA
  (iv) Summary Link to Autonomous System boundary router LSA
  (v)   External Link LSA

A short description of each LSA is given below :

(i) Router Link LSA

- The router generates router link advertisement for the area too which it belongs.

- This advertisement describes the collected states of router's links to the area.

- This advertisement also tells that whether the router is an area border router or it is an Autonomous System boundary router.

(ii) Network Link LSA

- This advertisement is originated by the designated router for the transit network.
- It gives the information about all OSPF routers fully adjacent to the designated router.

(iii) Summary Link LSA

- Summary link advertisement describes a single route to a destination.
- The destinations described are external to the area but internal to the autonomous system.
- Some condensing of the routing information occurs when creating these summary link state advertisements.

(iv) Summary Link to Autonomous System Boundary Router LSA

- These are similar to summary link advertisement except they describe routes to Autonomous system boundary routers.

(v) External Link LSA

- These describe the routes external to the autonomous systems.

4.18.3 Border Gateway Protocol (BGP)   (Nov. 16)

- It is an exterior/interdomain protocol.
- The purpose of exterior gateway protocol is to enable two different Autonomous Systems (AS) to exchange routing information so that IP traffic can flow across the autonomous system border.
- BGP was developed for use in conjunction with internets that employ the TCP/IP protocol suite.
- BGP is based on the routing method called as "Vector path routing".
- BGP is interdomain routing protocol that is used to exchange network reachability information among BGP routers (BGP speakers).
- Each BGP speaker establishes a connection with one or more BGP speakers (routers).
- Two routers are considered to be neighbours, if they are attached to the same subnetwork.
- If two routers are in different autonomous systems, they may wish to exchange routing information.

BGP Performs Three Functional Procedures

1. **Neighbour Acquisition :** These procedure are used to exchange the routing information between two routers in different autonomous systems.

   To perform neighbour acquisition, one router sends an open message to another.

   If target router accepts the message, it returns a keep alive message in response.

2. **Neighbour Reachability :** Once a neighbour relationship is established, the neighbour reachability procedure is used to maintain the relationship.

   Both sides' needs to be assured that the other side still exists and is still engaged in the neighbour relationship.

   For this purpose both routers send keepalive messages to each other.

3. **Network Reachability :** Both sides router maintains a database of the subnetworks that can reach and the preferred route for reaching the subnetworks.

   If the database changes, router issues an update message that is broadcast to all other routers implementing BGP.

   By broadcasting this update message, all the BGP routers can build up and maintain routing information.

   BGP connections inside an autonomous system are called as internal BGP and BGP connections between different autonomous systems are called as external BGP.

BGP Messages

- BGP has four different types of messages : Open, Update, Keepalive and Notification (Refer following Fig. 4.61).

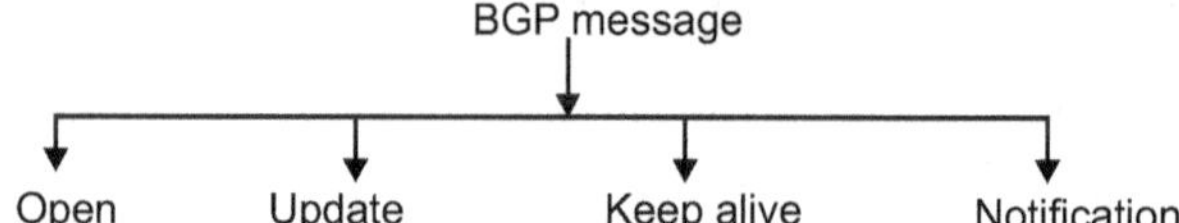

Fig. 4.61 : Types of BGP messages

1. **Open Message :** To create a neighbourhood relationship, a router running BGP opens a connection with a neighbour and sends an open message. If neighbour accepts the neighbourhood relationship, it responds with a *keep-alive* message, which means that a relationship has been established between the two routers.

2. **Update Message :** The update message is the heart of BGP protocol. It is used by a router to withdraw destinations that have been advertised previously, announce a route to a new destination, or do both.

Note that BGP can withdraw several destinations that were advertised before, but it can only advertise one new destination in a single update message.

3. **Keep Alive Message :** Routers running BGP protocols exchange keep-alive messages regularly to tell each other that they are alive.

4. **Notification Message :** A notification message is sent by a router whenever an error condition is detected or a router wants to close a connection.

Packet Format

All the BGP packets (messages) share the same common header. The fields of this header are as follows : (refer following Fig. 4.62)

- **Marker :** The 16 byte marker field is reserved for authentication. The sender may insert value in this field that would be used as part of an authentication mechanism to enable the recipient to verify the identity of the sender.

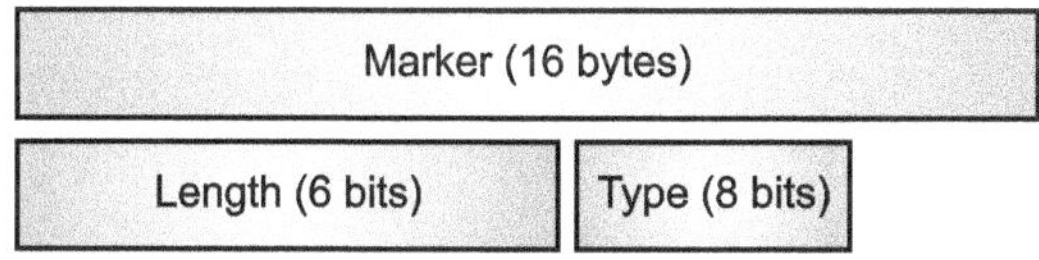

Fig. 4.62 : BGP header format

- **Length :** This 2 byte field defines the length of the total message including the header.

- **Type :** This field indicates the type of the message, which we have already seen.

BGP Operation

- To acquire a neighbour, a router first opens a TCP connection to the neighbour router of interest.

- It then sends the open message. The message identifies the AS to which the sender belongs and provides the IP address of the router.

- It also includes a hold timer parameter. The hold timer defines the maximum number of seconds that can elapse until one of the parties receive a keepalive or update message from the other.

- If a router does not receive one of these messages during the hold time period, it considers the other party dead.

- The KEEPALIVE messages are exchanged often enough as to not cause the hold timer to expire.

- The recommended time between successive KEEPALIVE messages is one-third of the hold time interval.

- This value ensures that, the KEEPALIVE message arrive at the receiving router almost always before the hold timer expires, even if the transmission delay of TCP is variable.

- If hold time is zero, then KEEPALIVE messages will not be send.

- When a BGP router detects an error, the router sends a NOTIFICATION message and then closes the TCP connection.

- After the connection is established, BGP peers exchange routing information by using the UPDATE message.

- The UPDATE message may contain three pieces of information   :
  1. Unfeasible Routes
  2. Path Attributes and
  3. Network Layer Reachability Information.

An UPDATE message can advertise a single route and withdraw a list of routes.

- UPDATE messages are used to construct a graph of AS connectivity.

- It also withdraws multiple unfeasible routes.

- A BGP router uses Network Layer Reachability Information (NLRI), the total path attribute length and the path attributes to advertise a route.

- The NLRI field contains a list of IP address prefixed that can be reached by the route.

EXERCISE

1. Explain the difference between circuit switching and packet switching.

2. Explain the IP header.

3. Explain the purpose of ICMP protocol.

4. Write a short note on ARP and RARP protocol.

5. Explain IGMP.

6. Name different protocols in network layer.

7. Explain the concept of Unicast routing.

8. What are the limitations of IPv4?

9. What are the advantages of IPv6 ?

10. Compare IPv4 and IPv6.

11. What is CIDR ?

12. How does CIDR work ? How does it differ from Classfull IP Addressing ?

13. What is routing information protocol (RIP) ?

14. Compare OSPF and RIF.

15. Explain different features of OSPF.

16. Write a short note on OSPF header.

17. Explain BGP protocol.

18. What are the different types of BGP messages ?

19. Explain the BGP operation.

20. What is multicast routing ?

21. Explain IGMP.

22. Write a short note on Multiprotocol Label Switching (MPLS) ?

23. What is Virtual LAN (VLAN) ?

24. Explain the concept of Bootstrap (BOOTP) protocol.

25. Why do we need BOOTP ?

26. Explain BOOTP packet format.

27. Explain the concept of Dynamic Host configuration protocol ? Why it is required ?

28. What are the different classes of IP address ? Also specify the range of each class of IP address.

29. Explain IPv6 header format.

30. What is IP subnetting ?

31. Write a short note on subnet masks.

32. What is supernetting ? Why it is required ?

33. What do you mean by private IP address ? Why they are required ?

34. Explain RIP message format.

35. Explain the types of RIP messages.

TRANSPORT LAYER

5.1 INTRODUCTION

- The network layer is concerned with getting packets from the source all the way to the destination. Getting the packets from source may require making many intermediate routes along the way to the destination. To achieve this goal, the network layer must know about the topology of the network and choose appropriate paths even for large networks.

- This unit is all about network layer. It covers the switching techniques such as packet switching, circuit switching, message switching etc. Later on we have covered details of IPv4 and IPv6 addressing scheme and subnetting. There are many protocols like distance vector, link state path vector which find optimal path from source to destination.

- All these protocols are covered here. Protocols like RIP, OSPF and BGP in the context of routing in Internet are also covered here. Later on congestion control and QoS concepts are discussed. After that MPLS and mobile IP topics are covered in detail and finally routing protocols in MANET like AODV and DSR are covered at the end of the unit.

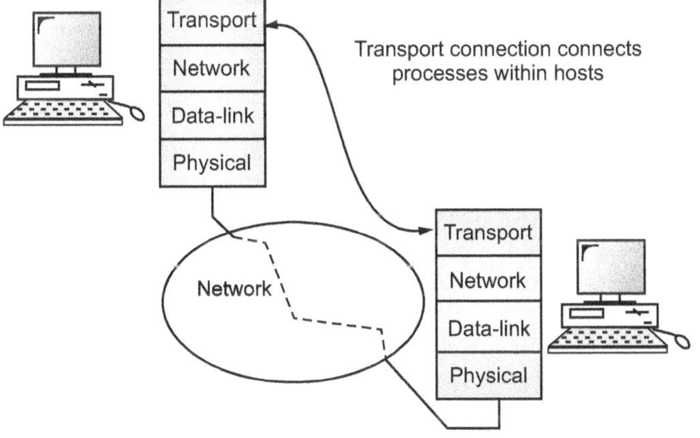

Fig. 5.1 : Transport layer

- The Network layer protocols are largely concerned with finding a path through the network from a source host to a destination host.

- The Network layer doesn't provide an interface for application processes.

- Above the Network layer is the Transport layer. The Transport layer provides an application process interface for actually connecting one process to another via the underlying network.

- The goal of the Transport layer is that application processes should communicate without needing to consider the network technology.

- The Transport layer is an important layer in the protocol stack and perhaps the only layer seen by the ordinary users of the network. It is like a outer skin of the network which hides all the organs and functions of the network.

- Its main task is to provide reliable connection from a network application running on a source computer to its corresponding application running on the destination computer.

- The connection could be both connectionless and connection-oriented. The OSI model specifies the Transport layer as providing only connection-oriented services. In TCP/IP standards, the Transport layer has provision for both connectionless and connection-oriented services.

5.2 TRANSPORT LAYER DUTIES

- The basic duties of the transport layer is to accept data/message from the application (such as email application) which is running on the application layer.

- Transport layer then splits the received data/message into smaller units if needed.

- It then passes these smaller data units (TPDU) to the network layer and ensures that the pieces all arrive correctly at the other end. But before sending the data units to network layer, transport layer adds transport layer header in front of each data unit.

- Transport layer header prominently contains source & destination port numbers as well as sequence numbers.

- The transport layer is a true end-to-end layer, all the way from the source to the destination.

- In other words, a program on the source machine carries on a conversation with a similar program on the destination machine, using the message headers and control messages.

- Transport layer provides end to end flow control and error control.

5.3 PROCESS TO PROCESS COMMUNICATION

- The ultimate goal of the transport layer is to provide efficient, reliable, and cost-effective service to its users.

- To achieve this goal, the transport layer takes help of the services provided by the network layer.

- The hardware and/or software within the transport layer that does the work is called the **transport entity**.

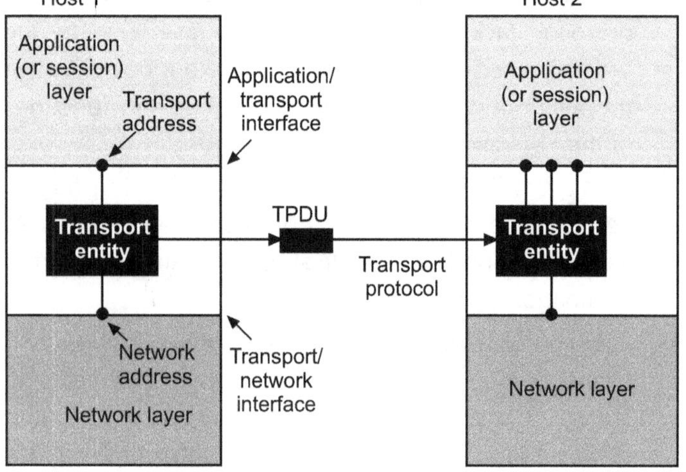

Fig. 5.2 : The relation between network, transport, and application layers

- The transport entity can be located in the operating system kernel, in a separate user process, in a library package bound into network applications, or conceivably on the network interface card.

- The (logical) relationship of the network, transport, and application layers is illustrated in Fig. 5.2.

- Similar to the network services, transport layer also contains two types of transport service.

    1. **Connectionless :** It is very much similar to the connectionless network service.

    2. **Connection Oriented :** The connection-oriented transport service is similar to the connection-oriented network service in many ways.

In both cases, connections have three phases :

    1. Establishment,

    2. Data transfer and

    3. Release.

- Addressing and flow control are also similar in both layers.

- If the transport layer service is so similar to the network layer service, why are there two distinct layers?

- The transport code runs completely on the users' machines, but the network layer mostly runs on the routers.

- The users have no real control over the network layer, so they cannot solve the problem of poor service by using better routers or putting more error handling in the data link layer.

- The only possibility is to put another layer on top of the network layer that improves the quality of the service.

- If, in a connection-oriented subnet, a transport entity is informed halfway through a long transmission that its network connection has been abruptly terminated, with no indication of what has happened to the data currently in transit, it can set up a new network connection to the remote transport entity.

- Using this new network connection, it can send a query to its peer asking which data arrived and which did not, and then pick up from where it left off.

- In essence, the existence of the transport layer makes it possible for the transport service to be more reliable than the underlying network service.

- Lost packets and garbled data can be detected and compensated for by the transport layer.

Advantage of Transport Layer

By using the transport layer, application programmers can write code according to a standard set of primitives and have these programs work on a wide variety of networks, without having to worry about dealing with different subnet interfaces and unreliable transmission.

5.4 USER DATAGRAM PROTOCOL (UDP)

(May 16, Nov. 16)

- It is a connectionless, unreliable transport protocol.

- UDP provides a way for applications to send encapsulated IP datagram's without having to establish a connection.

- UDP transmits segments consisting of an 8-byte header followed by the payload. The data segment sent using the UDP protocol is called as datagram.

- It serves as intermediary between the application programs and network operations.

- It helps in process to process communication; it uses port numbers to accomplish this task.

- Another responsibility is to provide control mechanism at transport layer.

- It performs very limited error checking.

- It does not contain flow control and acknowledgement mechanism for the received packets.

- It is very simple protocol with a minimum overhead. If process wants to send a small message and does not care much about the reliability, it can use UDP.

1. User Datagram

UDP packets, called user datagrams, have a fixed size header of 8 bytes. Following Fig. 5.3 shows the format of a user datagram header.

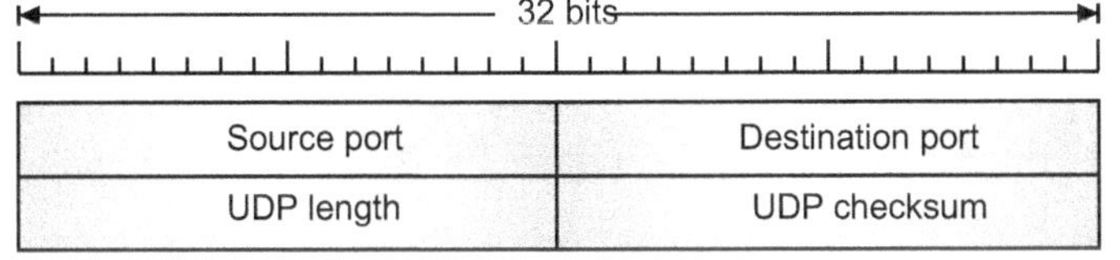

Fig. 5.3 : User datagram header format

The fields are as follows :

Source Port Number

- This is the port number used by the process running on source host.

- It is 16 bit long, which means that port number can range from 0 to 65,535.

- If the source host is the client (a client sending a request), the port number, in most cases is an ephemeral port number requested by the process and chosen by the UDP software running on the source host.

- If source host is server (a server sending a response), the port number, in most cases is a well-known port number.

Destination Port Number :

- This is a port number used by the process running on the destination host.

- It is also 16 bit long.

- If destination host is the server, the port number in most cases, is a well-known port.

- If destination host is a client, the port number in most cases is ephemeral port number. In this case, server copies the ephemeral port number it has received in the request packet.

UDP Length :

- This is a 2 byte (16 bit) field that defines the total length of user datagram (i.e. header + data).

- The 16 bit can define total length of 0 to 65,535 bytes. However, the total length needs to be much less because an UDP user datagram is stored in an IP datagram with the total length of 65535 bytes.

- The length field in a UDP user datagram is actually not necessary.

- A user datagram is encapsulated in an IP datagram. There is a field in the IP datagram that defines the total length. There is another field in the IP datagram that defines the length of the header. So after subtracting the value of second field from the first, we can find out the length of UDP datagram that is encapsulated in an IP datagram.

$$\text{UDP length} = \text{IP length} - \text{IP header's length}$$

- However, the designers of the UDP protocol felt that it was more efficient for the destination UDP to calculate the length of the data from the information provided in the UDP user datagram rather than asking the IP software to supply this information. Also, we should remember that when the IP software delivers the UDP user datagram to the UDP layer, it has already dropped the IP header.

UDP Checksum :

- This field is used to detect errors over the entire user datagram (header + data).

- UDP checksum calculation involves three sections : the pseudoheader, the UDP header and data coming from the application layer. (Pseudoheader is a part of the header of IP packet in which user datagram is to be encapsulated.)

2. Encapsulation and Decapsulation

- To send a message from one process to another, the UDP protocol encapsulates and decapsulates messages (Shown in Fig. 5.4).

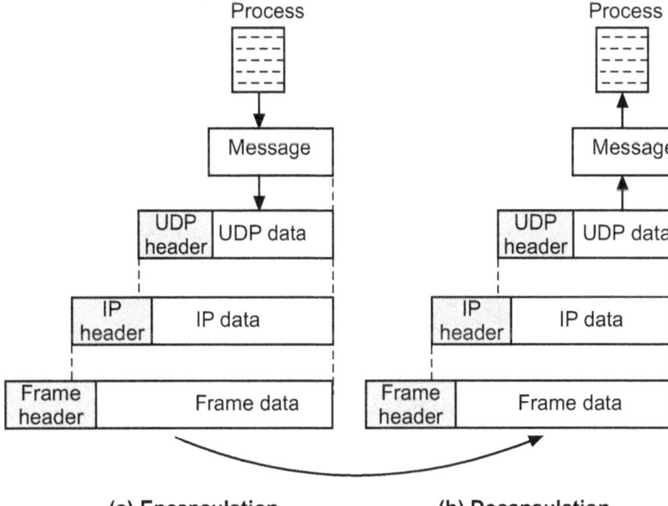

Fig. 5.4 : Encapsulation and Decapsulation

Encapsulation

- When a process has a message to send through UDP, it passes the message to UDP alongwith a pair of socket addresses and length of data.

- UDP receives the data and adds the UDP header.
- UDP then passes the user datagram to IP with socket addresses.
- IP adds its own header, using the value 17 in the protocol field, indicating that the data has come from the UDP protocol.
- The IP datagram is then passed to the data link layer.
- Data link layer then receives the IP datagram, adds its own header and passes it to the physical layer.
- The physical layer encodes the bits into electrical or optical signals and sends it to the remote machine.

Decapsulation :

- When message arrives at the destination host, the physical layer decodes the signals into bits and passes it to the data link layer.
- The data link layer uses the header (and trailer) to check the data.
- If there is no error, the header and trailer are dropped and datagram is passed to IP.
- The IP software does its own checking.
- If there is no error, the header is dropped and the user datagram is passed to UDP.
- UDP uses the checksum to check the entire user datagram. If there is no error, the header is dropped and the application data along with the sender socket address is passed to the process.
- The sender socket address is passed to the process in case it needs to respond to the message received.

5.5 TRANSMISSION CONTROL PROTOCOL (TCP) (Feb. 16)

- TCP is a connection oriented, reliable transport protocol.
- It lies between application layer and network layer and serves as intermediary between application programs and network operation.
- TCP provides a way for applications to send encapsulated IP packets by establishing a connection.
- It uses flow and error control mechanisms in the transport layer.
- It helps in process to process communication; it uses port numbers to accomplish this task.
- It is a full duplex protocol, meaning that each TCP connection supports a pair of byte streams, one flowing in each direction.
- TCP also implements a congestion-control mechanism.

5.5.1 Introduction to TCP

- TCP (Transmission Control Protocol) was specifically designed to provide a reliable end-to-end delivery of data over an unreliable internetwork.
- An internetwork differs from a single network because different parts may have wildly different topologies, protocols, bandwidths, delays, packet sizes, and other parameters.
- TCP was designed to dynamically adapt to properties of the internetwork and to be robust to face many kinds of failures.
- The reliability mechanism of TCP allows devices to deal with lost, delayed, or duplicate packets.
- The IP layer gives no guarantee that datagram's will be delivered properly, so by using the timeout mechanism; the lost packets are detected and then retransmitted.
- Datagrams may arrive in the wrong order; so it is up to TCP to reassemble them into messages in the proper sequence.
- Following are some of the services offered by TCP to the processes at the application layer
  - ➢ Stream delivery service
  - ➢ Full Duplex service
  - ➢ Connection oriented service
  - ➢ Reliable service

5.5.2 TCP Services

Stream Delivery Service

- TCP is stream-oriented protocol.
- It allows sending and receiving process to send and accept data as a stream of bytes.
- TCP creates an environment, in which an imaginary "tube" connects the sending and receiving process. This tube carries their data across the internet.
- This imaginary environment is shown in Fig. 5.5, where sender's process writes a stream of bytes on the tube at one end and the receiver's process reads the stream of bytes from the other end.

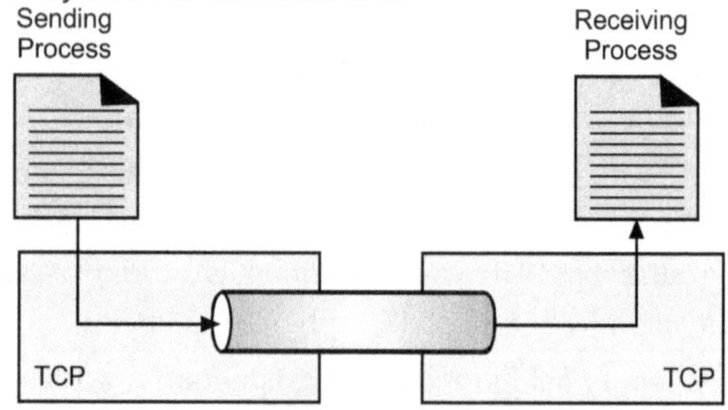

Fig. 5.5 : Stream delivery

- Because the sending and receiving processes may not read or write the data at the same speed, TCP needs buffer storage.

  Hence, TCP contains two buffers :

  1. Sending buffer and
  2. Receiving buffer

- Used at sender and receiver side process respectively.

- These buffers can be implemented by using a circular array of 1 byte locations as shown in Fig. 5.6. For simplicity, we have shown buffers of smaller size. Practically, the buffers are much larger than this.

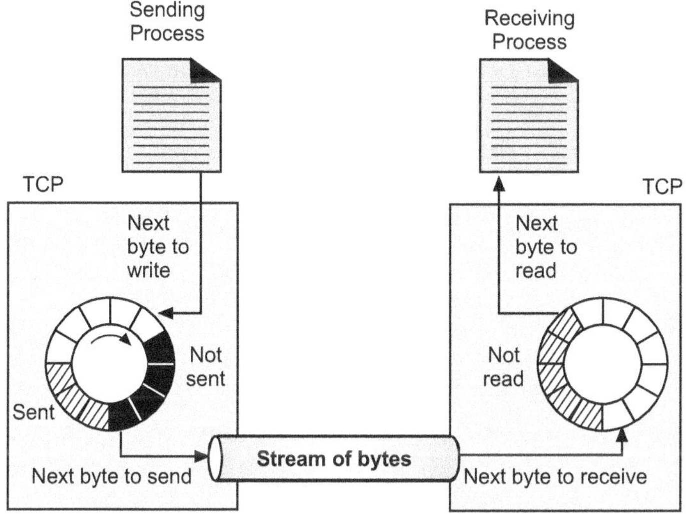

Fig. 5.6 : Sending and receiving buffer

- Fig. 5.6 shows data movement in one direction. At the sending site, the buffer contains three types of compartments.

- The white section contains empty compartments, which can be filled by the sending process.

- The area holds bytes that have been sent but not yet acknowledged.

- TCP keeps these bytes in the buffer until it receives the acknowledgement.

- The black area contains bytes to be sent by the sending TCP process.

- TCP may be able to send only part of this black section. This could be due to the slowness of the receiving process or due to the network congestion problem.

- When the bytes in the gray section are acknowledged, the chambers become free and get available for use to the sending process.

- At receiver side, the circular buffer is divided into two areas (white and gray).

- The white area contains empty compartments, to be filled by bytes, received from the network.

- The gray area contains received bytes that can be read by the receiving process. Once the byte is read by the receiving process, the compartment is freed and added to the pool of empty compartments.

- Buffering is used to handle the difference between the speed of data transmission and data reception. But only buffering is not enough, we need one more step before we can send the data.

- IP layer, as a service provider for TCP, needs to send data in packets, not as a stream of bytes.

- At transport layer, TCP groups the number of bytes together into a packet called as a segment; TCP adds a header to each segment and delivers the segment to the IP layer for transmission.

- The segments are encapsulated in an IP datagram and then transmitted.

Full Duplex Communication Service :

- TCP provides full duplex service, when data can flow in both directions at the same time.

- Each TCP then have a sending and receiving buffer and segments in both directions.

Connection Oriented Service :

- TCP is a connection oriented protocol.

- When one process at site A wants to send and receive data from another process at site B, following things take place :

  ➢ The two TCPs establish a connection between them.

  ➢ Data are exchanged in both the directions.

  ➢ The connection is terminated.

- Note that the connection is virtual and not the physical.

Reliable Service

- TCP is reliable transport protocol. It uses acknowledgement mechanism to check the safe and sound arrival of the data.

1. The TCP Service Model

- TCP service is obtained by both the sender and receiver creating end points, called sockets.

- Each socket has a socket number (address) consisting of the IP address of the host and a 16-bit number local to that host, called a port.

- A port is the TCP name for a TSAP. For obtaining TCP service, a connection must be explicitly established between a socket on the sending machine and a socket on the receiving machine.

- Two or more connections may terminate at the same socket. Connections are identified by the socket identifiers at both ends, that is, (*socket1*, *socket2*). No virtual circuit numbers or other identifiers are used.

- See the listing of socket calls in the table given below :

Table 5.1

Primitive	Meaning
SOCKET	Create a new communication end point
BIND	Attach a local address to a socket
LISTEN	Announce willingness to accept connections
ACCEPT	Block the caller until a connection attempt arrives
CONNECT	Actively attempt to establish a connection
SEND	Send some data over the connection
RECEIVE	Receive some data from the connection
CLOSE	Release the connection

- Port numbers below 1024 are called "well-known ports" and are reserved for standard services. For example, any process wishing to establish a connection to a host to transfer a file using FTP can connect to the destination host's port 21 to contact its FTP daemon.

- All TCP connections are full duplex and point-to-point. Full duplex means that traffic can go in both directions at the same time.

- Point-to-point means that each connection has exactly two end points.

- TCP does not support multicasting or broadcasting.

- A TCP connection is a byte stream, not a message stream.

- When an application passes data to TCP, TCP may send it immediately or buffer it.

- However, sometimes, the application really wants the data to be sent immediately. To force data out, applications can use the PUSH flag, which tells TCP not to delay the transmission.

2. Urgent Data :

- In this case, if the sending application puts some control information in the data stream and gives it to TCP along with the URGENT flag, this event causes TCP to stop accumulating data and transmit everything it has for that connection immediately.

- When the urgent data is reached at the destination, the receiving application is interrupted, so it can stop whatever it was doing and read the data stream to find the urgent data.

- The end of the urgent data is marked so the application knows when it is over.

- The start of the urgent data is not marked. It is up to the application to figure it out.

5.6 WORKING OF TCP PROTOCOL (Feb. 15)

- The sending and receiving TCP entities exchange data in the form of segments.

- A TCP segment consists of a fixed 20-byte header (plus an optional part) followed by zero or more data bytes.

- The TCP software decides how big segments should be.

- It can accumulate data from several writes into one segment or can split data from one write over multiple segments.

- Two limits restrict the segment size.

- Each segment, including the TCP header, must fit in the 65,515-byte IP payload.

- Each network has a maximum transfer unit, or MTU, and each segment must fit in the MTU.

- The basic protocol used by TCP entities is the sliding window protocol.

- When a sender transmits a segment, it also starts a timer. When the segment arrives at the destination, the receiving TCP entity sends back a segment (with data if any exist, otherwise without data) bearing an acknowledgement number equal to the next sequence number of the segment it expects to receive. If the sender's timer goes off before the acknowledgement is received, the sender transmits the segment again.

- Following Problems can Occur During Transmission of Segments :
  - ➢ Segments can arrive out of order.
  - ➢ Segments can be delayed in the transit.

- TCP must be prepared to deal with these problems and solve them in an efficient way. A considerable amount of effort has gone into optimizing the performance of TCP. A number of the algorithms used by many TCP implementations will be discussed later.

5.6.1 The TCP Segment Header

- The following Fig. 5.7 describe the TCP header.

- TCP protocol has fixed 20 byte header.

- This header is followed by the optional field, and which is further followed by the data part.

- Data part is also optional. Segments without any data are legal and are commonly used for acknowledgements and control messages.

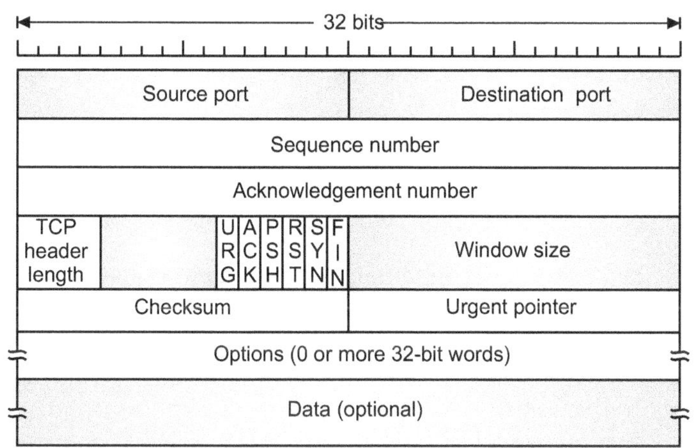

Fig. 5.7 : TCP header

Fields of TCP Header

- **SourcePort :** 16 bits.

- **DestinationPort :** 16 bits

- The Source port and Destination port fields identify the local end points of the connection.

- The source and destination end points together identify the connection.

- **Sequence Number :** 32 bits – has a dual role :

- If the SYN flag is set, then this is the initial sequence number. The sequence number of the actual first data byte (and the acknowledged number in the corresponding ACK) will then be this sequence number plus 1.

- If the SYN flag is clear, then this is the accumulated sequence number of the first data byte of this packet for the current session.

- **Acknowledgement Number :** It specifies the next data byte that is expected.

TCP Header length :

- The TCP header length tells how many 32-bit words are contained in the TCP header.

- This information is needed because the Options field which is the part of TCP header is of variable length, due to that the TCP header also becomes of variable length.

- Next 6 bit field is unused.

Bit Flags : There are 6 one bit flags which are given below.

(i) URG is set to 1 if the Urgent pointer field is in use. The Urgent pointer is used to indicate a byte offset from the current sequence number at which urgent data are to be found.

(ii) The ACK bit is set to 1 to indicate that the Acknowledgement number is valid. If ACK is 0, the segment does not contain an acknowledgement so the Acknowledgement number field is ignored.

(iii) The PSH bit indicates PUSHed data. This bit requests the receiver to deliver the data to the application upon arrival and not store the data in the buffer until a full buffer signal has been received.

(iv) The RST bit is used to reset a connection that has become confused due to a host crash or some other reason. In general, if you get a segment with the RST bit on, you have a problem in your hands.

(v) SYN bit is used to synchronize sequence numbers. Only the first packet sent from each end should have this flag set.

(vi) The FIN bit is used to release a connection. It specifies that the sender has no more data to transmit.

Window Size : The size of the window, which specifies the number of bytes that the sender/receiver is currently willing to send/receive.

Checksum : A Checksumis also provided for extra reliability. It checksums the header, the data, and the conceptual pseudo header as shown in Fig. 5.8.

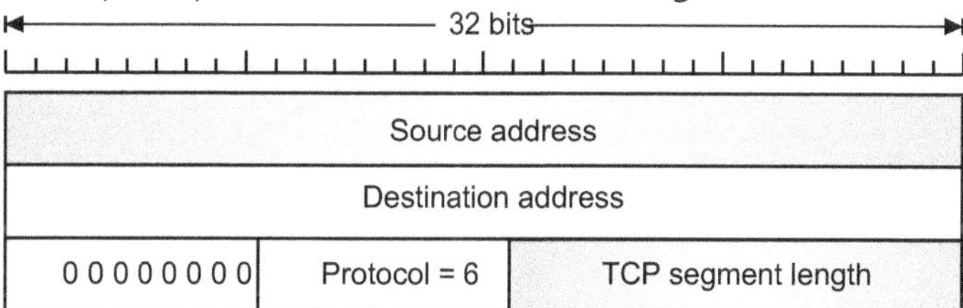

Fig. 5.8 : Pseudo header

Option : The Optionsfield provides a way to add extra facilities not covered by the regular header such as Maximum Segment Size, Window scale factor, Timestamp and many more. Now days some of the options are obsolete.

5.6.2 TCP Transmission Policy

- The following Fig. 5.9 describes the window management in TCP.

- When sender sends data, the receiver gives acknowledgement to the received data.

- But while giving the acknowledgement, it also tells the sender about the current size of receiver window (refer Fig. 5.9).

- Suppose the receiver has a 4096-byte buffer, as shown in Fig. 5.9.

- If the sender transmits a 2048-byte segment that is correctly received, the receiver will acknowledge the segment.

- However, since it now has only 2048 bytes of buffer space (until the application removes some data from the buffer), it will advertise a window size of 2048 for the next byte expected.

- Now the sender transmits another 2048 bytes. Now consider the scenario at the receiver side buffer.

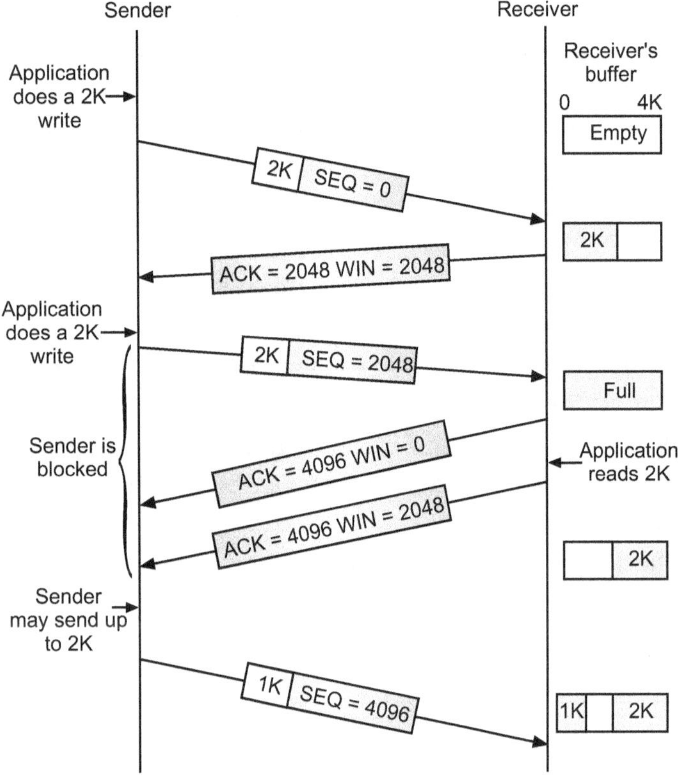

Fig. 5.9 : Window management in TCP

- Previously the buffer was having data of 2048 bytes, and now once again sender has sent the same amount of data. So, as the buffers capacity is of 4096 bytes, it gets full.

- In this situation, when the receiver sends the acknowledgement back to the server, that time (as the receiver side buffer if full) it advertises its window size as 0 bytes.

- Now the sender must stop until the application process on the receiving host removes some data from the buffer.

- When the receiver's application reads/consumes the 2048 byte of data, it advertises the window size as 2048 bytes (which is the free space of receiver buffer).

- That time server gets unblocked, and it starts sending the data once again, till it gets the receiver window having non-zero byte size.

- When the window is 0, the sender may not normally send segments.

- But sender can send the segments with two exceptions.

- Urgent data may be sent, for example, to allow the user to kill the process running on the remote machine.

- The sender may send a 1-byte segment to make the receiver re-announce the next byte expected and window size.

- The TCP standard explicitly provides this option to prevent deadlock if a window announcement ever gets lost.

- Senders are not required to transmit data as soon as it comes from the application.

- Neither are receivers required to send acknowledgements as soon as possible.

- For example, when the first 2 kb of data came in, sender transport entity, knowing that it had a 4 kb receiver window available, then it is completely correct to buffer the data until another 2 kb came in, so that the total size of transmitting segment will be 4 kb (as that of the size of receiving window). This freedom can be further used to improve performance.

- This is used to reduce the usage of the system. Another way to reduce the system uses has been stated in the Nagle's Algorithm.

5.6.3 Nagle's Algorithm

- When data comes into the sender one byte at a time, just send the first byte and buffer all the rest until the outstanding byte is acknowledged.

- Then send all the buffered characters in one TCP segment and start buffering again until they are all acknowledged.

- Nagle's algorithm is widely used by TCP implementations, but there are times when it is better to avoid it.

- For example, when an X Windows application is being run over the Internet, mouse movements have to be sent to the remote computer. (The X Window system is the windowing system used on most UNIX systems.) Gathering them up to send in bursts makes the mouse cursor move inconsistently, which can irritate the users.

- So, it is better to send each mouse movement separately, but that degrades the TCP performance.

5.6.4 Silly Window Syndrome

- Another problem that can degrade TCP performance is the silly window syndrome.

- The main reason of this problem is sender sends the data in larger blocks, but the receiver side application reads the data one byte at a time. Refer Fig. 5.10.

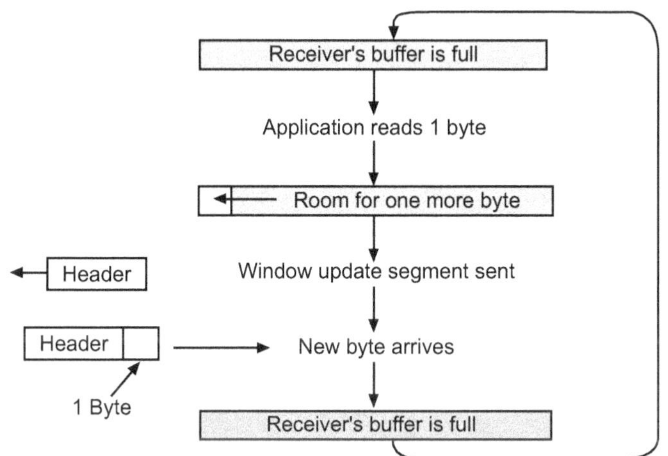

Fig. 5.10 : Silly window syndrome

- Initially, the buffer on the receiving side is full and the sender knows this as the receiver sends the window of size 0 with the acknowledgement after the successful data reception.

- Then the receiver's application reads one character from the buffer. Due to this receiver sends a window update to the sender saying that you can send 1 byte of data.

- Then the sender sends 1 byte.

- The buffer is now full, so the receiver acknowledges the 1-byte segment with advertising the window size equal to 0. This process gets repeated forever.

- Clark's solution is used to prevent the receiver from sending a window update for 1 byte.

- Instead it is forced to wait until it has a sufficient amount of buffer space available at the receiver side.

- Specifically, the receiver should not send a window update until it can handle the maximum segment size that it has advertised at the time of connection establishment or until its buffer is half empty, whichever is smaller.

- By not sending tiny segments, sender can also help to improve the performance. Instead, it should try to wait until it has accumulated enough space in the window to send a full segment or at least one containing half of the receiver's buffer size.

- Nagle's algorithm and Clark's solution to the silly window syndrome are opposite.

- Nagle was trying to solve the problem caused by the sending application delivering data to TCP a byte at a time.

- Clark was trying to solve the problem of the receiving application sucking the data up from TCP a byte at a time.

- Both solutions are valid and can work together. The goal is for the sender not to send small segments and the receiver not to ask for them.

5.7 TCP TIMER MANAGEMENT

(Feb. 16, May 17)

- TCP uses multiple timers (at least conceptually) to do its work.

- The most important of these is the retransmission timer.

- When a segment is sent, a retransmission timer is started.

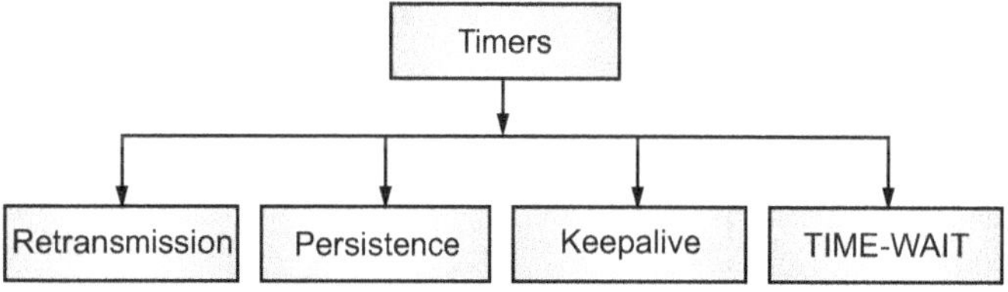

Fig. 5.11 : TCP timers

- If the segment is acknowledged before the timer expires, the timer is stopped.

- If, on the other hand, the timer goes off before the acknowledgement comes in, the segment is retransmitted (and the timer started again).

How long should the timeout interval be?

- This problem is much more difficult to solve in the transport layer than that of data link layer protocol.

- The timeout interval should be enough long such that as soon as the acknowledgement is received to the sender, the timer should be go off as shown in Fig. 5.12 (a).

- Normally the acknowledgements are rarely delayed in data link layer, the absence of an acknowledgement at the expected time generally means either the frame or the acknowledgement has been lost.

- Lots of factors decide the performance of the network, so it is very difficult to calculate the round trip time to the destination.

- If the timeout is set too short, say, T_1 in Fig. 5.12 (b), unnecessary retransmissions will occur, filling the Internet with useless packets.

- If it is set too long, (e.g. T_2), performance will suffer due to the long retransmission delay whenever a packet is lost.

- The best solution on this problem is to use a highly dynamic algorithm that constantly adjusts the timeout interval, based on continuous measurements of network performance.

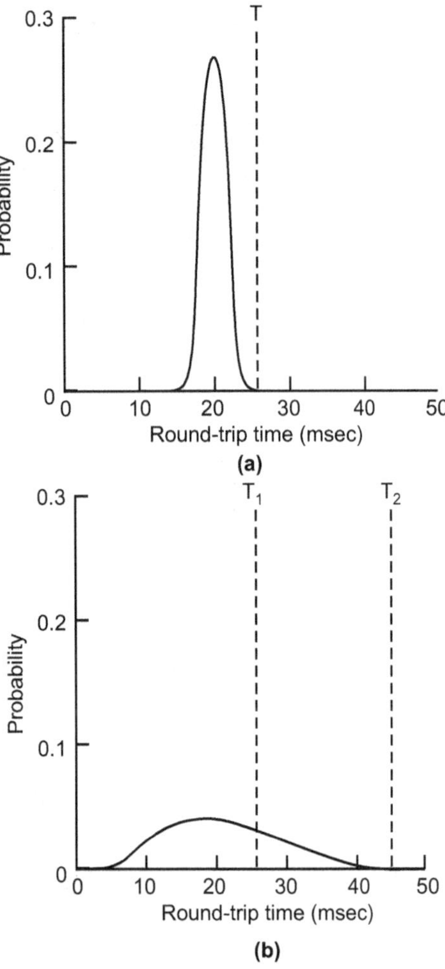

Fig. 5.12 : Probability density of acknowledgement arrival times in

5.7.1 Jacobson Algorithm of Timeout Interval

- For each connection, TCP maintains a variable called RTT that is the best current estimate of the round-trip time (RTT) to the destination.

- When a segment is sent, a timer is started, to see how long the acknowledgement takes and to trigger a retransmission if it takes too long.

- If the acknowledgement gets back before the timer goes off, TCP measures how long the acknowledgement took (say M). It then updates RTT according to the formula;

$$RTT = \alpha RTT + (1 - \alpha) M$$

where α is a smoothing factor. Typically $\alpha = 7/8$.

- For a given good value of RTT, choosing a suitable retransmission timeout is not an easy task.

- By studying all these facts, Jacobson proposed a new smoothing factor that has been given below;

$$D = \alpha D + (1 - \alpha) \, | \, RTT - M \, |$$

And the corresponding timeout is calculated by

$$Time\ out = 4D + RTT$$

- One problem that occurs with the dynamic estimation of RTT is what to do when a segment times out and is sent again.

- The solution to this problem is "don't update RTT on any segments that have been retransmitted. Instead of this, the timeout is doubled on each failure until the segments get through the first time. This solution is called as Karn's algorithm. Most TCP implementations use it."

5.7.2 Karn's Algorithm

- Suppose that a segment is not acknowledged during the retransmission period and is therefore retransmitted.

- When the sending TCP receives an acknowledgement for this segment, it does not know if the acknowledgement is for the original segment or for the retransmitted one.

- The value of the new RTT is based on the departure of the segment.

- However, if the original segment was lost and acknowledgement is for retransmitted one, the value of the current RTT must be calculated from the time the segment was retransmitted.

- This problem has been solved by Karn.

- Karn's solution is very simple.

- Do not consider Round Trip Time of a retransmitted segment in the calculation of the new RTT. Do not update the value of RTT until you send a segment and receive acknowledgement without the need of retransmission.

5.7.3 Other Types of TCP Timers

1. **Persistence Timer :**

- To deal with a zero-window-size advertisement, TCP needs another timer.

- If the receiving TCP announces a window size of zero, the sending TCP stops transmitting the segments until the receiving TCP sends an acknowledgement segment announcing a non-zero window size.

- This acknowledgement segment can be lost.

- Remember that acknowledgement segments are not acknowledged in TCP.

- If this acknowledgement is lost, the receiving TCP thinks that it has done its job and waits for sending TCP to send more segments.

- There is no retransmission timer for the segment containing only acknowledgement.

- The sending TCP has not received an acknowledgement and waits for the other TCP to send an acknowledgement advertising the size of the window.

- Both TCPs can continue to wait for each other forever, which creates a deadlock.

- To correct this deadlock, TCP uses a persistence timer for each connection.

- When the sending TCP receives an acknowledgement with a window size of zero, it starts a persistence timer.

- When persistence timer goes off, the sending TCP sends a special segment called as probe.

- This segment (probe) contains only one byte of data.

- It has a sequence number; but this sequence number is never acknowledged.

- The probe alerts the receiving TCP that the acknowledgement was lost and must be resent.

- The value of persistence timer is set to the value of retransmission time.

- However, if response is not received from the receiver, another probe segment is sent and the value of the persistence timer is doubled and reset.

- The sender continues sending the probe segments and doubling and resetting the value of the persistent timer until the value reaches a threshold (usually 60 sec).

- After that the sender sends one probe segment every 60 sec until the window is reopened.

2. Keepalive Timer :

- It is used in some implementations to prevent a long idle connection between two TCPs.

- Suppose that a client opens a TCP connection to a server, transfers some data, and becomes silent. Perhaps the client has crashed. In this case, the connection remains open forever.

- To remedy this situation, most implementations equip a server with a keepalive timer.

- Each time the server hears from a client, it resets this timer.

- The timeout is usually 2 hours.

- If the server does not hear from the client after two hours, it send a probe segment.

- If there is no response after 10 probes, each of which is 75 sec apart, it assumes that the client is down and terminates the connection

3. TIME-WAIT Timer :

- It is used during connection termination. This timer is set to a time equal to twice the maximum packet lifetime to ensure that after closing a connection all the packets created by it die off.

5.8 STREAM CONTROL TRANSMISSION PROTOCOL

- Stream Control Transmission Protocol (SCTP) is a new reliable, message-oriented transport-layer protocol.

- Figure shows the relationship of SCTP to the other protocols in the Internet protocol suite.

- SCTP lies between the application layer and the network layer and serves as the intermediary between the application programs and thenetwork operations.

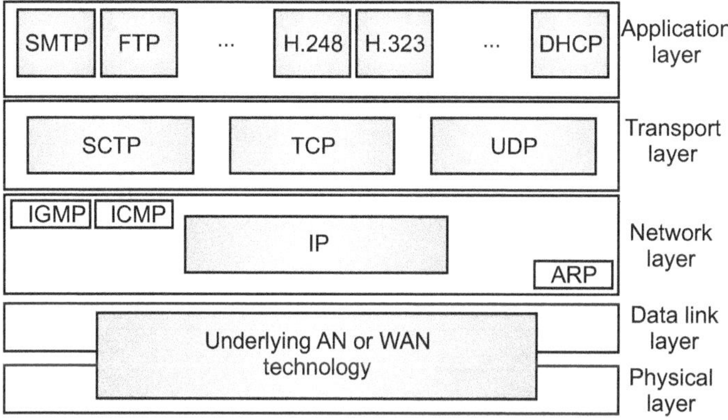

Fig. 5.13 : Relation of SCTP with other protocols

- SCTP, however, is mostly designed for Internet applications that have recently been introduced. These new applications, such as IUA (ISDN over IP), M2UA and M3UA (telephony signaling), H.248 (media gateway control), H.323 (IP telephony), and SIP (IP telephony), need a more sophisticated service than TCP can provide.

- SCTP provides this enhanced performance and reliability.

- We briefly compare UDP, TCP, and SCTP:

UDP

- UDP is a message-oriented protocol.

- A process delivers a message to UDP, which is encapsulated in a user datagram and sent over the network. UDP conserves the message boundaries; each message is independent from any other message.

- This is a desirable feature when we are dealing with applications such as IP telephony and transmission of real-time data.

- However, UDP is unreliable; the sender cannot know the destiny of messages sent. A message can be lost, duplicated, or received out of order.

- UDP also lacks some other features, such as congestion control and flow control, needed for a friendly transport-layer protocol.

TCP

- TCP is a byte-oriented protocol.

- It receives a message or messages from a process, stores them as a stream of bytes, and sends them in segments. There is no preservation of the message boundaries.

- However, TCP is a reliable protocol.

- The duplicate segments are detected, the lost segments are resent, and the bytes are delivered to the end process in order. TCP also has congestion control and flow control mechanisms.

- SCTP combines the best features of UDP and TCP.

- SCTP is a reliable message oriented protocol.

- It preserves the message boundaries and at the same time detects lost data, duplicate data, and out-of-order data. It also has congestion control and flow control mechanisms.

5.8.1 SCTP Services

1. Process-to-Process Communication

SCTP uses all well-known ports in the TCP space. Table 5.2 lists some extra port numbers used by SCTP.

Table 5.2

Protocol	Port Number	Description
IUA	9990	ISDN over IP
M2UA	2904	SS7 telephony signaling
M3UA	2905	SS7 telephony signaling
H.248	2945	Media gateway control
H.323	1718, 1719, 11720	IP telephony
SIP	5060	IP telephony

2. Multiple Streams

- TCP is a stream-oriented protocol. Each connection between a TCP client and a TCP server involves one single stream.

- The problem with this approach is that a loss at any point in the stream blocks the delivery of the rest of the data.

- This can be acceptable when we are transferring text; it is not when we are sending real-time data such as audio or video.

- SCTP allows multistream service in each connection, which is called association in SCTP terminology. If one of the streams is blocked, the other streams can still deliver their data.

- The idea is similar to multiple lanes on a highway. Each lane can be used for a different type of traffic. For example, one lane can be used for regular traffic, another for car pools. If the traffic is blocked for regular vehicles, car pool vehicles can still reach their destinations. Fig. 5.14 shows the idea of multiple-stream delivery.

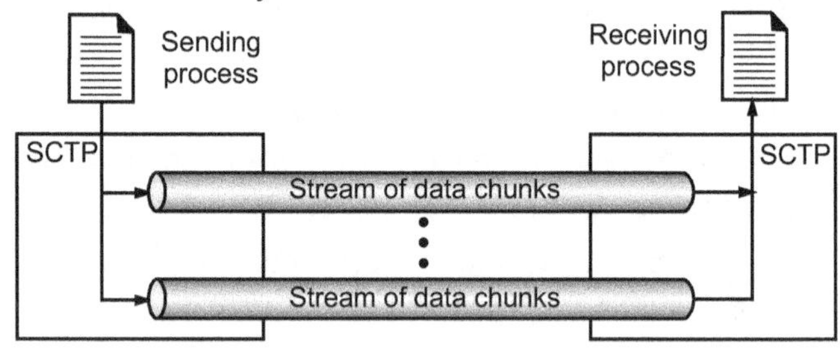

Fig. 5.14 : Multiple stream delivery

3. Multihoming

- A TCP connection involves one source and one destination IP address. This means that even if the sender or receiver is a multihomed host (connected to more than one physical address with multiple IP addresses), only one of these IP addresses per end can be utilized during the connection.

- An SCTP association, on the other hand, supports multihoming service. The sending and receiving host can define multiple IP addresses in each end for an association.

- In this fault-tolerant approach, when one path fails, another interface can be used for data delivery without interruption. This fault-tolerant feature is very helpful when we are sending and receiving a real-time payload such as Internet telephony. Fig. 5.15 shows the idea of multihoming.

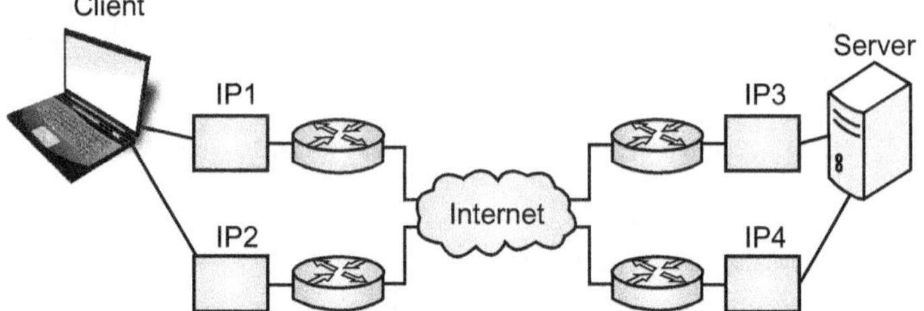

Fig. 5.15 : Multihoming

- In the figure, the client is connected to two local networks with two IP addresses.

- The server is also connected to two networks with two IP addresses.

- The client and the server can make an association using four different pairs of IP addresses.
- However, note that in the current implementations of SCTP, only one pair of IP addresses can be chosen for normal communication; the alternative is used if the main choice fails.
- In other words, at present, SCTP does not allow load sharing between different paths.

4. Full-Duplex Communication

- Like TCP, SCTP offers full-duplex service, where data can flow in both directions at the same time. Each SCTP then has a sending and receiving buffer and packets are sent in both directions.

5. Connection-Oriented Service

- Like TCP, SCTP is a connection-oriented protocol. However, in SCTP, a connection is called an association.
- When a process at site A wants to send and receive data from another process at site B, the following occurs:
  - ➢ The two SCTPs establish an association between each other.
  - ➢ Data are exchanged in both directions.
  - ➢ The association is terminated.

6. Reliable Service

- SCTP, like TCP, is a reliable transport protocol. It uses an acknowledgment mechanism to check the safe and sound arrival of data.

5.8.2 SCTP Features

Transmission Sequence Number (TSN)

- The unit of data in TCP is a byte. Data transfer in TCP is controlled by numbering bytes using a sequence number.
- On the other hand, the unit of data in SCTP is a data chunk, which may or may not have a one-to-one relationship with the message coming from the process because of fragmentation (discussed later). Data transfer in SCTP is controlled by numbering the data chunks.
- SCTP uses a transmission sequence number (TSN) to number the data chunks. In other words, the TSN in SCTP plays the analogous role as the sequence number in TCP.
- TSNs are 32 bits long and randomly initialized between 0 and 232 − 1. Each data chunk must carry the corresponding TSN in its header.

Stream Identifier (SI)

- In TCP, there is only one stream in each connection.
- In SCTP, there may be several streams in each association.
- Each stream in SCTP needs to be identified using a stream identifier (SI). Each data chunk must carry the SI in its header so that when it arrives at the destination, it can be properly placed in its stream.
- The SI is a 16-bit number starting from 0.

Stream Sequence Number (SSN)

- When a data chunk arrives at the destination SCTP, it is delivered to the appropriate stream and in the proper order.
- This means that, in addition to an SI, SCTP defines each data chunk in each stream with a stream sequence number (SSN).

Packets

- In TCP, a segment carries data and control information. Data are carried as a collection of bytes; control information is defined by six control flags in the header.
- The design of SCTP is totally different: data are carried as data chunks, control information as control chunks. Several control chunks and data chunks can be packed together in a packet.
- A packet in SCTP plays the same role as a segment in TCP. Figure compares a segment in TCP and a packet in SCTP.

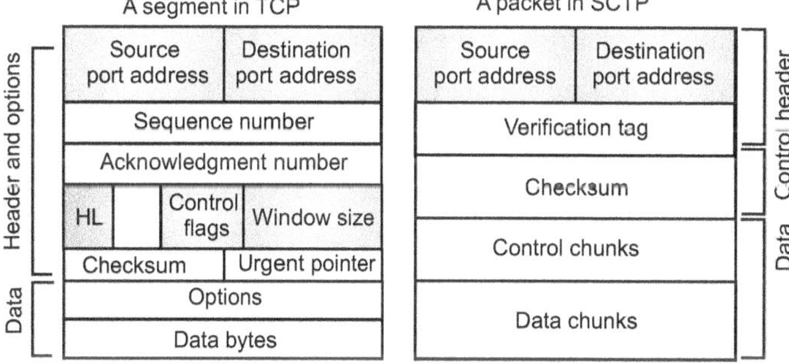

Fig. 5.16 : Segment in TCP V/s packet in SCTP

- The control information in TCP is part of the header; the control information inSCTP is included in the control chunks. There are several types of control chunks; each is used for a different purpose.
- The data in a TCP segment treated as one entity; an SCTP packet can carry several data chunks; each can belong to a different stream.
- The options section, which can be part of a TCP segment, does not exist in an SCTP packet. Options in SCTP are handled by defining new chunk types.

- The mandatory part of the TCP header is 20 bytes, while the general header in SCTP is only 12 bytes.

The SCTP Header is Shorter due to the Following:

- An SCTP sequence number (TSN) belongs to each data chunk, and hence is located in the chunk's header.

- The acknowledgment number and window size are part of each control chunk.

- There is no need for a header length field (shown as HL in the TCP segment) because there are no options to make the length of the header variable; the SCTP header length is fixed (12 bytes).

- There is no need for an urgent pointer in SCTP, as we will see later.

- The checksum in TCP is 16 bits; in SCTP, it is 32 bits.

- The verification tag in SCTP is an association identifier, which does not exist in TCP. In TCP, the combination of IP and port addresses define a connection; in SCTP we may have multihoming using different IP addresses. A unique verificationtag is needed to define each association.

- TCP includes one sequence number in the header, which defines the number of thefirst byte in the data section. An SCTP packet can include several different data chunks. TSNs, ISs, and SSNs define each data chunk.

- Some segments in TCP that carry control information (such as SYN and FIN), need to consume one sequence number; control chunks in SCTP never use a TSN, IS, or SSN number. These three identifiers belong only to data chunks, not to the whole packet.

- In SCTP, we have data chunks, streams, and packets.

- An association may send many packets, a packet may contain several chunks, and chunks may belong to different streams.

- To make the definitions of these terms clear, let us suppose that process A needs to send 11 messages to process B in three streams.

- The first four messages are in the first stream, the second three messages are in the second stream, and the last four messages are in the third stream.

- Although a message, if long, can be carried by several data chunks, we assume that each message fits into one data chunk. Therefore, we have 11 data chunks in three streams.

- The application process delivers 11 messages to SCTP, where each message is earmarked for the appropriate stream.

- Although the process could deliver one message from the first stream and then another from the second, we assume that it delivers all messages belonging to the first stream first, all messages belonging to the second stream next, and finally, all messages belonging to the last stream.

- We also assume that the network allows only 3 data chunks per packet, which means that we need 4 packets as shown in Figure.

- Data chunks in stream 0 are carried in the first and part of the second packet; those in stream 1 are carried in the second and the third packet; those in stream 2 are carried in the third and fourth packet.

- Note that each data chunk needs three identifiers: TSN, SI, and SSN. TSN is a cumulative number and used for flow control and error control. SI defines the stream to which the chunk belongs. SSN defines the chunk's order in a particular stream. In our example, SSN starts from 0 for each stream.

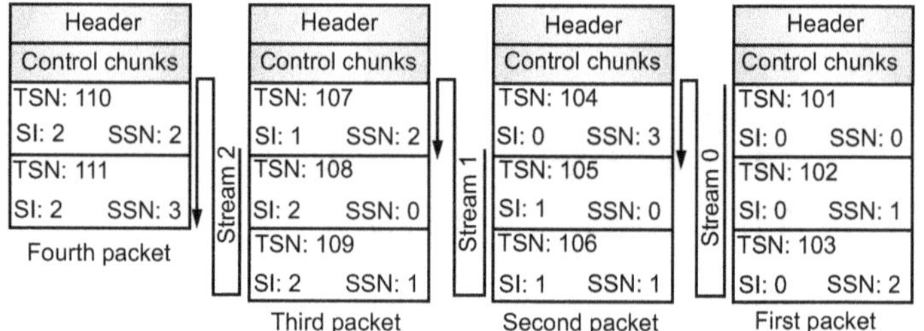

Fig. 5.17 : Packet, data chunks, streams

Acknowledgment Number

- TCP acknowledgment numbers are byte-oriented and refer to the sequence numbers.

- SCTP acknowledgment numbers are chunk-oriented. They refer to the TSN.

- A second difference between TCP and SCTP acknowledgments is the control information. Recall that this information is part of the segment header in TCP.

- To acknowledge segments that carry only control information, TCP uses a sequence number and acknowledgment number (for example, a SYN segment needs to be acknowledged by an ACK segment).

- In SCTP, however, the control information is carried by control chunks, which do not need a TSN.

- These control chunks are acknowledged by another control chunk of the appropriate type (some need no acknowledgment). For example, an INIT control chunk is acknowledged by an INIT-ACK chunk.

- There is no need for a sequence number or an acknowledgment number.

Flow Control

- Like TCP, SCTP implements flow control to avoid overwhelming the receiver.

Error Control

- Like TCP, SCTP implements error control to provide reliability. TSN numbers and acknowledgment numbers are used for error control.

Congestion Control

- Like TCP, SCTP implements congestion control to determine how many data chunks can be injected into the network.

5.9 CONGESTION CONTROL   (FEB. 15)

- Congestion occurs, when the load offered to any network is more than it can handle. The basic principal of congestion control is, do not inject more packets into the network until the old one is delivered properly.

- There are mostly two approaches to solve the problem of congestion :

  1. **Proactive Approach :** In this technique the congestion control algorithm tries to avoid the congestion from occurring.

  2. **Reactive Approach :** In this technique once the congestion takes place in the network, then efforts are taken to reduce it.

- In this section we will discuss algorithms that have been developed to deal with congestion.

- Along with the network layer, transport layer also tries to manage congestion; because the real solution to congestion is to slow down the data rate and that can only be done by the transport layer.

- The first step in managing congestion is detecting it. In the old days, detecting congestion was difficult.

- Before discussing how TCP reacts to congestion, we will first see how to try to prevent congestion from occurring in the first place (congestion avoidance mechanism).

- When a connection is established, a suitable window size has to be chosen. The receiver can specify a window based on its buffer size. If the sender sticks to this window size, problems will not occur due to buffer overflow at the receiving end, but they may still occur due to internal congestion within the network.

- For solving the problem of congestion on the internet, it is important to understand the two potential problems :
  1. Network capacity and
  2. Receiver capacity

- And to deal each of the problem separately.

- To do so, each sender maintains two windows : the window the receiver has granted and a second window, the congestion window. In Transmission Control Protocol (TCP), the congestion window, also called the TCP receive window, determines the number of bytes that can be outstanding at any time.

5.9.1 Slow Start Algorithm

- When a connection is established, the sender initializes the congestion window to the size of the maximum segment in use on the connection.

- It then sends one maximum segment. If this segment is acknowledged before the timer goes off, it adds one segment's worth of bytes to the congestion window to make it two maximum size segments and sends two segments.

- As each of these segments is acknowledged, the congestion window is increased by one maximum segment size.

- When the congestion window is n segments, if all n are acknowledged on time, the congestion window is increased by the byte count corresponding to n segments. In effect, each burst acknowledged doubles the congestion window.

- Until timeout occurs or the receiver's window is reached, the size of congestion window increases exponentially.

- In short, it tells that if bursts of size, say, 512, 1024, 2048, and 4096 bytes work fine but a burst of 8192 bytes gives a timeout, the congestion window should be set to 4096 to avoid congestion.

- As long as the congestion window remains at 4096, no traffic bursts longer than 4096 will be sent, no matter how much the size of receiver window is.

- Even though this algorithm is called slow start, but it is not slow at all.

- It is exponential in nature.

5.9.2 Internet Congestion Control Algorithm

Now let's see the effect of one more parameter known as Threshold on the congestion mechanism.

- Assume that initially threshold value is of 64 kB. When a timeout occurs during data transmission, the threshold is set to half of the current congestion window, and the congestion window is reset to one maximum segment.

- Slow start is then used to determine what the network can handle, except that exponential growth stops when the threshold is hit.

- From that point on, successful transmissions grow the congestion window linearly (by one maximum segment for each burst).

- In effect, this algorithm is guessing that it is probably acceptable to cut the congestion window in half, and then it gradually works its way up from there.

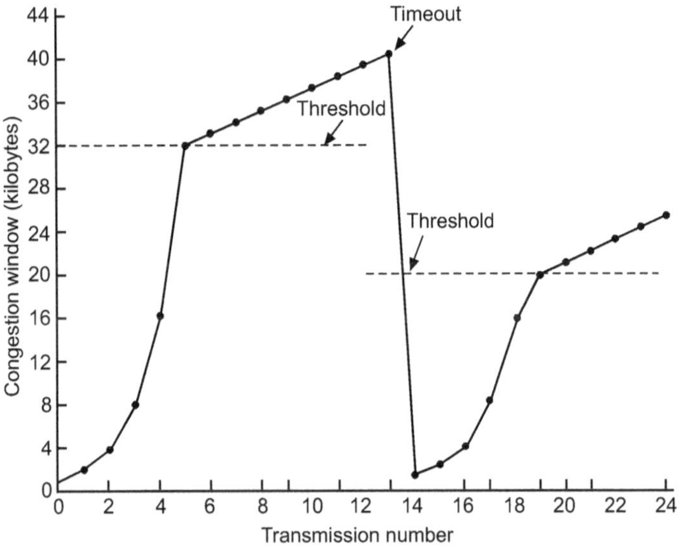

Fig. 5.18 : An example of the internet congestion algorithm

- See Fig. 5.18. Initially, the congestion window was 64 kB, but a timeout occurred, so the threshold is set to 32 kB and the congestion window to 1 kB for transmission number 0.

- The congestion window then grows exponentially until it hits the threshold (32 kB). From there, it starts growing linearly.

- At Transmission number 13 timeout occurred once again. The threshold is set to half the current window (by now 40 kB, so half is 20 kB), and slow start is initiated all over again.

- When the acknowledgements from transmission 14 start coming in, then for first four acknowledgements congestion window gets doubled, but after that, growth becomes linear again as it reaches to the threshold value (of 20 kB).

- If no timeout occurs in the future, the congestion window will continue to grow till the size of receiver's window, once the size of receiver's window is reached, the congestion window stops growing.

5.10 QUALITY OF SERVICE (QoS)

- Quality of service provided by Network layer is out of our control.

- Network layer involves routers across the Internet, which we do not own (the carriers own them).

- If routers or the routing mechanism is faulty, it can result in introducing errors in the data and packet lost.

- To deal with such poor service, we can't change Network layer as we don't own those machines.

- Our only option is to put a further layer on top of Network layer and try to improve quality of service.

- Transport layer tries to provide reliable service to Application layer.

- Transport service can be more reliable than underlying Network service. Unreliability of Network service is hidden from higher layer.

- For improving quality of service, following parameters play an important role

- **Connection Establishment Delay :** Generally for establishing the connection, the source does a request to the destination host; destination host gives acknowledgement to the source. If the acknowledgement is positive, the connection gets established. So, some time goes from the moment of requesting the connection till the instant at which connection is confirmed. This time difference is called as connection establishment delay. This delay should be as small as possible.

- **Transit Delay :** This is the time difference between, the instant at which the message is sent by the transport user on the source machine and the instant when it is received by the transport user at the destination machine.

- **Protection :** Normally, intruders monitor the network traffic. So, it is better to send the data securely. For sending the data in secure manner, some data protection mechanisms are required.

- **Residual Error Ratio :** It is the ratio of number of lost or garbled messages to the total message sent out on the network.

- **Priority :** Priority is associated with every network connection, which shows that some connections are more important than the other. It plays very important role in case of congestion control. During congestion, the higher priority connections get service before the lower one.

- **Throughput :** It is nothing but number of data bytes transferred per second. Every connection has different throughput value associated with it.

- **Resilience :** Due to some hardware/software problem or due to congestion, the transport layer spontaneously terminates a connection. The resilience parameter gives the probability of such a termination.

  - The techniques we looked at previous sections are designed to reduce congestion and improve network performance.

  - However, with the growth of multimedia networking, often these ad-hoc measures are not enough.

  - Serious attempts at guaranteeing quality of service through the network and protocol design are needed.

  - A stream of packets from a source to destination is called a flow. In connection oriented network, all the packets belonging to a flow follow the same route; in connectionless network, they may follow different routes.

  - The needs of each flow can be characterized by four primary parameters: reliability, delay, jitter and bandwidth. Together these parameters determine quality of service. QoS of some applications is shown in the table below.

Table 5.3

Application	Reliability	Delay	Jitter	Bandwidth
E-mail	High	Low	Low	Low
File Transfer	High	Low	Low	Medium
Web access	High	Medium	Low	Medium
Remote Login	High	Medium	Medium	Low
Audio on Demand	Low	Low	High	Medium
Video on Demand	Low	Low	High	High
Telephony	Low	High	High	Low
Videoconferencing	Low	High	High	High

- **Reliability :**

  - Reliability is a characteristic that a flow needs.

  - Lack of reliability means losing a packet or acknowledgement, which entails retransmission.

  - However, the sensitivity of application programs to reliability is not the same.

  - For example, it is more important that electronic mail, file transfer, and internet access have reliable transmission than telephony or audio conferencing.

- **Delay :**

  - Source to destination delay is another flow characteristic.

  - Again, application can tolerate delay in different degrees.

  - In this case, telephony, audio conferencing, video conferencing, and remote login need minimum delay, while delay in file transfer or email is less important.

- **Jitter :**

  - Jitter is the variation in delay for packets belonging to the same flow.

  - Real time audio and video cannot tolerate high jitter. e.g. a real time video broadcast is useless if there is a 2 ms delay for the first and second packet and 16 ms delay for the third and fourth.

  - On the other hand, it does not matter if packet carrying information in a file has different delays. The transport layer at the destination waits until all packets arrive before delivery to the application layer.

- **Bandwidth :**

  - Different applications need different bandwidths. In video conferencing we need to send millions of bits per second to refresh a color screen while the total number of bits in an email may not reach even a million.

5.11 CONGESTION CONTROL ALGORITHMS

Congestion control in the network can be controlled by employing traffic shaping mechanism. Traffic shaping is a mechanism to control the amount and the rate of the traffic sent to the network. Two techniques can shape traffic: leaky bucket and token bucket.

5.11.1 Leaky Bucket

- If a bucket has a small hole at the bottom, the water leaks from the bucket at a constant rate as long as there is water in the bucket.

- The rate at which the water leaks does not depend on the rate at which the water is in put to the bucket unless the bucket is empty.

- The input rate can vary, but the output rate remains constant. Similarly, in networking, a technique called leaky bucket can smooth out bursty traffic.

- Bursty chunks are stored in the bucket and sent out at an average rate. Fig. 5.19 shows aleaky bucket and its effects.

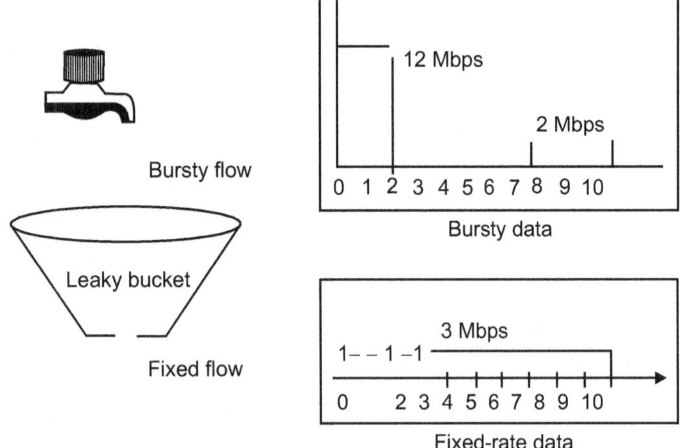

Fig. 5.19 : Leaky Bucket

- In the Fig. 5.19, we assume that the network has committed a bandwidth of 3Mbps for a host. The use of the leaky bucket shapes the input traffic to make it conform to this commitment. In Fig. 5.19 the host sends a burst of data at a rate of 12 Mbps for 2s, for a total of 24 Mbits of data.

- The host is silent for 5 s and then sends data at a rate of 2Mbps for 3s, for a total of 6Mbits of data. Inall, the host has sent 30 Mbits of data in IOs. The leaky bucket smooths the traffic by sending out data at a rate of 3 Mbps during the same 10s. Without the leaky bucket, the beginning burst may have hurt the network by consuming more bandwidth than is set aside for this host. We can also see that the leaky bucket may prevent congestion. As an analogy, consider the free way during rush hour (bursty traffic). If, instead, commuters could stagger their working hours, congestion o' nour freeways could be avoided.

- A simple leaky bucket implementation is shown in Fig. 5.19. A FIFO queue holds the packets. If the traffic consists of fixed-size packets (e.g., cells in ATM networks),the process removes a fixed number of packets from the queue at each tick of the clock.

- If the traffic consists of variable length packets, the fixed output rate must be base do the number of by orbits.

- The following is an algorithm for variable length packets:

  1. Initialize a counter to n at the tick of the clock.

  2. If n is greater than the size of the packet, send the packet and decrement the counter by the packet size. Repeat this step until n is smaller than the packet size.

  3. Reset the counter and goto step1.

5.11.2 Token Bucket

- The leaky bucket is very restrictive. It does not credit an idle host. For example, if a host is not sending for a while, its bucket becomes empty. Now if the host has bursty data, the leaky bucket allows only an average rate. The time when the host was idle is not taken into account.

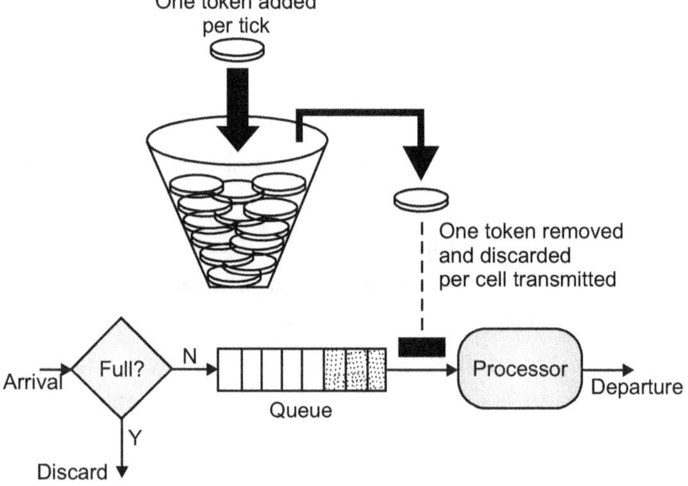

Fig. 5.20 : Token bucket

- On the other hand, the token bucket algorithm allow sidle hosts to accumulate credit for the future in the form of tokens.

- For each tick of the clock, the system sends n tokens to the bucket. The system removes one token for every cell (or byte) of data sent. For example, if n is 100 and the host is idle for 100 ticks, the bucket collects 10,000 tokens. Now the host can consume all these tokens in one tick with 10,000 cells, or the host takes 1000 ticks with 10 cells per tick.

- In other words, the host can send bursty data as long as the bucket is not empty Fig. 5.20 shows the idea.

- The token bucket can easily be implemented with a counter. The token is initialized to zero. Each time a token is added, the counter is incremented by 1. Each time a unit of data is sent, the counter is decremented by 1. When the counter is zero, the host can not send data.

5.12 CONGESTION AVOIDANCE

As long as non-duplicate ACKs are received, the congestion window is additively increased by one MSS every round trip time. When a packet is lost, the likelihood of duplicate ACKs being received is very high (it's possible though unlikely that the stream just underwent extreme packet reordering, which would also prompt duplicate ACKs). The behavior of Tahoe and Reno differ in how they detect and react to packet loss:

5.12.1 RTT Estimation

- For each connection, TCP maintains a variable called RTT that is the best current estimate of the round-trip time (RTT) to the destination.

- When a segment is sent, a timer is started, to see how long the acknowledgement takes and to trigger a retransmission if it takes too long.

- If the acknowledgement gets back before the timer goes off, TCP measures how long the acknowledgement took (say *M*). It then updates RTT according to the formula;

$$RTT = \alpha RTT + (1 - \alpha) M$$

where a is a smoothing factor. Typically a = 7/8.

- For a given good value of RTT, choosing a suitable retransmission timeout is not an easy task.

- By studying all these facts, Jacobson proposed a new smoothing factor that has been given below;

$$D = \alpha D + (1 - \alpha) | RTT - M |$$

And the corresponding timeout is calculated by

$$Time\ out = 4D + RTT$$

- One problem that occurs with the dynamic estimation of RTT is what to do when a segment times out and is sent again ?

- The solution to this problem is "don't update RTT on any segments that have been retransmitted. Instead of this the timeout is doubled on each failure until the segments get through the first time. This solution is called as Karn's algorithm. Most TCP implementations use it."

5.12.2 Retransmission

- The heart of the error control mechanism is the retransmission of segments. When a segment is corrupted, lost, or delayed, it is retransmitted. In modern implementations, a segment is retransmitted on two occasions: when are transmission timer expires or when the send er receives three duplicate ACKs.

- In modern implementations, are transmission occurs if there transmission timer expires or three duplicate ACK segments have arrived.

- Note that more transmission occurs for segments that do not consume sequence numbers. In particular, there is no transmission for an ACK segment.

No retransmission timer is set for an ACK segment.

1. Retransmission After RTO

- A recent implementation of TCP maintains one retransmission time-out(RTO) timer for all outstanding (sent, but not acknowledged segments. When the timer matures, the earliest outstanding segment is retransmitted even though lack of a received ACK can be due to a delayed segment, a delayed ACK, or a lost acknowledgment. Note that no time-out timer is set for a segment that carries only an acknowledgment, which means that no such segment is resent.

- The value of RTO is dynamic in TCP and is updated based on the round-trip time (RTT) of segments. An RTI is the time needed for a segment to reach a destination and for an acknowledgment to be received.

2. Retransmission After Three Duplicate ACK Segments

- The previous rule about retransmission of a segment is sufficient if the value of RTO is not very large. Sometimes, however, one segment is lost and the receiver receives so many out-of-order segments that they can not be saved (limited buffer size).

- To alleviate this situation, most implementations today follow the three duplicate-ACKs rule and retransmit the missing segment immediately. This feature is referred to as fast retransmission, which we will see in an example shortly.

3. Fast Retransmission

- When the receiver receives the fourth, fifth, and sixth segments, it triggers an acknowledgment. The sender receives four acknowledgments with the same value (three duplicates).

- Although the timer for segment has not matured yet, the fast transmission requires that segment 3, the segment that is expected by all these acknowledgments, be resent immediately. Note that only one segment is retransmitted although four segments are not acknowledged. When the sender receives the retransmitted ACK, it knows that the four

Segments are safe and sound because acknowledgment is cumulative.

5.12.3 TCP Tahoe

- Triple duplicate ACKS are treated the same as a timeout. Tahoe will perform "fast retransmit", set the slow start threshold to half the current congestion window, reduce congestion window to 1 MSS, and reset to slow-start state.

5.12.4 Fast Recovery

- There is a variation to the slow-start algorithm known as Fast Recovery, which uses fast retransmit followed by Congestion Avoidance.

- In the Fast Recovery algorithm, during Congestion Avoidance mode, when packets are not received (detected through three duplicate ACKs), the congestion window size is reduced to the slow-start threshold, rather than the smaller initial value.

EXERCISE

1. Explain TCP with its header format.
2. Explain the multiplexing technique used in transport layer.
3. Explain how TCP provide flow control.
4. What is silly window syndrome ?
5. Explain the three-way handshake algorithm for TCP connection establishment.
6. How will you differentiate a stream socket from a raw socket ? How data transmissions happen in a datagram mode?
7. List and discuss the performance issues of the transport layer.
8. Why does UDP exist ? Would it not have been enough to just let user processes send raw IP packet ?
9. In a network assume that maximum TPDU size of 128 bytes, a maximum TPDU lifetime of 30 seconds, and an 8-bit sequence number. Calculate the maximum data rate per connection.
10. Describe any two flow control mechanisms. How timer plays the role in the flow control ?
11. What is a socket ? Explain various socket primitives used in client-server interaction.
12. Explain with a suitable diagram, the parameters involved in process to process communication. Give the different types of parts with their ranges.
13. Explain how TCP provide flow control mechanism.
14. List and discuss the performance issues of the transport layer.
15. What do you mean by flow control ? What are the different methods to achieve it ?
16. Give two functions of four different timers used in TCP.
17. Explain how TCP provide a flow control mechanism.
18. What is silly-window syndrome ? Explain at least two methods to overcome it.
19. List and discuss performance issues of transport layer.
20. List and discuss performance issues of the transport layer.
21. Explain UDP Header? The following is a dump of UDP header in hexadecimal format. 06 32 00 0D 00 IC E2 17
    (i) What is source port number ?
    (ii) What is destination port number ?
    (iii) What is the total length of the user datagram ?
    (iv) What is the length of the data ?
    (v) Is the packet directed from a client to a server or vice versa ?
    (vi) What is the client process ?
22. Draw and explain UDP header in brief.
23. Differentiate between TCP and UDP protocol.
24. Why TCP need four different timers? Explain function of each.

◈ ◈ ◈

APPLICATION LAYER

6.1 INTRODUCTION (Feb. 16)

- In the Open System Interconnection (OSI) communication model, the application layer provides services for an application program to ensure that effective communication with another application program in a network is possible.

- In this unit, first of all we will learn the Domain Name System (DNS) topic which makes sure that the other party is identified and can be reached.

- Later on we will cover Hyper Text Transfer Protocol (HTTP) which is responsible for transferring the data in the form of text to the other devices on the network.

- Later we will study different protocols in the context of Email such as SMTP, MIME, POP3 web mail. After that we will study Dynamic Host Configuration Protocol and Simple Network Management Protocol.

6.2 DOMAIN NAME SYSTEM (DNS)

- DNS provides a protocol that allows client and servers to communicate with each other.

- To identify a computer on the internet, IP address is used by the TCP/IP protocol suit.

- However, it is difficult to remember so many IP addresses, for that people used to prefer names (such as www.puneatoz.net) instead of the numeric IP addresses.

- Therefore we need a system that can map a name to an IP address or an IP address to a name.

- Domain Name System provides this facility.

6.2.1 Working of DNS

- To map a URL name with IP address, the application program calls a library procedure called as 'Resolver'.

- The URL name is passed to the Resolver as an argument.

- This Resolver sends a UDP packet to local DNS server which goes through its database, finds out the corresponding IP address associated with the URL name, and returns this IP address to the application program.

- After getting the actual IP address, the application program establish the connection with the destination.

6.2.2 Domain Name Space

- Domain name space is hierarchical, which is similar to UNIX file system.

- In this design, names are defined in an inverted tree structure with the root at the top.

- Domain names are case insensitive (i.e. com and COM are same thing).

- Every node has a label of maximum 63 characters long.

- The root label is null string.

- A domain name that ends with a period is called as absolute domain name or fully qualified domain name.

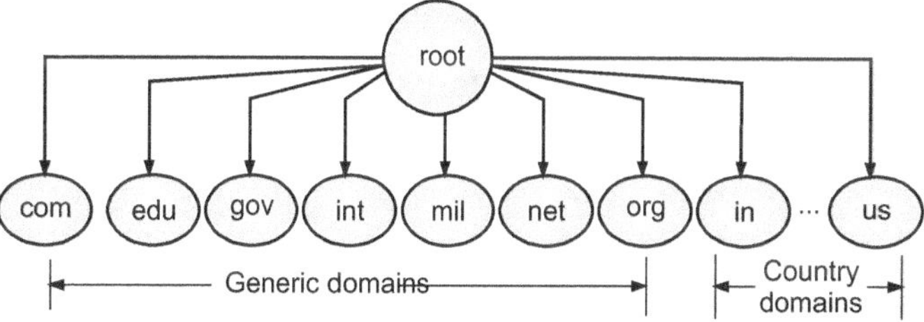

Fig. 6.1 : Hierarchical organization of DNS

- Top level domains are divided into two areas; i.e. generic and countries as shown in Fig. 6.1.

- Three character domains are called as generic domains

- Two character domains are called as country domains (e.g. in for India, us for United States, nz for New Zealand).

- The generic domains .GOV and .MIL are restricted to the United States only.

- Each domain is named by the path upward from it to the root. The components are separated by periods. This is called as hierarchical routing.

- As we know that root label is null, this means that a full domain name always ends in null label; this means the last character is dot because the null string is nothing.

- Following Fig. 6.2 shows some domain names

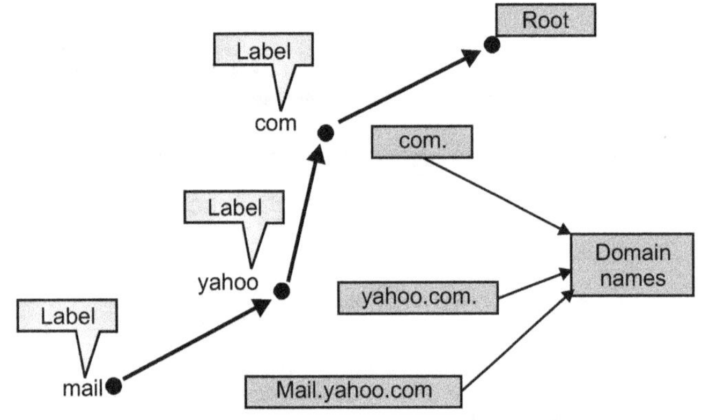

Fig. 6.2 : Domain names and labels

6.2.3 Fully Qualified Domain Names

- If a label is terminated by a null string, it is called as a Fully Qualified Domain Name (FQDN).
- A fully qualified domain name contains full name of the host.e.g. mail.yahoo.com.
- It is also called as absolute domain name. It always ends with a period (a dot).

6.2.4 Partially Qualified Domain Names

- If a label is not terminated by a null string, it is called as a partially qualified domain name (PQDN).
- PQDN starts from a node but never reach to the root node.

  e.g. mail.yahoo.com

- It is also called as relative domain name. It never ends with a period (a dot).

6.2.5 Domain

- A domain is a subtree of domain name space.
- The name of the domain is the domain name of the node at the top of the subtree.
- Following Fig. 6.3 shows .com and .edu domains.

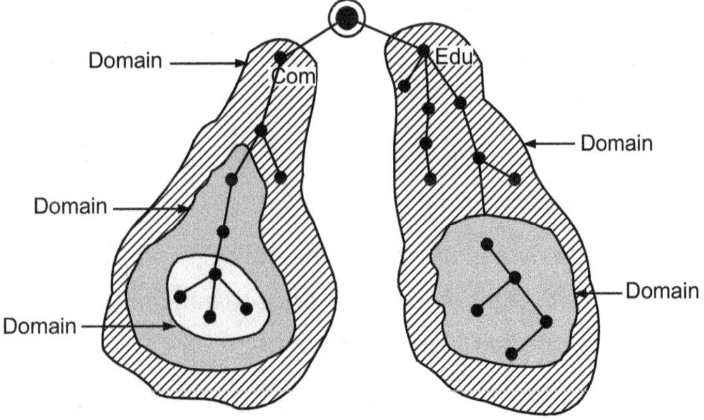

Fig. 6.3 : Domains

- Note that a domain may itself be divided into domains or subdomains as they are sometimes called.

6.2.6 Domain Name Servers

Need

- The information contained in domain name space must be stored.
- As the information is tremendous (nowadays the websites becomes countless), so it is not possible to store such a huge amount of information only on one computer.
- Even if we put the information on single computer, still it is very inefficient because for a single computer, responding to requests from all over the world will not be possible.
- Also, from the point of reliability, it can cause disaster because of machine failure. (The entire data on that machine will be inaccessible).
- So, the solution to these problems is to distribute the information among many computers called DNS servers.
- Firstly, the whole space is divided into many first level domains. The root server stands alone and can create as many first level domains as required.
- The first level domains are further divided into smaller domains called as subdomains.
- Each server can be responsible for either a large or small domain.
- In short, we have hierarchy of servers similar to the hierarchy of names.
- The whole DNS namespace is divided into non overlapping zones. We can define a zone as a contiguous part of the entire tree.
- The zones are explained below.

6.2.7 Zone

- Since the complete domain name hierarchy can not be stored on a single server, it is divided among many servers.
- Server is responsible for a zone.
- We can define a zone as a contiguous part of the entire tree.
- If a server accepts responsibility of a domain and does not divide the domain into smaller domains, the domain and the zone refer to the same thing.
- A server makes a database called a "zone file" and keeps all the information for every node under that domain.
- However, if the server divides its domain into subdomains and pass on a part of its authority to other

servers then domain and zone will be different from each other. This is shown in the following Fig. 6.4.

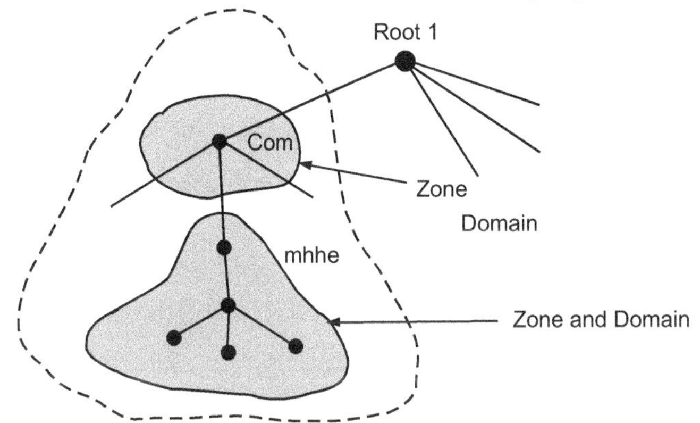

Fig. 6.4 : Zones and domains

- The information about the nodes in the subdomains is stored in the servers at the lower levels, with the original server keeping some sort of reference to these lower-level servers.

6.2.8 Root Server

- Root server is a server whose zone consists of the whole tree.

- Root server does not store any information about domains but pass on its authority to other servers.

- There are several root servers distributed all over the world.

6.2.9 Primary and Secondary Servers

- There are two types of DNS servers

  (i) Primary Server

  (ii) Secondary Server

- Primary server stores a file about the zone for which it is an authority.

- Primary server is responsible for creating, maintaining and updating a zone file. It stores the zone file on a local disk.

- A secondary server is a server that transfers the complete information about a zone from another server (primary and secondary) and stores the file on its local disk.

- The secondary server neither create nor update the zone files, updating is done only by the primary server, which further sends the updated zone file to the secondary.

- A primary server loads all information from the disk file; the secondary server loads all information from the primary server.

- **Zone Transfer :** When the secondary, downloads the information from the primary server, it is called as zone transfer.

6.2.10 Resolution

The mechanism of mapping a name to an address or an address to a name is called as name-address resolution.

Resolver

- DNS is a client-server application.

- A resolver program is called by the client when a host wants to map an address to a name or a name to address.

- For this purpose, the nearest DNS server is accessed by the resolver.

- If this server has information, it satisfies the resolver; otherwise, it refers the resolver to other servers or asks other servers to provide the information.

- After receiving the correct mapping, resolver delivers the result to the process that requested it.

Mapping Names to Addresses

- Many times, the resolver gives a domain name to the DNS server and asks for the corresponding address.

- Here, server checks the generic or country domains to find mapping.

- If the domain name is from the generic domain area, the resolver receives the domain name such as "mail.viit.edu.".

- The query is sent by the resolver to the local DNS for resolution.

- If the local DNS server seems unable to solve the query, it either refers the resolver to other servers or it itself contact to other DNS servers.

Mapping Addresses to Names

- In this case, client sends an IP address to a server to be mapped to a domain name. This type of query is called as PTR query.

- To solve such queries, DNS uses the inverse domain.

- For example, if the resolver receives an IP address 129.132.12.37, the resolver first inverts the address and then adds the two labels before sending.

- The domain name sent is "129.132.12.37.in-addr.arpa", which is received by the local DNS and resolved.

Recursive Resolution

- The client can ask for a recursive answer from a name server.

- This means that the resolver expects the server to supply the final answer.

- If the server is the authority for the domain name, it checks its database and responds.

- If the server is not the authority, it sends the request to another server and waits for the response.

- If the parent is the authority, it responds; otherwise it sends the query to yet another server.

- When the query is finally resolved, the response travels back until it finally reaches the requesting client.

- Refer following Fig. 6.5.

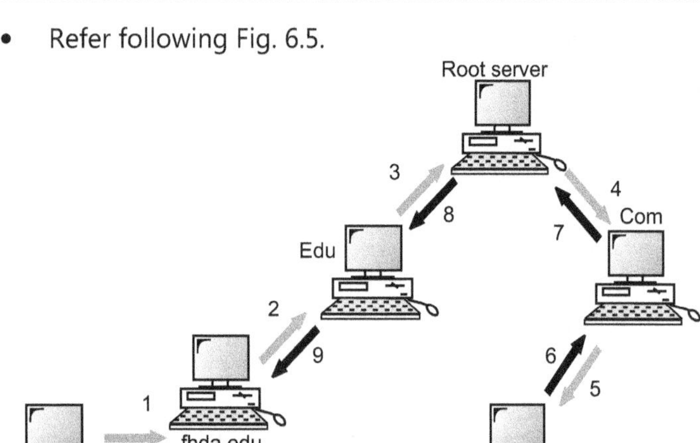

Fig. 6.5 : Recursive resolution

Iterative Resolution

- If the client does not ask for a recursive answer, the mapping can be done iteratively.

- If the server is an authority for the name, it sends the answer. If it is not, it returns the IP address of the server that it thinks can resolve the query.

- The client has to repeat the query to the second server. If the new server can resolve the problem, it answers the query with the IP address, otherwise it returns the IP address of the new server to the client.

- Now the client once again repeats the query to the third server. The process becomes iterative as the client is repeating the same query to multiple servers.

- Refer the following Fig. 6.6 for the same.

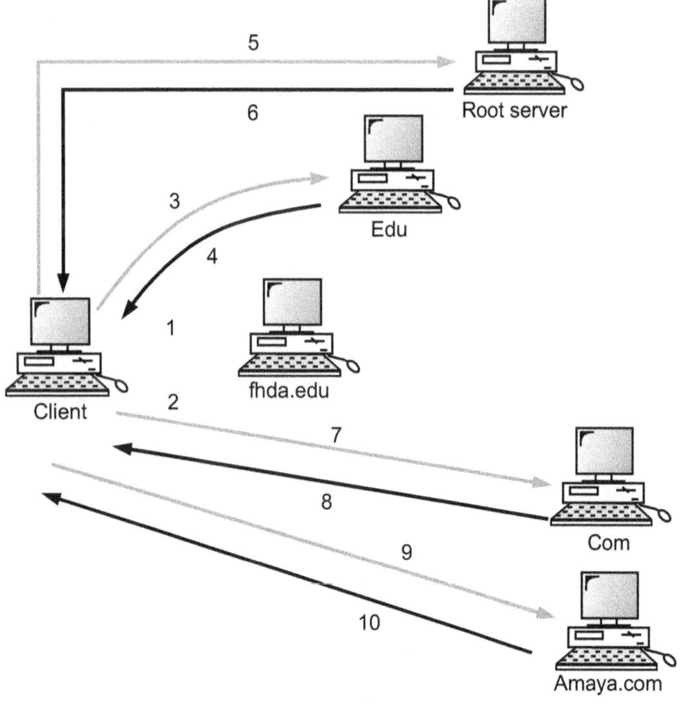

Fig. 6.6 : Iterative resolution

Caching :

- After getting the query, server does searching in the database.

- Searching mechanism requires some time. Reduction of this search time can increase the server efficiency.

- For reducing the search time, DNS uses a mechanism known as caching.

- When a server asks for mapping from another server and receives the response, it stores this information in its cache memory before sending it to client.

- If the same or another client asks for the same mapping, it can check the cache memory and resolve the problem, so it reduces the time that is required for searching in the database.

- Caching speeds up the resolution, but it can also be problematic. If the server caches a mapping for a long time, it may send outdated mapping to the client.

- There are two solutions on this problem. First, the authority server always adds information to the mapping called time-to-live (TTL).

- It defines the time in seconds that the receiving server can cache the information. After that time, the mapping is invalid and any query must be sent again to the authority server.

- In the second solution, DNS server requires that each server keep a TTL counter for each mapping it caches. The cache memory must be searched periodically and those mapping with an expired TTL must be washed out.

6.2.11 DNS Messages

- DNS has two types of messages : query and response. See the following Fig. 6.7.

  1. Query message

  2. Response message

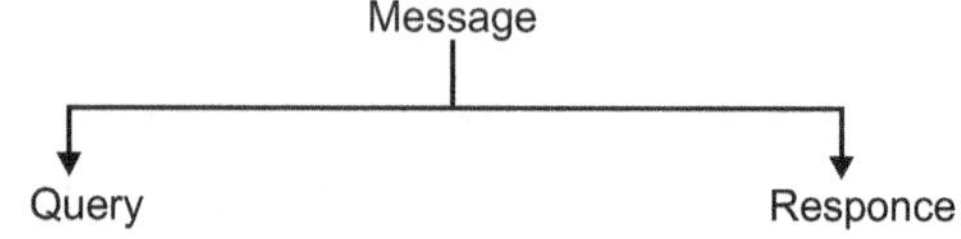

Fig. 6.7 : DNS messages

- Both types have same format. The query message consists of a header question record; the response message consists of a header, question records, authoritative records and additional records. Refer the following Fig. 6.8.

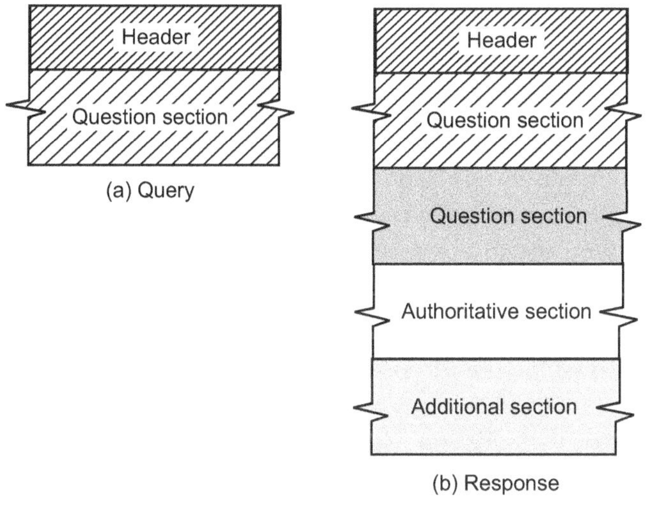

Fig. 6.8 : Query and response messages

Header :

- Both the query and response messages have the same header format with some fields set to zero for the query message.

- The header is of 12 bytes. The format of the header is shown in the following Fig. 6.9.

Identification	Flags
Number of question records	Number of answer records (all 0s in query message)
No. of authoritative records (all 0s in query message)	No. of additional records (all 0s in query message)

Fig. 6.9 : Header format

The header fields are as follows :

- **Identification :** This is 16 bit field used by the client to match the response with the query.

- **Flags :** This is 16 bit field, which is a collection of subfields that defines the type of the message, type of the answers requested, type of the desired resolution and so on.

- **Number of Question Records :** It contains the number of queries in the question section of the message.

- **Number of Answer Records :** It contains the number of answer records in the answer section of the response message. Its value is zero in the query message.

- **Number of Authoritative Records :** It contains the number of authoritative records in the authoritative section of the response message. Its value is zero in the query message.

- **Number of Additional Records :** It contains the number of additional records in the additional section of the response message. Its value is zero in the query message.

- **Question Section :** This section consists of one or more question records. It is present on both query and response messages.

- **Answer Section :** This is a section consisting of one or more resource records. It is present only in response message.

- **Authoritative Section :** This section consists of one or more resource records. It is present only on response message. This section gives information about one or more authoritative servers for the query.

- **Additional Information Section :** This is a section consisting of one or more resource records. It is present only on response message. This section provides additional information that may help the resolver.

6.3 DDNS

- In the initial stages of the Internet (ARPANET) addressing of hosts on the network was achieved by static translation tables that mapped hostnames to IP addresses. The tables were maintained manually in form of the host file. The Domain Name System brought a method of distributing the same address information automatically online through recursive queries to remote databases configured for each network, or domain. Even this DNS facility still used static lookup tables at each participating node.

- IP addresses, once assigned to a particular host, rarely changed and the mechanism was initially sufficient. However, the rapid growth of the Internet and the proliferation of personal computers in the workplace and in homes created the substantial burden for administrators of keeping track of assigned IP addresses and managing their address space.

- The Dynamic Host Configuration Protocol (DHCP) allowed enterprises and Internet service providers (ISPs) to assign addresses to computers automatically as they powered up. In addition, this helped conserve the address space available, since not all devices might be actively used at all times and addresses could be assigned as needed. This feature required that DNS servers be kept current automatically as well.

- The first implementations of dynamic DNS fulfilled this purpose: Host computers gained the feature to notify their respective DNS server of the address they had received from a DHCP server or through self-configuration. This protocol-based DNS update method was documented and standardized in IETF publication RFC 2136 in 1997 and has become a standard part of the DNS protocol (see also nsupdate program).

- The explosive growth and proliferation of the Internet into homes brought a growing shortage of available IP addresses. DHCP became an important tool for ISPs as well to manage their address spaces for connecting home and small-business end-users with a single IP address each by implementing network address translation (NAT) at the customer-premises router.

- The private network behind these routers uses address space set aside for these purposes (RFC 1918), masqueraded by the NAT device. This, however, broke the end-to-end principle of Internet architecture and methods were required to allow private networks, with frequently changing external IP addresses, to discover their public address and insert it into the Domain Name System in order to participate in Internet communications properly. Today, numerous providers, called Dynamic DNSservice providers, offer such technology and services on the Internet.

6.3.1 Domain Name System

- DNS is based on a distributed database that takes some time to update globally. When DNS was first introduced, the database was small and could be easily maintained by hand. As the system grew this task became difficult for any one site to handle, and a new management structure was introduced to spread out the updates among many domain name registrars. Registrars today offer end-user updating to their account information, typically using a web-based form, and the registrar then pushes out update information to other DNS servers.

- Due to the distributed nature of the DNS systems and its registrars, updates to the global DNS system may take hours to distribute. Thus DNS is only suitable for services that do not change their IP address very often, as is the case for most large services like Wikipedia. Smaller services, however, are generally much more likely to move from host to host over shorter periods of time. Servers being run on certain types of Internet service provider, cable modems in particular, are likely to change their IP address over very short periods of time, on the order of days or hours.

DDNS

- Dynamic DNS is a system that addresses the problem of rapid updates. The term is used in two ways, which, while technically similar, have very different purposes and user populations. The first is "standards-based DNS updates", which uses an extension of the DNS protocol to ask for an update; this is often used for company laptops to register their address. The second is usually a web-based protocol, normally a single HTTP fetch with username and password which then updates some DNS records (by some unspecified method); this is commonly used for a domestic computer to register itself by a publicly-known name in order to be found by a wider group, for example as a games server or webcam.

- End users of Internet access receive an allocation of IP addresses, often only a single address, by their Internet service provider. The assigned addresses may either be fixed (i.e. static), or may change from time to time, a situation called dynamic. Dynamic addresses are generally given only to residential customers and small businesses, as most enterprises specifically require static addresses.

- Dynamic IP addresses present a problem if the customer wants to provide a service to other users on the Internet, such as a web service. As the IP address may change frequently, corresponding domain names must be quickly re-mapped in the DNS, to maintain accessibility using a well-known URL.

- Many providers offer commercial or free Dynamic DNS service for this scenario. The automatic reconfiguration is generally implemented in the user's router or computer, which runs software to update the DDNS service. The communication between the user's equipment and the provider is not standardized, although a few standard web-based methods of updating have emerged over time.

6.3.2 Standards-Based Dynamic DNS Update

- The standardized method of dynamically updating domain name server records is prescribed by RFC 2136, commonly known as dynamic DNS update. The method described by RFC 2136 is a network protocol for use with managed DNS servers, and it includes a security mechanism. RFC 2136 supports all DNS record types, but often it is used only as an extension of the DHCP system, and in which the authorized DHCP servers register the client records in the DNS. This form of support for RFC 2136 is provided by a plethora of client and server software, including those that are components of most current operating systems. Support for RFC 2136 is also an integral part of many directory services, including LDAP and Windows' Active Directory domains.

Function

- In Microsoft Windows networks, dynamic DNS is an integral part of Active Directory, because domain controllers register their network service types in DNS so that other computers in the domain (or forest) can access them.

- Increasing efforts to secure Internet communications today involve encryption of all dynamic updates via the public Internet, as these public dynamic DNS services have been abused increasingly to design security breaches. Standards-based methods within the DNSSEC protocol suite, such as TSIG, have been developed to secure DNS updates, but are not widely in use. Microsoft developed alternative technology (GSS-TSIG) based on Kerberos authentication.

- Some free DNS server software systems, such as dnsmasq, support a dynamic update procedure that directly involves a built-in DHCP server. This server automatically updates or adds the DNS records as it assigns addresses, relieving the administrator of the task of specifically configuring dynamic updates.

6.3.3 DDNS for Internet Access Devices

- Dynamic DNS providers offer a software client program that automates the discovery and registration of the client system's public IP addresses. The client program is executed on a computer or device in the private network.

- It connects to the DDNS provider's systems with a unique login name; the provider uses the name to link the discovered public IP address of the home network with a hostname in the domain name system. Depending on the provider, the hostname is registered within a domain owned by the provider, or within the customer's own domain name. These services can function by a number of mechanisms. Often they use an HTTP service request since even restrictive environments usually allow HTTP service. The provider might use RFC 2136 to update the DNS servers.

- Many home networking modem/routers include client applications in their firmware, compatible with a variety of DDNS providers.

6.3.4 DDNS for Security Appliance Manufacturers

- Dynamic DNS is an expected feature or even requirement for IP-based security appliances like DVRs and IP cameras.[citation needed] Many options are available for today's manufacturer, and these include the use of existing DDNS services or the use of custom services hosted by the manufacturer themselves.

- In almost all cases, a simple HTTP based update API is used as it allows for easy integration of a DDNS client into a device's firmware. There are several pre-made tools that can help to ease the burden of server and client development, like MintDNS, cURL and Inadyn.

- Most web-based DDNS services use a standard user name and password security schema. This requires that a user first create an account at the DDNS server website and then configure their device to send updates to the DDNS server whenever an IP address change is detected.

- Some device manufacturers go a step further by only allowing their DDNS Service to be used by the devices they manufacture, and also eliminate the need for user names and passwords altogether. Generally this is accomplished by encrypting the device's MAC address using an cryptographic algorithm kept secret on both the DDNS server and within the device's firmware. The resulting decryption or decryption failure is used to secure or deny updates. Resources for the development of custom DDNS services are generally limited and involve a full software development cycle to design and field a secure and robust DDNS server.

6.4 TELNET

6.4.1 Introduction to TELNET

- Where the TCP protocol makes it possible to connect the remote computers, the TELNET protocol makes it possible to use them. The TELNET protocol offers a user the possibility to connect and log on to any other hosts in the network from user's own computer by offering a remote log on capability.

- Historically, TELNET was the first TCP/IP application and still is widely used as a terminal emulator. Today, while the applications are more and more equipped with the graphical user interface, the terminal-based applications are becoming minority among the applications.

- The TELNET has found its future as a toolkit lying below several client/server software. For example : FTP, SMTP, SNMP, NNTP and HTTP are more or less dependent on the TELNET protocol.

6.4.2 TELNET Model

- For the connections, TELNET uses the TCP protocol. The TELNET service is offered in the host machine's TCP port 23. The user at the terminal interacts with the local telnet client.

- The TELNET client acts as a terminal accepting any keystrokes from the keyboard, interpreting them and displaying the output on the screen.

- The client on the computer makes the TCP connection to the host machine's port 23 where the TELNET server answers. The TELNET server interacts with applications in the host machine and assists in the terminal emulation.

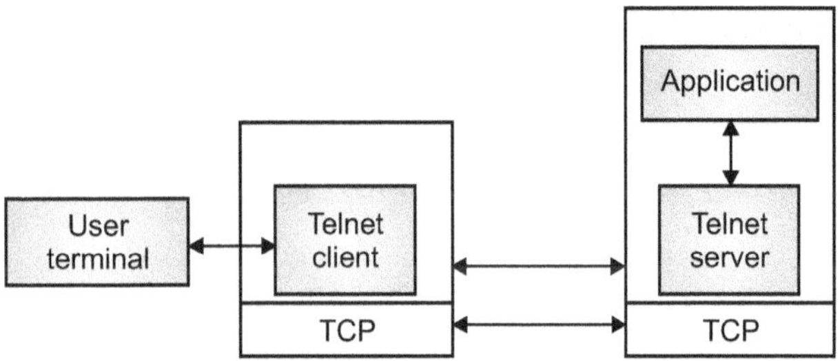

Fig. 6.10 : TELNET Protocol Model

- As the connection is setup, the both ends of the TELNET connection are assumed to be originated and terminated at the Network Virtual Terminal (NVT). The NVT is a network wide terminal which is host independent so that both the server and the client in the connection may not need to keep any information about each other's terminal's characteristics as both sees each other as a NVT terminal.

- As there are several types of terminals, which may be able to provide additional services from those provided by the NVT, the TELNET protocol contains a negotiation method for the user and the server to negotiate changes to the terminal provided in the NVT. Typically, the client and the server stays in the NVT just as long as it takes to negotiate some terminal type to be emulated.

6.4.3 Options

The TELNET has a set of options and these options can be negotiated through a simple protocol inside the TELNET. The negotiation protocol contains commands DO, WILL, WON'T and DON'T. Following examples present the accepted command sequences

- DO (sender wants receiver to enable the option)
- WILL (receiver acknowledges)
- DO (sender wants receiver to enable the option)
- WON'T (receiver will not acknowledge the request)

- WILL (sender wants to enable the option)
- DO (receiver gives permission)
- WILL (sender wants to enable the option)
- DON'T (receiver does not give permission to do so)
- WON'T (sender wants to disable option)
- DON'T (receiver has to answer OK)
- DON'T (sender wants receiver to disable option)
- WON'T (receiver must say OK)

- Mostly, the options are used in the beginning of the connection to setup a desired set of options for the TELNET. In some cases, the options are changed during the session. The key option to be negotiated is the terminal type.

- The symmetry in the negotiation protocol indicates that some loops are possible. Without any further restrictions, the following sequence could take place:
  - ➢ DO TERMINAL TYPE VT100
  - ➢ WILL TERMINAL TYPE VT100
  - ➢ DO TERMINAL TYPE VT100

- For this situation and similar situations, the protocol introduces set of rules
  - ➢ Parties may only request a change in the option (this rule overcomes previous problem).
  - ➢ If a party receives a request enter some mode that it is already in, the request should not be acknowledged.
  - ➢ If the option affects the way, how the data is processed, the command must be inserted in the data stream exactly in the place where it is desired to take effect.
  - ➢ The rejected request should not be repeated until something changes in the operating environment. For example, the process runs other program or other user commands is executed.

- The options are set through TELNET commands. To indicate that the next byte is a command byte, the IAC (interpret as command) byte (0xFF) is sent. The data byte 0xFF is sent as two consecutive 0xFF bytes. The option negotiation requires three bytes: IAC, request (WILL, WON'T, DO, DON'T) and the option ID byte to be enabled or disabled. The negotiations are either symmetrical or non-symmetrical.

- With a symmetrical option both sides may start the negotiation sequence. A non-symmetrical option is always requested by the other part. An example of a non-symmetrical option can be the line-mode option, which can only be requested by the client. The negotiation may require sub option negotiations. These negotiations take place when the option does not have only two modes: enable and disable. An example for such sub negotiation is a terminal type negotiation. The Terminal type option is first enabled with a normal 3 byte negotiation

IAC, WILL, 24 (24 = terminal type)

- The server responds hopefully

IAC, DO, 24

- The server then asks the terminal type of the client

IAC, SB, 24, 1, IAC, SE

(SB = suboption, 24 = suboption terminal type, 1 = sent your terminal type, SE = suboption end)

Client responds

IAC, SB, 24, 0, 'V', 'T', '1', '0', '0', IAC, SE

(0 = my terminal type, string VT100)

6.4.4 TELNET Commands

- Typical for the terminals directly connected to a computer are that the keystrokes by the user are immediately interpreted by the computer's operating system. For the purpose certain keystroke combinations were invented.

- For example, by pressing ctrl+'z', the processes were suspended or ctrl+'c' was used for killing current process. TELNET cannot transmit such codes as the codes are commands containing two keystrokes and do not map to the 7-bit ASCII chart used in the NVT.

- This requires that the client has to translate the terminal's control codes to the TELNET commands and transmit the commands to the server host's operating system.

- As explained earlier, the TELNET commands are presented with IAC byte with a followed command and parameters. All the TELNET commands are presented in the Table 6.1. The TELNET command IP can be used for send, what would be ctrl + 'c' in the keyboard and the command EC for what would be the backspace.

- The IP command is not always the right choice for interrupting the process because it acts similarly with the command sent from real terminal, which is interpreted right away while the IP signal may take some time to be transmitted over the connection and through the buffered data. For overcoming this problem, the data mark (DM) command is introduced.

- With the DM command, the client gets server's attention faster and the client tells the server to throw away all the data but the commands.

- The client uses the Synch Signal TCP segment for the purpose. The Synch signal is marked as an urgent data. The server will throw away everything except the commands until DM is reached. The DM marks the spot where the data is no longer discarded and TELNET resumes the normal operating mode.

Table 6.1 : TELNET Commands

Name	Code	Description
EOF	236	End of file
SUSP	237	Suspend process
ABORT	238	Abort process
EOR	239	End of record
SE	240	Suboption end
NOP	241	No operation
DM	242	Data mark
BRK	243	Break
IP	244	Interrupt process
AO	245	Abort output
AYT	246	Are you there
EC	247	Escape character
EL	248	Erase line
GA	249	Go ahead
SB	250	Suboption
WILL	251	Option negotiation
WONT	252	Option negotiation
DO	253	Option negotiation
DONT	254	Option negotiation
IAC	255	Interpret as command

6.4.5 TELNET for Client/Server Applications

- TELNET can be used as a tool set to build the client/server applications. As a basis for the application, TELNET rarely requires the terminal extensions and the negotiations but TELNET operates in the basic NVT mode.

- The NVT is bi-directional half-duplex device, which contains a terminal and a keyboard. Basically, the keyboard is the user keyboard. The keyboard produces client's outgoing data sent over the TELNET connection to the server. The printer is the user's display where the TELNET server sends the characters. The NVT is half duplex.

- This means that in the TELNET protocol, either the client or the server has the control. The control from the client to the server is changed by the CR/LF. The server uses the CR/LF for changing the line, not for returning the control to the client. The client receives the control after receiving the data from the server and accepting the GO AHEAD control code.

6.4.6 Telnet and Security

- Since the eavesdropping and the snooping are easy to implement to any machine connected to a LAN and the fact that the password and the user IDs are sent through the TELNET connection which is unencrypted, if not otherwise required, the TELNET protocol is a security risk.

- Therefore, the TELNET protocol defines an option for the authentication. The actual authentication is exchanged in the authentication subnegotiation.

- TELNET's authentication options support such authentication standards like Kerberos, SPX, RSA, LOKI and SSA.

6.5 ELECTRONIC MAIL

6.5.1 Simple Mail Transfer Protocol   (Feb. 15)

E-mail system is implemented with the help of Message Transfer Agents (MTA).

- There are normally two MTAs in each mailing system : One for sending e-mails and another for receiving e-mails.

- The formal protocol that defines the MTA client and server in the internet is called Simple Mail Transfer Protocol (SMTP).

- Fig. 6.11 shows the range of SMTP protocol.

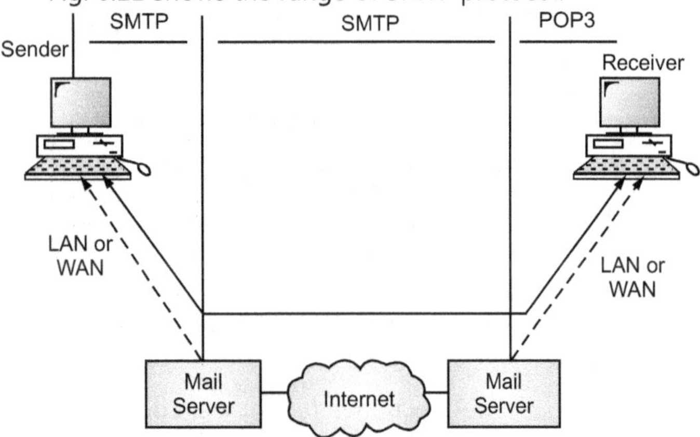

Fig. 6.11 : SMTPRange

- By referring the above diagram, we can say that SMTP is used two times. That is between the sender and sender's mail server and between the sender's mail server and receiver's mail server.

Another protocol is used between the receiver's mail server and receiver.

- SMTP simply defines how commands and responses must be sent back and forth.

- SMTP is a simple ASCII protocol.

- It establishes a TCP connection between a sender and port number 25 of the receiver.

  No checksums are generally required because TCP provides a reliable byte stream.

- After exchanging all the e-mail, the connection is released.

Commands and Responses

- SMTP uses commands and responses to transfer messages between an MTA client and MTA server.

- Each command or reply is terminated by a two character (carriage return and line feed) end-of-line token.

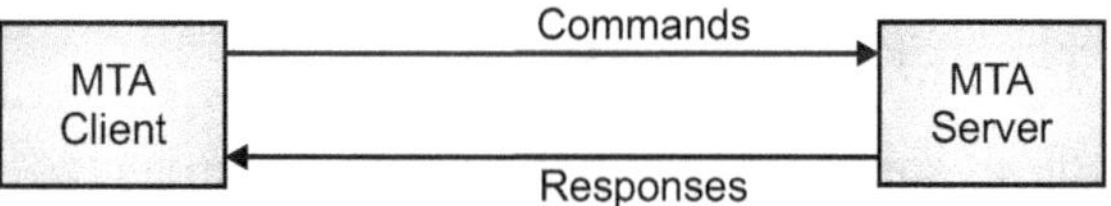

Fig. 6.12 : Commands and Responses

Commands

- Client sends commands to the server. SMTP defines 14 commands.

- Out of that first 5 commands are mandatory; every implementation must support these commands.

- The next three commands are often used and are highly recommended.

- Last six commands are hardly used.

- Following table 6.2 shows FTP commands.

Table 6.2

Keyword	Description
HELO	Used by the client to identify itself. The argument is domain name of the client host. The format is : HELO : mail.viit.ac.in
MAIL FROM	Used by the client to identify the sender of the message. The argument is email address of the sender. The format is MAIL FROM : amol.dhumane@gmail.com
RCPT TO	Used by the client to identify the intended recipient of the message. The argument is the email address of the recipient. The format is: RCPT TO : ashwini.dhumane@yahoo.co.in

Conti...

DATA	This command is used to send the actual message. The lines following DATA command are treated as mail message. The format is: DATA There is an important meeting on this Sunday. So please be present for it. Regards; ABC
QUIT	It terminates the message. The format is : QUIT
RSET	It aborts the current email transaction. The stored information of the sender and receiver is deleted after executing the command. The connection gets reset. The format is : RSET
VRFY	This command is used to verify the address of the recipient. In this sender asks the receiver to confirm that a name identifies a valid recipient. Its format is : VRFY : sai@puneatoz.net
NOOP	By using this command the client checks the status of the recipient. It requires an answer from recipient. Its format is : NOOP
TURN	It reverses the role of sender and receiver.
EXPN	It asks the receiving host to expand the mailing list.
HELP	It sends system specific documentation. The format is : HELP : mail
SEND FROM	This command specifies that the mail is to be delivered to the terminal of the recipient, and not the mailbox. If the recipient is not logged in, the mail is bounced back. The argument is the address of the sender. The format is : SEND FROM : amol.dhumane@gmail.com
SMOL FROM	This command specifies to send the mail to the terminal if possible, otherwise to the mailbox.
SMAL FROM	It sends mail to the terminal and mail box.

SMTP Working

SMTP works in the following three stages :

1. Connection Establishment
2. Message Transfer
3. Connection Termination

These are explained below :

1. Connection Establishment

- Once the TCP connection is made on port no. 25, SMTP server starts the connection phase.

- This phase involves following three steps which are explained in the Fig. 6.13.

- The server tells the client that it is ready to receive mail by using the code 220. If the server is not ready, it sends code 421 which tells that service is not available.

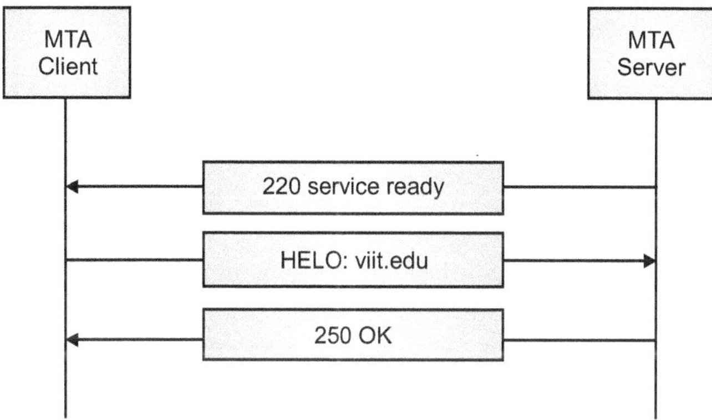

Fig. 6.13 : Connection establishment

- Once the server becomes ready to receive the mails, client sends HELO message to identify itself using the domain name address. This is important step which informs the server of the domain name of the client. Remember that during TCP connection establishment, the sender and receiver know each other through their IP addresses.

- Server responds with code 250 which tells that the request command is completed.

2. Message Transfer

- Once the connection has been established, SMTP server sends messages to SMTP receiver.

- The messages are transferred in three stages

  1. A MAIL command identifies the message originator.
  2. RCPT command identifies the receiver of the message.
  3. DATA command transfers the message text.

3. Connection Closing

- After the message is transferred successfully, the client terminates the connection.

- The connection is terminated in two steps

  1. The client sends the quit command.
  2. The server responds with code 221 or some other appropriate code.

After the connection termination phase, the TCP connection must be closed.

6.5.2 MIME   (Feb. 16)

- Short for Multipurpose Internet Mail Extensions, a specification for formatting non-ASCII messages so that they can be sent over the Internet.

- Many e-mail clients now support MIME, which enables them to send and receive graphics, audio, and video files via the Internet mail system. In addition, MIME supports messages in character sets other than ASCII.

- There are many predefined MIME types, such as GIF graphics files and PostScript files. It is also possible to define your own MIME types.

- In addition to e-mail applications, Web browsers also support various MIME types. This enables the browser to display or output files that are not in HTML format.

MIME-Version

The presence of this header indicates the message is MIME-formatted. The value is typically "1.0" so this header appears as

MIME-Version: 1.0

Content-Type

- This header indicates the media type of the message content, consisting of a *type* and subtype, for example

  Content-Type: text/plain

- Through the use of the *multipart* type, MIME allows mail messages to have parts arranged in a tree structure where the leaf nodes are any non-multipart content type and the non-leaf nodes are any of a variety of multipart types. This mechanism supports:

- Simple text messages using text/plain (the default value for "Content-Type: ")

- Text plus attachments (multipart/mixed with a text/plain part and other non-text parts). A MIME message including an attached file generally indicates the file's original name with the "Content-disposition:" header, so the type of file is indicated both by the MIME content-type and the (usually OS-specific) filename extension

- Reply with original attached (multipart/mixed with a text/plain part and the original message as a message/rfc822 part)

- Alternative content, such as a message sent in both plain text and another format such as HTML (multipart/alternative with the same content in text/plain and text/html forms)

- Image, audio, video and application (for example, image/jpeg, audio/mp3, video/mp4, and application/MS Word and so on)

- Many other message constructs

6.5.3 POP 3   (Feb. 15, Nov. 15)

- POP Stands for "Post Office Protocol." POP3, sometimes referred to as just "POP," is a simple, standardized method of delivering e-mail messages. A POP3 mail server receives e-mails and filters them into the appropriate user folders. When a user connects to the mail server to retrieve his mail, the messages are downloaded from mail server to the user's hard disk.

- When you configure your e-mail client, such as Outlook (Windows) or Mail (Mac OS X), you will need to enter the type of mail server your e-mail account uses. This will typically be either a POP3 or IMAP server. IMAP mail servers are a bit more complex than POP3 servers and allow e-mail messages to be read and stored on the server. Many "webmail" interfaces use IMAP mail servers so that users can manage all their mail online.

- Still, most mail servers use the POP3 mail protocol because it is simple and well-supported. You may have to check with your ISP or whoever manages your mail account to find out what settings to use for configuring your mail program. If your e-mail account is on a POP3 mail server, you will need to enter the correct POP3 server address in your e-mail program settings. Typically, this is something like "mail.servername.com" or "pop.servername.com." Of course, to successfully retrieve your mail, you will have to enter a valid username and password too.

- POP, or Post Office Protocol, is a way of retrieving email information that dates back to a very different Internet than we use today. Computers only had limited, low bandwidth access to remote computers, so engineers created POP in an effort to create a dead simple way to download copies of emails for offline reading, then remove those mails from the remote server.

Mail Server Functionality

- POP3 has become increasingly sophisticated so that some administrators can configure the protocol to "store" email on the server for a certain period of time, which would allow an individual to download it as many times as they wished within that given time frame. However, this method is not practical for the vast majority of email recipients.

- While mail servers can use alternate protocol retrieval programs, such as IMAP, POP3 is extremely common among most mail servers because of its simplicity and high rate of success. Although the newer version of

POP offers more "features," at its basic level, POP3 is preferred because it does the job with a minimum of errors.

Working With Email Applications

- Because POP3 is a basic method of storing and retrieving email, it can work with virtually any email program, as long as the email program is configured to host the protocol. Many popular email programs, including Eudora and Microsoft Outlook, are automatically designed to work with POP3.

- Each POP3 mail server has a different address, which is usually provided to an individual by their web hosting company.

- This address must be entered into the email program in order for the program to connect effectively with the protocol. Generally, most email applications use the 110 port to connect to POP3.

- Those individuals who are configuring their email program to receive POP3 email will also need to input their username and password in order to successfully receive email.

- POP3 and IMAP are two different protocols (methods) used to access email.

- Both of the two, IMAP is the better option - and the recommended option - when you need to check your emails from multiple devices, such as a work laptop, a home computer, or a tablet, smartphone, or other mobile device. Tap into your synced (updated) account from any device with IMAP.

- POP3 downloads email from a server to a single computer, then deletes it from the server. Because your messages get downloaded to a single computer or device and then deleted from the server, it can appear that mail is missing or disappearing from your Inbox if you try to check your mail from a different computer.

- Here are the differences between POP3 and IMAP.

Table 6.3

POP3 - Post Office Protocol	IMAP - Internet Messaging Access Protocol
You can use only one computer to check your email (no other devices)	You can use multiple computers and devices to check your email
Your mails are stored on the computer that you use	Your mails are stored on the server
Sent mail is stored locally on your PC, not on a mail server	Sent mail stays on the server so you can see it from any device.

6.5.4 Webmail

- Webmail is a way of sending and receiving emails from a web browser, instead of from an email client.

- All email travels over the internet, and is stored on servers. Those servers can belong to email providers (like Gmail), internet service providers, or web hosting providers. This server is where email is collected and stored, until you delete it.

- To access your webmail provider's server, you connect to the internet, and log in to a site that connects to your email account. When you use webmail, you are directly accessing the email, from your provider's server. This lets you send and receive mail from anywhere in the world, from any device, as long as you have an internet-connected web browser.

- Your mail always remains on your provider's server, so if you do not have an internet connection, or the provider's servers are down, you will not be able to access your email using webmail. Also, webmail interfaces may not provide as much functionality as a more robust email client.

- Webmail accounts don't require you to install or use an email application software such as Microsoft Outlook on your computer. You can read, send, reply, forward, organize your email into folders and save attachments.

- The webmail service stores of all of your emails on their computers and storage systems, and gives you a web page to use for accessing your email account. You login to your webmail account from the webmail service's webpage using your email address and password.

- You might already have a webmail account, and common webmail providers include Google's Gmail, AOL Mail, Microsoft's Hotmail/Live/Outlook.com, Yahoo! Mail, and Apple's iCloud email. Many of the webmail services that are offered on the Internet are free but can include ads that display on the screen. An upgrade to a paid email plans will remove the ads.

- Your Internet service provider may also provide you with access to webmail when you subscribe to their service so your existing email account may already be webmail capable. You might have to check with them to see if your account can be used as a webmail account, or, you may be able to change your account to a webmail account and keep the same email address.

Advantages of Web Mail

- One of the biggest advantages of webmail is that you can access it from anywhere with almost any device that can connect to the Internet. You don't need to carry your laptop with you just to check your email. Instead, you could check your webmail from your smartphone or even from an Internet café or public library. All your emails are stored on the providers' servers until you choose to delete them.

- For those who do not regularly check their email on the same computer, do not have a computer or who frequently travel, webmail is a good option to access their email on public computers. Most webmail systems so you only need you to go to the provider's website, and then enter your username and password to access your account.

- Often webmail accounts are referred to as IMAP accounts, and while most webmail accounts use IMAP, not all IMAP accounts can be used with a Web browser. IMAP stands for Internet Message Access Protocol, which is a type of email account service. Many webmail accounts use this type of email protocol, which keeps a master copy of all email on the email service's computers. Webmail is the ability to access any email account using a Web browser, regardless of the type of software running the email service.

- While webmail works with a Web browser, it's also possible to setup your email account with a smartphone using an app or on a computer using a local application. Why would you do this you might ask since the point of webmail is to use it on the Web, right?

- Sometimes the apps and applications included with smartphones and computers are designed to make using email easier and faster, and sometimes they add additional capabilities. For example, the Mail app provided on Apple iPhones, iPads, and the Mail app provided on Android phones are all designed to work with touch gestures.

- While you could use your webmail account using the Web browser on your device, you would need to navigate around the screen using mouse clicks and keyboard commands designed for when you use a desktop or laptop. When you use your email account with the built-in apps on a smartphone or tablet, your email experience is optimized for a touch-screen device.

- Whether it's a Microsoft Windows model or an Apple Mac, the built-in email application provides additional features and functions, and the built-in email applications usually work more like other applications on the computer, which can make them easier to use for many people.

- For both smartphones and computers, using the built-in email application on the device with your webmail account lets you store a local copy of your email, so you can read and write emails while you're not connected to the Internet. When you connect to the Internet again, your new email messages will arrive in your Inbox, and your pending messages will be sent to their recipient.

- One other advantage of using webmail with more than one device is that your email and folders stay up to date on all of your devices since a master copy is kept by the webmail service. Each device checks in and sends and receives updates whenever the device is connected to the Internet.

Disadvantages of Web Mail

- If you only use webmail using a Web browser, in order to access your email to view or create an email, you must be connected to the Internet. Your email will not be saved locally on your computer. This means would also be unable to review your email later unless you have access to the Internet. Since Internet access is available with smartphones almost everywhere, this might not be a disadvantage unless you travel to places where cellular reception or WiFi access is not available.

- Webmail users may miss some of the features that may be in the local email application software used on smartphones and computers. Users may be subjected to ads that are included by the webmail service with their emails, in addition to having to deal with the service's spam-filters that may be too efficient and mark some emails from people you know as spam.

6.6 FILE TRANSFER PROTOCOL (FTP)

(May 15, 16, Feb. 16)

- The main purpose of computer networks is data and resource sharing.

  File transfer is a very common operation on the computer network.

- We require two types of protocols for transferring the files on the network :

  1. FTP　　　2. TFTP

- FTP is a standard mechanism provided by TCP/IP for copying a file from one host to another.

- There are some problems associated with file transfer mechanism from one machine to another, which can be as follows :

  1. Two systems may have different ways to represent text and data.

  2. These systems may have different directory structure.

- It is necessary for the FTP to solve all such sort of problems for transferring a file.

- For transferring a file, FTP establishes two connections between the hosts. One connection is used for data transfer, and other for control information (commands and responses).

- Separation of commands and data transfer makes FTP more efficient.

- Data connection uses very complex rules for transmission of data (due to variety of data types transferred); on the other hand, control connection uses very simple rules of communication.

Points to be Remembered

- FTP uses the service of TCP.

- It needs two TCP connections.

- Port 21 is used for control connection.

- Port 20 is used for data connection.

Following Fig. 6.14 explains the basic architecture of FTP. The server has two major components and the client has two major components as given in Fig. 6.14.

- The control connection is made between the control processes at server and client side while the data connection is made between the data transfer processes.

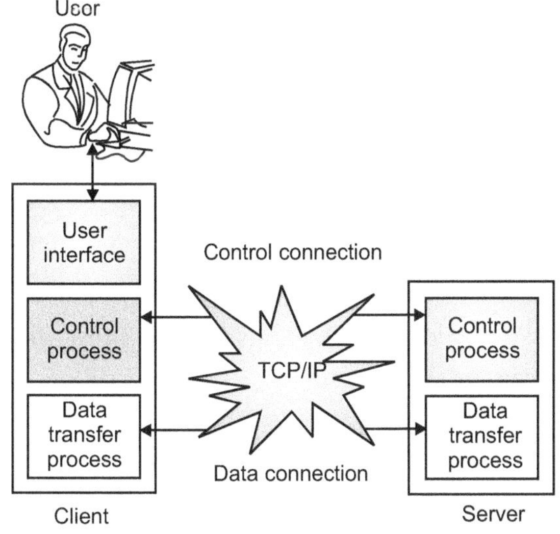

Fig. 6.14 : FTP

- One more thing that is very important about the control and data connection is that, the control connection remains open during the entire FTP interactive session, while the data connection is opened when the user wants to transmit a file and then it is closed after the file transfer. In short, the data connection is opened and closed for each file transferred.

6.6.1 Control Connection

- The process of opening the control connection is shown in Fig. 6.15. There are two steps:

  1. The server issues a passive open on well known port 21 and waits for a client.

  2. The client uses a temporary (ephemeral) port and issues an active open.

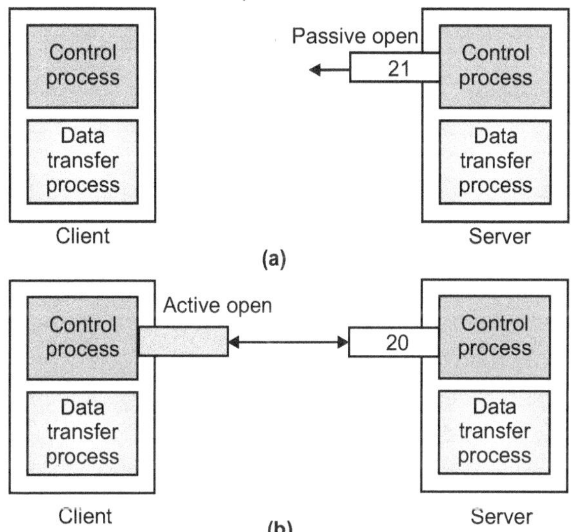

Fig. 6.15 : Opening the control connection

- The control connections remain alive during the entire process.

- The IP uses minimize delay type service because this is an interactive connection between a user and a server.

6.6.2 Data Connections

- Data connection uses port 20 at server side.

- Following steps are required for creating data connections :

  - The client issues a passive open using a temporary (ephemeral) port.

  - The client sends this port number to the server using PORT command.

  - The server receives the port number and issues an active open using port 20 and the received ephemeral port number.

These steps for creating the initial data connection are shown in Fig. 6.16.

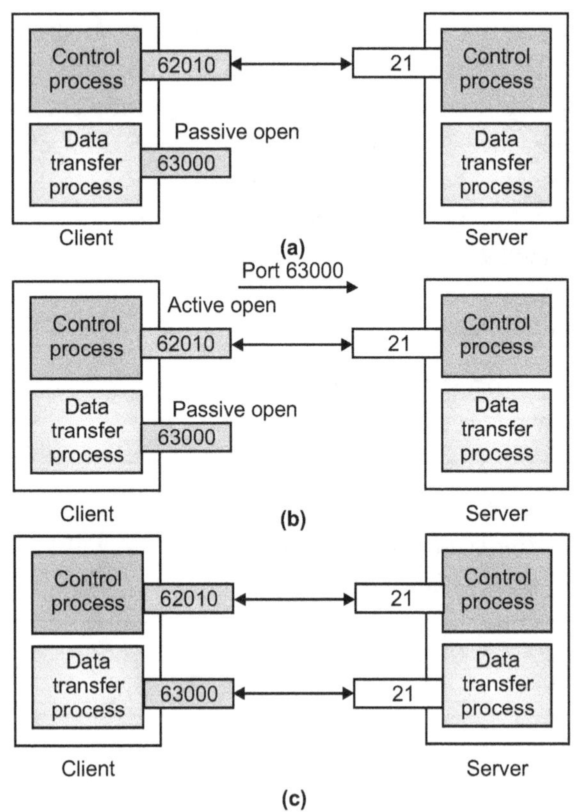

Fig. 6.16 : Creating data connections

6.6.3 Communication Over Control Connection

- Similar to the TELNET and SMTP, FTP communicates across the control connection.
- It uses NVT ASCII character set for communication as shown in the following Fig. 6.17.

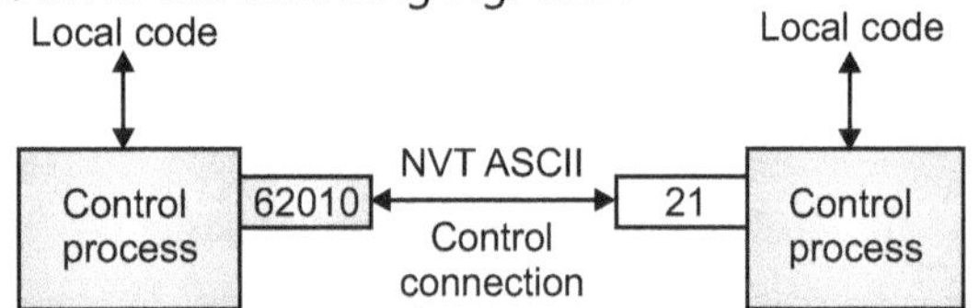

Fig. 6.17 : Using the control connection

- Communication is normally done using commands and responses.
  Only one command or response is sent at a time during the control connection.
- Each command or response is only one short line, so we need not worry about the file format or file structure.
- Every line is terminated with two characters i.e. carriage return and line feed.

6.6.4 Communication Over Data Connection

- Data connection is used to transfer files.
- Before transmission of the files, the client must specify following entities are :
  - ➢ The type of the file to be transferred
  - ➢ The structure of the data
  - ➢ Mode of data transmission.

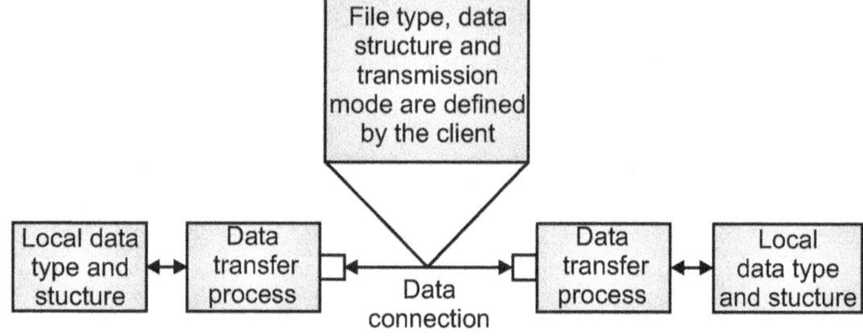

Fig. 6.18 : Using the data connection

- Before sending the file using data connection, we have to send proper control signals through the control connection.
- The problem of heterogeneity is solved by defining the above three communication attributes i.e. file type, data structure and transmission mode. These attributes are explained below. Refer Fig. 6.18.

File Type : FTP can transfer ASCII file, EBCDIC file or Image file.

- **ASCII File :** Text files are by default transferred using ASCII format.
- **EBCDIC File :** If one or both ends of the connection use EBCDIC encoding, the file can be transferred using EBCDIC encoding.
- **Image File :** This is default format of transferring binary files.
- **Data Structure :** FTP can use the following data structures :
- **File Structure :** The file has no structure; it is continuous stream of bytes.
- **Record Structure :** In this case, the text files are divided into records.
- **Page Structure :** The file is divided into pages, in which each page is having a page number and page header. The pages can be accessed randomly or sequentially.

Transmission Modes : FTP uses stream, block and compressed mode for file transfer.

- **Stream Mode :** In this case, data is transferred in continuous stream of bytes.
- **Block Mode :** In this case, data is transferred in the form of blocks.
- **Compressed Mode :** In this method, by using run length encoding data is compressed.

6.6.5 File Transfer

File transfer takes place over the data connection in association with the control connection. File transfer in FTP means one of the following :

1. **Retrieving a File :** A file is copied from server to the client.

2. **Storing a File :** A file can be copied from client to the server.

3. A Server sends a list of directory or file names to the client. FTP treats such a list of directory as a file. The file transfer has been shown in the following Fig. 6.19.

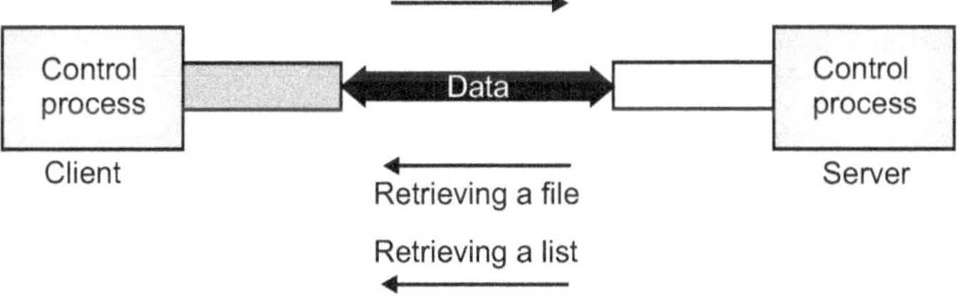

Fig. 6.19 : File transfer

6.6.6 FTP Commands

- FTP creates a control connection between a client control process and server control process.

- During this communication, commands are sent from the client to the server and responses are sent from the server to the client.

- FTP commands are roughly divided into six groups : (1) access, (2) file management, (3) data formatting, (4) port defining, (5) file transferring and (6) miscellaneous commands.

- These commands are given in the following Table 6.4.

Table 6.4

Sr. No.	Command	Description
Access Commands		
1.	USER	User information
2.	PASS	Password
3.	ACCT	Account information
4.	REIN	Reinitialize
5.	QUIT	Log out of the system
6.	ABOR	Abort the previous command
Data Formatting Commands		
1.	TYPE	Defines the file type and if necessary the print format
2.	STRU	Defines the organization of data
3.	MODE	Defines the transmission mode

File Management Commands		
1.	CWD	Change to another directory
2.	CDUP	Change to parent directory
3.	DELE	Delete a File
4.	LIST	List subdirectories or files
5.	MKD	Create (make) new directory
6.	PWD	Display name of current (present working) directory
7.	RMD	Delete (remove) directory
8.	RNFR	Identify a file to be renamed (rename from)
9.	RNTO	Rename the file (rename to)
Port Defining Commands		
1.	PORT	Client chooses a port
2.	PASV	Server chooses a port
File Transfer Commands		
1.	RETR	Retrieve files
2.	STOR	Store files: files are transferred from client to server
3.	APPE	Similar to STOR except if file exists, data must be appended to it
4.	STOU	Similar to STOR except that the file name will be unique in the directory
5.	ALLO	Allocate the storage space for the files at the server
6.	REST	Position a file marker at a specified data point
7.	STAT	Return the status of the files
Miscellaneous Commands		
1.	HELP	Ask information about the server
2.	NOOP	Check if server is alive
3.	SITE	Specify the site specific commands
4.	SYST	Ask about the operating system used by the server

6.6.7 Trivial File Transfer Protocol (TFTP)

- Trivial File Transfer Protocol (TFTP) is a file transfer protocol, with the functionality of a very basic form of File Transfer Protocol (FTP); it was first defined in 1980.

- Due to its simple design, TFTP could be implemented using a very small amount of memory.

- It is therefore useful for booting computers such as routers which did not have any data storage devices.

- It is still used to transfer small amounts of data between hosts on a network when a remote X Window System Terminal or any other thin client boots from a network host or server.
- The initial stages of some network based installation systems (such as Solaris, Symantec Ghost and Windows NT's Remote Installation Services) use TFTP to load a basic kernel that performs the actual installation.
- Trivial File Transfer Protocol (TFTP) is a simple protocol to transfer files.
- It has been implemented on top of the User Datagram Protocol (UDP) using port number 69.
- TFTP is designed to be small and easy to implement, Therefore, lacks most of the features of a regular FTP.
- TFTP only reads and writes files (or mail) from/to a remote server.
- It cannot list directories, and currently has no provisions for user authentication.

1.　TFTP Operation

- In TFTP, any transfer begins with a request to read or write a file, which also serves to request a connection.
- If the server grants the request, the connection is opened and the file is sent in fixed length blocks of 512 bytes.
- Each data packet contains one block of data, and must be acknowledged by an acknowledgement packet before the next packet can be sent.
- A data packet of less than 512 bytes signals termination of a transfer.
- If a packet gets lost in the network, the intended recipient will timeout and may retransmit his last packet (which may be data or an acknowledgement), thus causing the sender of the lost packet to retransmit that lost packet.
- The sender has to keep just one packet on hand for retransmission, since the lock step acknowledgement guarantees that all older packets have been received.
- One sends data and receives acknowledgements; the other sends acknowledgements and receives data.

2.　Modes of TFTP

- Three modes of transfer are currently supported by TFTP :
  - **(i)　Netascii :** That it is 8 bit ascii.
  - **(ii)　Octet :** This replaces the "binary" mode of previous versions of this document.
  - **(iii)　Mail :** Netascii characters sent to a user rather than a file.

3.　Uses

- TFTP is used to read files from, or write files to, a remote server.
- Due to the lack of security, it is dangerous over the open Internet. Thus TFTP is generally only used on private, local networks.

4.　Disadvantages of TFTP

- TFTP cannot list directory contents.
- TFTP has no authentication or encryption mechanisms.
- TFTP allows big data packets which may burst and cause delay in transmission.
- TFTP cannot download files larger than 1 Terabyte.

6.6.8 Comparison between FTP and TFTP

Table 6.5

Sr. No.	Parameter	FTP	TFTP
1.	Operation	Transferring Files	Transferring Files
2.	Authentication	Yes	No
3.	Control and Data	Separated	Not separated
4.	Protocol	TCP	UDP
5.	Ports	21-Control, 20-Data	Port 3214, 69, 4012
6.	Data Transfer	Reliable	Unreliable
7.	Number of connections	Two connections	One connection
8.	Commands	Provides many commands	Provides only 5 commands

6.7 WORLD WIDE WEB (WWW)

- It is an information repository spread all over the world and linked together.
- The WWW is distributed client-server service, in which a client using a browser can access a service provided by the server.
- This service is distributed to many locations using the websites.

Hypertext and Hypermedia :

- In hypertext, set of documents containing information are linked using the concept of pointers.
- An item can be associated with another document by a pointer.
- The reader, who is browsing through the document, can move to other documents by clicking the items that are linked to other documents. (e.g. if user clicks on Pune city, then he/she will get more information on Pune city. If user clicks on Maharashtra then he/she will get some more information about Maharashtra state)

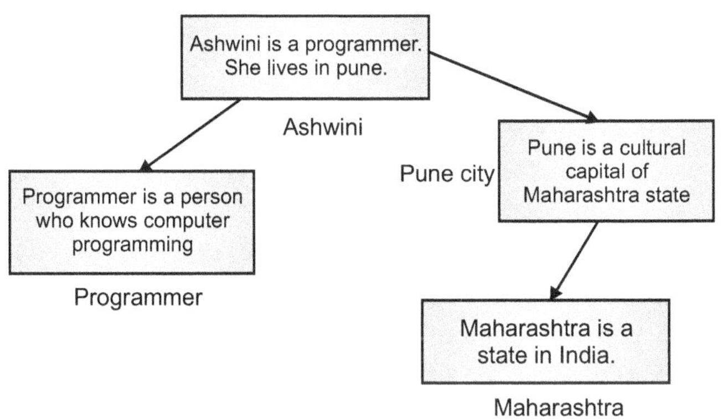

Fig. 6.20 : Hypertext

- Fig. 6.20 shows the concept of hypertext.

- Hypertext documents contain only text.

- Hypermedia documents can contain pictures, graphics and sounds.

- The hypertext and hypermedia available on the web in the form of webpages.

- The main page of the website is called as homepage.

6.7.1 Browser Architecture

- Refer the following Fig. 6.21.

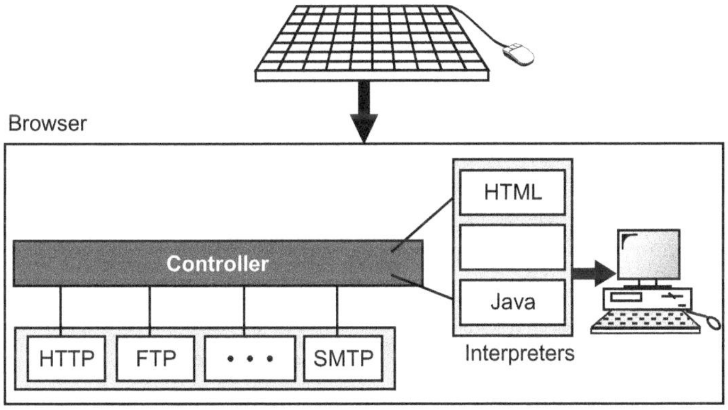

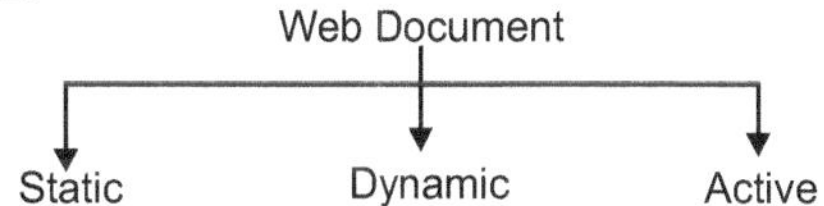

Fig. 6.21 : Browser Architecture

- Different vendors provide different browsers (such as internet explorer, Firefox mozilla, Netscape etc.).

- Although the browsers are different which have been developed by different vendors, they are having nearly same architecture.

- Each browser usually has three parts :

　(i)　A controller

　(ii)　A client program

　(iii)　Interpreters

- A controller receives input from the keyboard or mouse and uses client program to access the document.

- After the document has been accessed, the controller uses one of the interpreter to display the document on the screen.

- The client programs can be one of the protocols, such as HTTP, FTP or SMTP.

- The interpreter can be HTML or JAVA depending on the type of the document.

6.7.2 Categories of Web Documents

- There are three types of web document as shown in Fig. 6.22.

Fig. 6.22 : Categories of Web Documents

1.　Static Documents :

- The contents of the static documents are fixed and that are created and stored in the server.

- The client can get a copy of a document when it does the request to the server.

- User cannot change the content of the document, but it is possible to change the content of the document at the server side. (See Fig. 6.23)

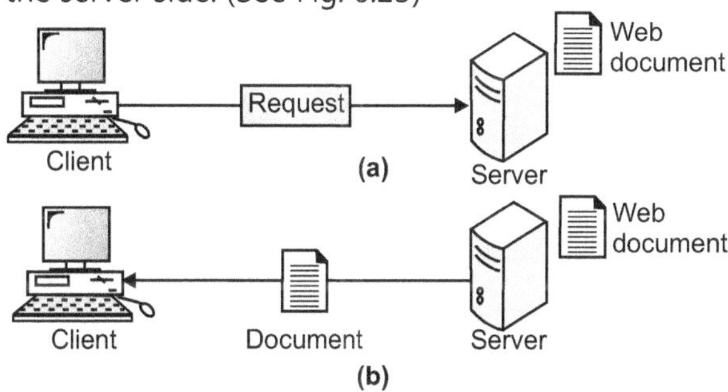

Fig. 6.23 : Static Document

2.　Dynamic Documents

- Dynamic documents are created by the server whenever it gets request from the client's browser.

- This document does not exist in predefined format.

- Whenever, the request comes from the client, the web server runs the application program which creates the dynamic document. The server returns the output of the program as a response to the browser that requested the document.

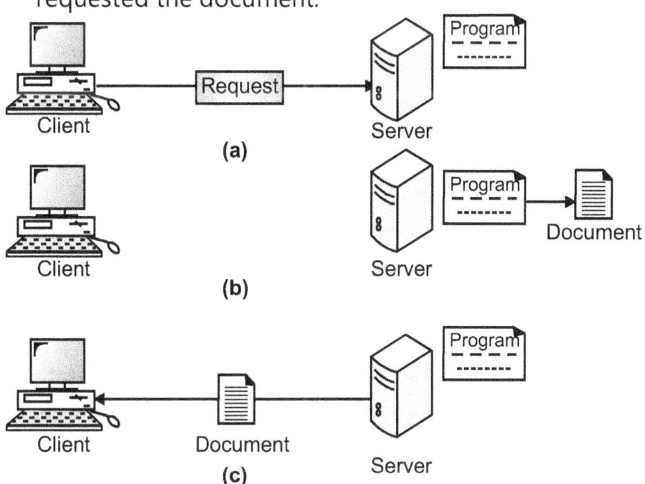

Fig. 6.24 : Dynamic Document

- Because the fresh document is created for each request, the contents of the dynamic document may vary from one request to another.

- A very simple example of dynamic document is getting the time and date from the server. (Time and date changes from moment to moment.)

- Observe Fig. 6.25.

- Server follows the following steps to handle the dynamic documents :

  (a) The server examines the URL to find if it defines a dynamic document.

  (b) If the URL defines a dynamic document, the server executes the program.

  (c) It sends the output of the program to the client.

Common Gateway Interface (CGI)

- CGI technology creates and handles dynamic documents.

- CGI is a set of standards that defines how a dynamic document should be written, how input data should be supplied to the program and how the output result should be used.

- Remember that, CGI is not a new language; instead it allows programmers to use any of several languages such as C, C++, Perl etc.

- The only thing is that CGI defines a set of rules and terms that the programmer should follow.

- The word 'common' in CGI shows that this standard defines some rules which are common to any language.

- The term 'Gateway' here means that a CGI program is a gateway that can be used to access the resources such as databases and graphics packages.

- The term 'interface' means that there is a set of predefined terms, calls, variables etc. which can be used in any CGI program.

3. Active Documents

- Active documents are the programs that run on the client side.

- For example, imagine we want to run a program that creates animated graphics on the screen or interacts with the user.

- When the browser requests an active document, the server sends a copy of the document in the form of **byte code**. The document is then run on the client browser.

- Active document does not create an overhead to the server because it is stored in the form of **binary code** in the server and server does not run it.

- When the client receives the document, it can also store the document in its own storage area; due to that client can run that document again and again without making another request to the server. (e.g. Applets in Java programming can be an example of active document.)

- Server sends an active document to the client in the binary form.

- While sending the document, it is compressed at server side and is decompressed at the client side, which saves both, bandwidth and transmission time.

Creation, Compilation and Execution of Active Document :

- Programmer writes a source code and deploys it on the server side.

- At server side, the program is compiled and binary code is created, which is stored in a file.

- The client requests a copy of binary code, which is transmitted in the compressed form from server to the client.

- The client converts the received program from binary code to executable code using its own browser software.

- The client runs the program and creates the result that can include animation or interaction with the user.

6.8 HYPERTEXT MARKUP LANGUAGE

- By using this language the web pages are created.

- Book publishing agencies normally use the word markup language. (Before a book is typeset and printed, a copy editor reads the manuscript and puts a lot of marks on it. These marks tell the designer how to format the text.)

- For example : in case of HTML, if we want to give a heading in the page, then we must include the beginning and ending tags (marks) in the text as given below :

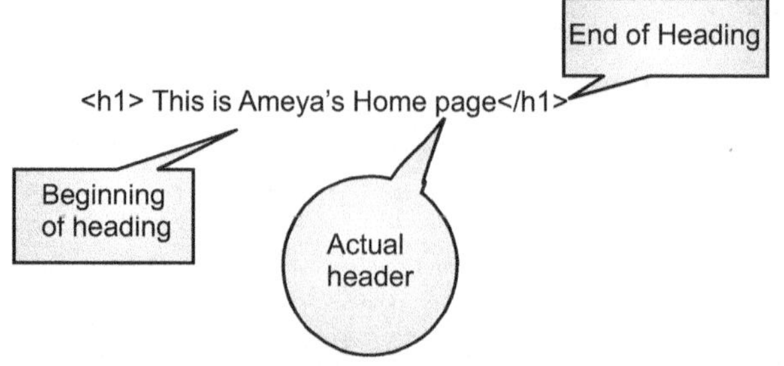

Fig. 6.25 : Heading Tags

- The two tags <h1> and </h1> are instructions for the browser. When the browser sees these two marks, it knows that the text is a heading in the document.
- As HTML allows embedding the formatting instructions inside the file itself, it is called as markup language.
- With the help of these embedded instructions, the browser at client side can read these instructions and format the text accordingly.
- HTML lets the user to use only ASCII characters for both the main text and formatting instructions.
- In this way, every computer can receive the whole document as an ASCII document. The main text is used as a data and formatting instructions can be used by the browser to format the data.

6.8.1 Web Page Structure

Web page is made up of two parts :

1. Head and
2. Body

1. **Head :** The head is the first part of the webpage. The head contains the title of the page and other parameters that the browser will use.

2. **Body :**

- The actual content of the page are in the body, which includes the text and the formatting tags.
- The text is the actual information contained in a page and tags define the appearance of the document.
- A tag can have a list of attributes, each of which is followed by an equal sign and a value associated with the attribute. Fig. 6.26 shows the format of the tags.

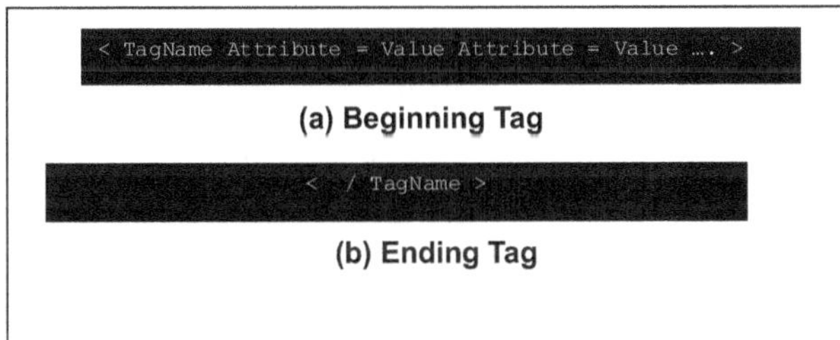

Fig. 6.26 : Beginning and Ending Tags

6.8.2 Some Common HTML Tags

Beginning Tag	Ending Tag	Meaning
Skeletal Tags		
<HTML>	</HTML>	Defines a HTML document
<HEAD>	</HEAD>	Defines Head of the Document
<BODY>	</BODY>	Defines the body of the document
Title and Header Tags		
<TITLE>	</TITLE>	Defines the title of the document
Text Formatting Tags		
<B>	</B>	Boldface
<I>	</I>	Italic
<U>	</U>	Underlined
Data Flow Tags		
<CENTER>	</CENTER>	Centered
 		Line Break
Image Tags		
<IMG>		Defines an Image
Hyperlink Tags		
<A>	</A>	Defines an address

6.9 HYPER TEXT TRANSFER PROTOCOL (HTTP)   (Feb. 15, 16, May 17)

Now a days Internet has become very essential for us. If we want to access the information on any topic we just search the topic in World Wide Web. Now the public has become aware of the power of internet through WWW.

- HTTP protocol is used to access the data on World Wide Web.
- This protocol normally transfers the data in the form of plain text, hypertext, audio, video and so on.
- It is called as Hypertext Transfer Protocol because it is used in an environment where there are rapid jumps from one document to another.
- HTTP functions like a combination of FTP and SMTP.
- It is said to be similar to FTP because it transfers files.
- It is said to be similar to SMTP because the data transferred between the client and server are similar to the SMTP messages.
- However, HTTP differs from SMTP in the way the messages are sent from client to server and from server to client. In case of SMTP, messages are only transferred from server to client.

How does HTTP Work?

- The working of HTTP is very simple.
- A client sends a request and server sends a reply (response) to the client. See Fig. 6.27.
- HTTP uses the services of TCP on well known port 80.

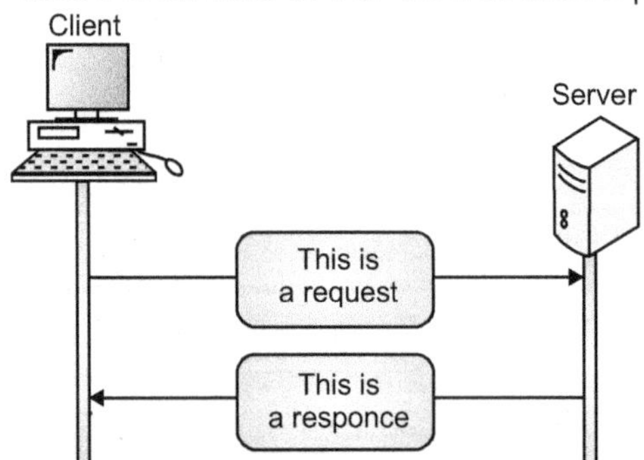

Fig. 6.27 : HTTP transaction

- Although HTTP uses the services of TCP, HTTP itself is a "stateless protocol". The client initializes the transaction by sending a request message. The server replies by sending a response.

6.9.1 Types of HTTP Messages

There are two types of HTTP messages

1. Request Message
2. Response Message

1. Request Message

Request message consists of a request line, headers and sometimes a body. See the following Fig. 6.28.

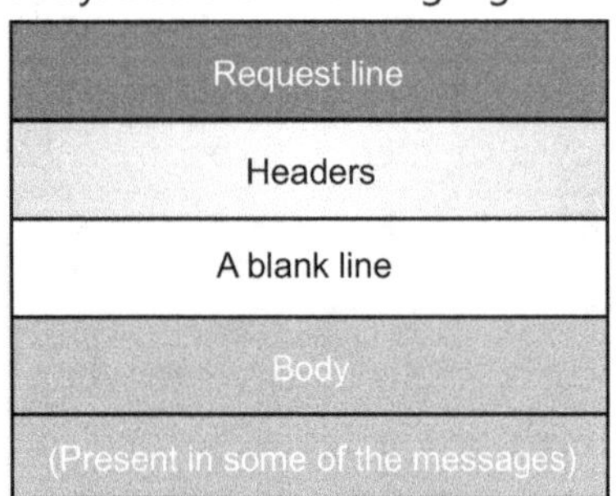

Fig. 6.28 : Request message

Fields of request message are described as below :

- **Request Line :** The request line defines the request type, Uniform resource locator (URL) and HTTP version as shown in Fig. 6.29.

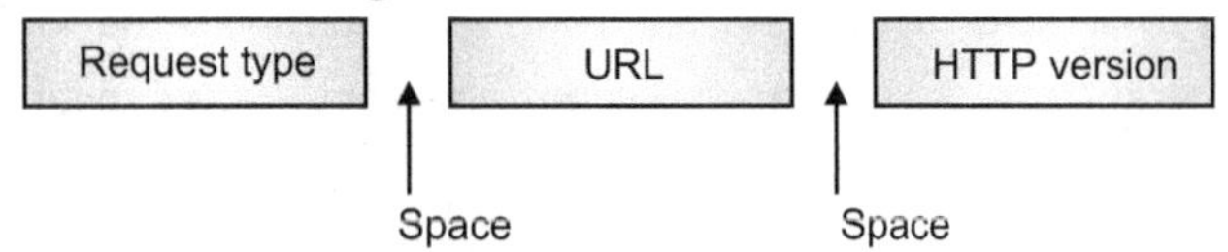

Fig. 6.29 : Request line

- **Request Type :** It categorizes the request message into several methods, which will be discussed later.

- **Uniform Resource Locator (URL) :** A client that wants to access a web page needs an address. HTTP uses URL to facilitate the access of any document distributed over the world. The URL defines four things : method, host computer, port and path as shown in Fig. 6.30.

URL

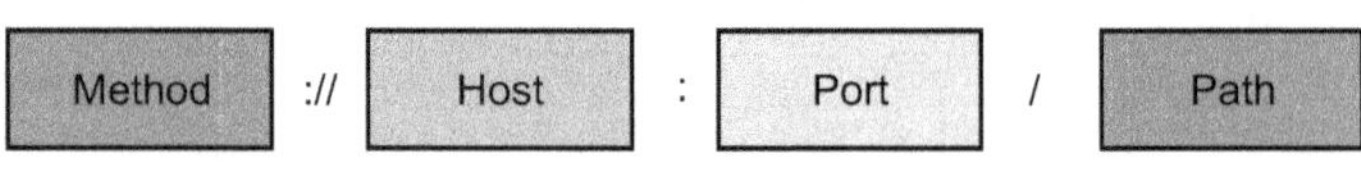

Fig. 6.30 : URL

- **Method :** It is a protocol such as FTP, HTTP that is used to retrieve the document from the web.

- **Host :** Host is a computer where the URL specific information is located. This information can be in terms of webpages or audio video files or other types of files such as doc files or pdf files etc.

- **Port :** URL can originally contain the port number of the server. If the port number is included, it is inserted between the host and the path, and it should be separated from the host by a colon.

- **Path :** It is the pathname of the file where the information is located.

- **Version :** The latest version of HTTP is 1.1 but the versions 0.9 and 1 are also used.

Methods in the URL Field

The request method is the actual request that a client issues to the server. Following are some of the important methods :

- **GET :** The GET method is used when the client wants to retrieve a document from the server. The address of the document is defined in the URL; this is the main method for retrieving a document. The server usually responds with the contents of the document in the body of the response message unless there is an error.

- **HEAD :** The HEAD method is used when the client wants some information about a document but not the document itself. It is similar to GET, but the response from the server, does not contain a body.

- **POST :** The POST method is used by the client to provide some information to the server. e.g. it can be used to send input to a server.

- **PUT :** The PUT method is used by the client to provide a new or replacement document to be stored on the server. The document is included in the body of the request and stored in the location defined by the URL.

- **PATCH :** PATCH is similar to PUT, except that the request contains a list of differences that should be implemented in the existing file.

- **COPY :** The COPY method copies a file to another location. The location of the source file is given in the request line (URL). The location of the destination is given in the entity header.

- **MOVE :** The MOVE method moves a file to another location. The location of the source file is given in the request line (URL); the location of the destination is given in the entity header.

- **DELETE :** The DELETE method removes a document on the server.

- **LINK :** The LINK method creates a link or links from a document to another location. The location of the file is given in the request line (URL); the location of the destination is given in the entity header.

- **UNLINK :** The UNLINK method deletes links created by the LINK method.

- **OPTION :** The OPTION method is used by the client to ask the server about the available options.

2. Response Message

A response message consists of a status line, a header and sometimes a body. Refer Fig. 6.31 (a).

Status Line :

The status line defines the status of the response message. It consists of the HTTP version, a status code and a status phrase as shown in Fig. 6.31 (b).

- **HTTP Version :** This field is the same as the corresponding field in the request line.

- **Status Code :** The status code field is similar to those in the FTP and the SMTP protocol. It consists of three digits.

- **Status Phrase :** This field explains the status code in text form.

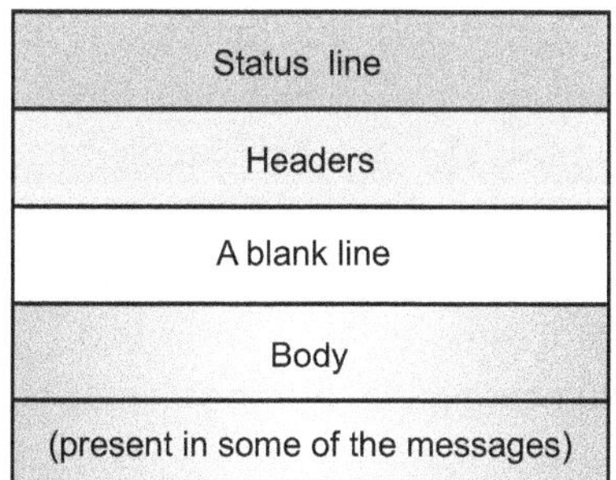

Fig. 6.31 (a) : Response Message

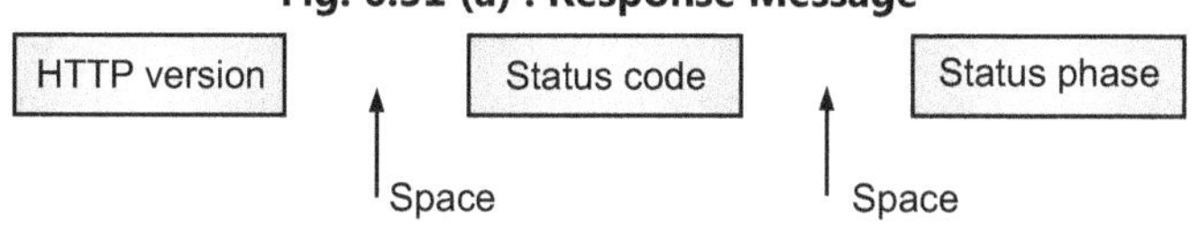

Fig. 6.31 (b) : Status Line

Headers :

- The headers exchange additional information between the client and the server.

- The header can be of one or more lines. Each header line is made of a header name, a colon, a space and a header value as shown in the following Fig. 6.32.

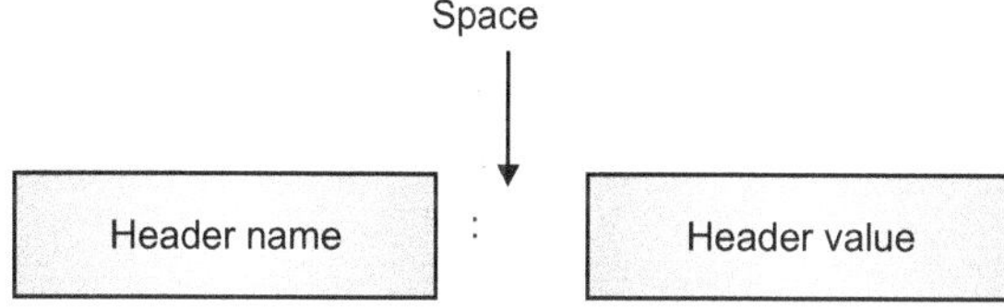

Fig. 6.32 : Header Format

- The header lines can be of four categories :
  (i)   General Header,
  (ii) Request Header,
  (iii) Response Header,
  (iv) Entity Header

(i) General Header : It gives general information about the message. It is present in request as well as response message.

(ii) Request Header : It can be present only in request message, which is used to specify client's configuration and client's preferred file format.

(iii) Response Header : It can be present only in the response message. It is used for specifying the server's configuration.

(iv) Entity Header : It is used for giving the information about the body of the document. It can be present in the request message as well as the response message.

- A request message can contain general request and entity headers while response message can contain general response and entity headers.

A comparison of request message and response message is shown in Fig. 6.33.

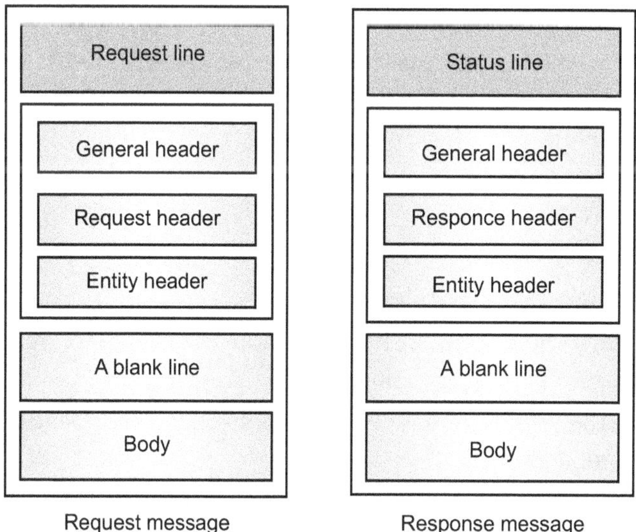

Fig. 6.33 : Request and Response Message Headers

6.9.2 HTTP is a Stateless Protocol

- In this protocol, the server does not store any information about the state of current transaction.

- So, when the client does a request for some files, server sends these files to client without storing any state information about the client.

- If same client asks for the same information again and again to the server, so the server would not understand that it has already transferred this information to the same client, so server resends this information back to the client again and again as and when the client requests for those files.

- As HTTP server does not maintain any information about the state of client, it is called as "stateless protocol".

6.9.3 HTTP Connections Types

HTTP connections are of two types

1. Persistent Connections
2. Non-Persistent Connections.

These are explained in detail as follows :

1.   Persistent Connections

- It is specified in HTTP version 1.1.

- In a persistent connection, the server leaves the connection open for more requests after sending a response.

- The server can close the connection at the request of a client or if timeout has been reached.

- The sender usually sends a length of the data with each response.

- However, there are certain situations where the sender does not know the length of the data. In such situation the document is created dynamically.

- In such cases, the server informs the client that the length of the data is not known and closes the connection after sending the data so the client knows that the end of the data has been reached.

  HTTP version 1.1 by default specifies a persistent connection.

2.   Non-Persistent Connection

In this type one TCP connection is made for each request/response by using the following steps :

- The client opens a TCP connection and sends a request.

- The server sends a response and closes the connection.

- The client reads the data until it encounters an end-of-file marker; the client then closes the connection.

6.9.4 Cookies

- As we have seen repeatedly, the HTTP is basically stateless. There is no concept of a login session. The browser sends a request to a server and gets back a file. Then the server forgets that it has ever seen that particular client.

- At first, when the Web was just used for retrieving publicly available documents, this model was perfectly adequate.

- But as the Web started to acquire other functions, it caused problems. For example, some Web sites require clients to register (and possibly pay money) to use them. This raises the question of how servers can distinguish between requests from registered users and everyone else. A second example is from e-commerce. If a user wanders around an electronic store, tossing items into her shopping cart from time to time, how does the server keep track of the contents of the cart? A third example is customized Web portals such as Yahoo. Users can set up a detailed initial page with only the information they want (e.g. their stocks and their favorite sports teams), but how can the server display the correct page if it does not know who the user is.

- To solve this problem, Netscape devised a much criticized technique called **cookies.** The name derives from ancient programmer slang in which a program calls a procedure and gets something back that it may need to present later to get some work done. In this sense, a UNIX file descriptor or a Windows object handle can be considered as a cookie.

- When a client requests a Web page, the server can supply additional information along with the requested page. This information may include a cookie, which is a small (at most 4 kB) file (or string). Browsers store offered cookies in a cookie directory on the client's hard disk unless the user has disabled cookies. Cookies are just files or strings, not executable programs. In principle, a cookie could contain a virus, but since cookies are treated as data, there is no official way for the virus to actually run and do damage. However, it is always possible for some hacker to exploit a browser bug to cause activation.

- A cookie may contain up to five fields, as shown in the following table. The *Domain* tells where the cookie came from. Browsers are supposed to check that servers are not lying about their domain. Each domain may store not more than 20 cookies per client. The

Path is a path in the server's directory structure that identifies which parts of the server's file tree may use the cookie. It is often /, which means the whole tree.

- The *Content* field takes the form *name = value*. Both *name* and *value* can be anything the server wants. This field is where the cookie's content is stored.

- The Expires field specifies when the cookie expires. If this field is absent, the browser discards the cookie when it exits. Such a cookie is called a non-persistent cookie. If a time and date are supplied, the cookie is said to be persistent and is kept until it expires. Expiration times are given in Greenwich Mean Time. To remove a cookie from a client's hard disk, a server just sends it again, but with an expiration time in the past.

Table 6.6

Domain	Path	Content	Expires	Secure
toms-casino.com	/	CustomerID=4977 93521	15-10-02 17:00	Yes
joes-store.com	/	Cart=1.00501;1-07031;2-13721	11-10-02 14:22	No
aportal.com	/	Prefs=Stk:SUNW+ORCL;Spl:Jets	31-12-10 23:59	No
sneaky.com	/	UserID=36272391 01	31-12-12 23:59	No

- Finally, the Securefield can be set to indicate that the browser may only return the cookie to a secure server. This feature is used for e-commerce, banking, and other secure applications.

- We have now seen how cookies are acquired, but how are they used? Just before a browser sends a request for a page to some web site, it checks its cookie directory to see if any cookies there were placed by the domain the request is going to. If so, all the cookies placed by that domain are included in the request message. When the server gets them, it can interpret them any way it wants to.

- To maintain some privacy, some users configure. their browsers to reject all cookies. However, this can give problems with legitimate Web sites that use cookies. To solve this problem, users sometimes install cookie-eating software. These are special programs that inspect each incoming cookie upon arrival and accept or discard it depending on choices the user has given it (e.g. about which Web sites can be trusted). This gives the user fine-grained control over which cookies are accepted and which are rejected. Modern browsers, such as Mozilla (www.mozilla.org), have elaborate user controls over cookies built in.

6.10 SIMPLE NETWORK MANAGEMENT PROTOCOL (SNMP)

- SNMP stands for simple network management protocol. It is a way that servers can share information about their current state, and also a channel through which an administer can modify pre-defined values. While the protocol itself is very simple, the structure of programs that implement SNMP can be very complex.

- In this guide, we will introduce you to the basics of the SNMP protocol. We will go over its uses, the way that the protocol is typically used in a network, the differences in its protocol versions, and more.

Basic Concepts

- SNMP is a protocol that is implemented on the application layer of the networking stack (click here to learn about networking layers). The protocol was created as a way of gathering information from very different systems in a consistent manner. Although it can be used in connection to a diverse array of systems, the method of querying information and the paths to the relevant information are standardized.

- There are multiple versions of the SNMP protocol, and many networked hardware devices implement some form of SNMP access. The most widely used version is SNMPv1, but it is in many ways insecure. Its popularity largely stems from its ubiquity and long time in the wild. Unless you have a strong reason not to, we recommend you use SNMPv3, which provides more advanced security features.

- In general, a network being profiled by SNMP will mainly consist of devices containing SNMP **agents**. An agent is a program that can gather information about a piece of hardware, organize it into predefined entries, and respond to queries using the SNMP protocol.

- The component of this model that queries agents for information is called an SNMP **manager**. These machines generally have data about all of the SNMP-enabled devices in their network and can issue requests to gather information and set certain properties.

SNMP Managers

- An SNMP manager is a computer that is configured to poll SNMP agent for information. The management component, when only discussing its core functionality, is actually a lot less complex than the client configuration, because the management component simply requests data.

- The manager can be any machine that can send query requests to SNMP agents with the correct credentials. Sometimes, this is implemented as part of a monitoring suite, while other times this is an administrator using some simple utilities to craft a quick request.

- Almost all of the commands defined in the SNMP protocol (we will go over these in detail later) are designed to be *sent* by a manager component. These include GetRequest, GetNextRequest, GetBulkRequest, SetRequest, InformRequest, and Response. In addition to these, a manager is also designed to respond to Trap, and Response messages.

SNMP Agents

- SNMP agents do the bulk of the work. They are responsible for gathering information about the local system and storing them in a format that can be queried.updating a database called the "management information base", or **MIB**.

- The MIB is a hierarchical, pre-defined structure that stores information that can be queried or set. This is available to well-formed SNMP requests originating from a host that has authenticated with the correct credentials (an SNMP manager).

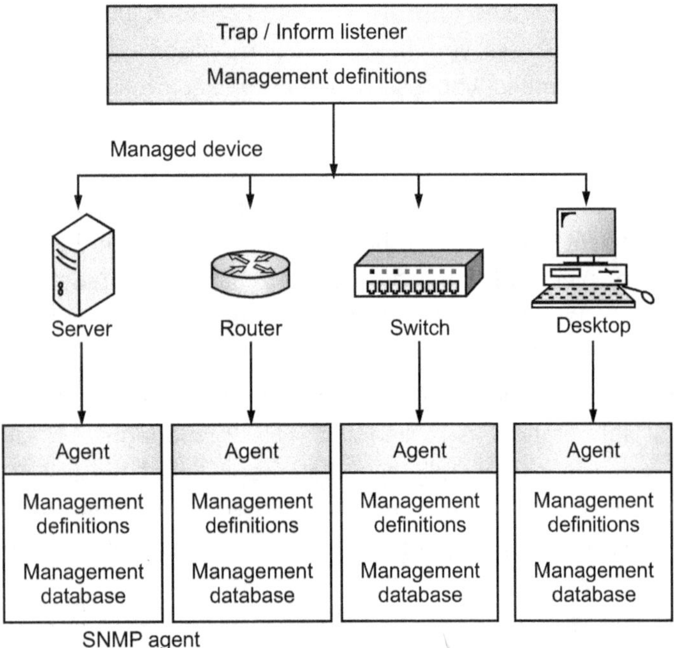

Fig. 6.34

- The agent computer configures which managers should have access to its information. It can also act as an intermediary to report information on devices it can connect to that are not configured for SNMP traffic.

This provides a lot of flexibility in getting your components online and SNMP accessible.

- SNMP agents respond to most of the commands defined by the protocol. These include GetRequest, GetNextRequest, GetBulkRequest, SetRequest and InformRequest. In addition, an agent is designed to send Trap messages.

Understanding the Management Information Base

- The most difficult part of the SNMP system to understand is probably the **MIB**, or management information base. The MIB is a database that follows a standard that the manager and agents adhere to. It is a hierarchical structure that, in many areas, is globally standardized, but also flexible enough to allow vendor-specific additions.

- The MIB structure is best understood as a top-down hierarchical tree. Each branch that forks off is labeled with both an identifying number (starting with 1) and an identifying string that are unique for that level of the hierarchy. You can use the strings and numbers interchangeably.

- To refer to a specific node of the tree, you must trace the path from the unnamed root of the tree to the node in question. The lineage of its parent IDs (numbers or strings) are strung together, starting with the most general, to form an address. Each junction in the hierarchy is represented by a dot in this notation, so that the address ends up being a series of ID strings or numbers separated by dots. This entire address is known as an object identifier, or **OID**.

- Hardware vendors that embed SNMP agents in their devices sometimes implement custom branches with their own fields and data points. However, there are standard MIB branches that are well defined and can be used by any device.

- The standard branches we will be discussing will all be under the same parent branch structure. This branch defines information that adheres to the MIB-2 specification, which is a revised standard for compliant devices.

- The base path to this branch is:
  1.3.6.1.2.1

- This can also be represented in strings like:
  iso.org.dod.internet.mgmt.mib-2

- The section 1.3.6.1 or iso.org.dod.internet is the OID that defines internet resources. The 2 or mgmtthat follows in our base path is for a management subcategory. The 1 or mib-2 under that defines the MIB-2 specification.

- Basically, if we want to query our devices for information, most of the paths will begin with 1.3.6.1.2.1. You can browse the tree interfaces to learn what kind of information is available to query and set.

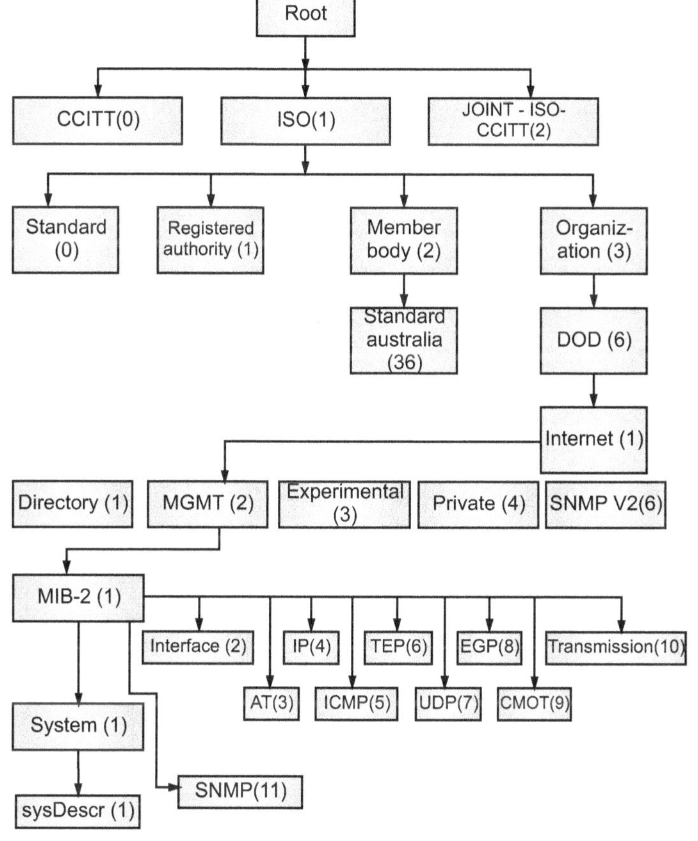

Fig. 6.35

SNMP Protocol Commands

One of the reasons that SNMP has seen such heavy adoption is the simplicity of the commands available. There are very few operations to implement or remember, but they are flexible enough to address the utility requirements of the protocol.

The following PDUs, or protocol data units, describe the exact messaging types that are allowed by the protocol:

- **Get**: A Get message is sent by a manager to an agent to request the value of a specific OID. This request is answered with a Response message that is sent back to the manager with the data.

- **GetNext**: A GetNext message allows a manager to request the next sequential object in the MIB. This is a way that you can traverse the structure of the MIB without worrying about what OIDs to query.

- **Set**: A Set message is sent by a manager to an agent in order to change the value held by a variable on the agent. This can be used to control configuration information or otherwise modify the state of remote hosts. This is the only write operation defined by the protocol.

- **GetBulk**: This manager to agent request functions as if multiple GetNext requests were made. The reply back to the manager will contain as much data as possible (within the constraints set by the request) as the packet allows.

- **Response**: This message, sent by an agent, is used to send any requested information back to the manager. It serves as both a transport for the data requested, as well as an acknowledgement of receipt of the request. If the requested data cannot be returned, the response contains error fields that can be set with further information. A response message must be returned for any of the above requests, as well as Inform messages.

- **Trap**: A trap message is generally sent by an agent to a manager. Traps are asynchronous notifications in that they are unsolicited by the manager receiving them. They are mainly used by agents to inform managers of events that are happening on their managed devices.

- **Inform**: To confirm the receipt of a trap, a manager sends an Inform message back to the agent. If the agent does not receive this message, it may continue to resend the trap message.

With these seven data unit types, SNMP is capable of querying for and sending information about your networked devices.

Protocol Versions

- The SNMP protocol has gone through many changes since it was first introduced. The initial spec was formulated with RFC 1065, 1066, and 1067 in 1988. By the simple fact that it has been around so long, this version is still widely supported. However, there are many security issues with the protocol, including authenticating in plain text, so its use is highly discouraged, especially when used on unprotected networks.

- Work on version 2 of the protocol was initiated in 1993 and offers some substantial improvements on the earlier standard. Included in this version was a new "party-based" security model meant to address the security issues inherent with the prior revision. However, the new model was not very popular because it was difficult to understand and implement.

- Because of this, a few "spin-offs" of version 2 were created, each of which kept the bulk of the version 2 improvements, but swapped out the security model. In

SNMPv2c, community-based authentication, the same model used in v1, was reintroduced. This was the most popular version of the v2 protocol. Another implementation, called SNMPv2u, uses user-based security, although this was never very popular. This allowed for per-user authentication settings.

- In 1998, the third (and current) version of the SNMP protocol entered as a spec proposal. From a user's perspective, the most relevant change was the adoption of a user-based security system. It allows you to set a user's authentication requirements as one of these models:

  NoAuthNoPriv: Users connecting with this level have no authentication in place and no privacy of the messages they send and receive.

  AuthNoPriv: Connections using this model must authenticate, but messages are sent without any encryption.

  AuthPriv: Authentication is required and messages are encrypted.

- In addition to authentication, an access control mechanism was implemented to provide granular control over which branches a user can access. Version 3 also has the ability to leverage the security provided by the transport protocols, such as SSH or TLS.

6.11 BLUETOOTH

Bluetooth is a new technology that comes as an alternative to cables for connecting portable and fixed electronic devices and uses short range (10 meter) frequency hopping radio links for communication. It operates within the unlicensed ISM (Industrial scientific and medical) band at 2.4 GHz.

Bluetooth Features

- **Robustness :** It uses a fast acknowledgement and frequency hopping scheme to make a radio link robust.
- **Low Complexity :** The necessary transceiver components present in the devices are simple.
- **Low Cost :** A wireless device with this technology is available at an affordable price.

Applications of Bluetooth

There are a variety of applications of Bluetooth such as

- Allows a transfer of images (or) word documents (or) applications (or) audio and video files between devices without the help of cables.
- Can be used for remote sales technology allowing wireless access to vending machines and other commercial enterprises.

- Provides inter accessibility of PDAs, palmtops and desktops for file and data exchanges.
- It can be used to setup a personal area network (PAN) or a wireless personal area network (WPAN).

Bluetooth Standards Documents

The documentation on Bluetooth can be split into two sections:

1. Bluetooth specification
2. Bluetooth profiles

The specification section deals with the protocol architecture required for implementing bluetooth, while the profiles section describes in what ways we can utilize the Bluetooth technology.

Bluetooth Architecture

The protocol architecture of Bluetooth is given below:

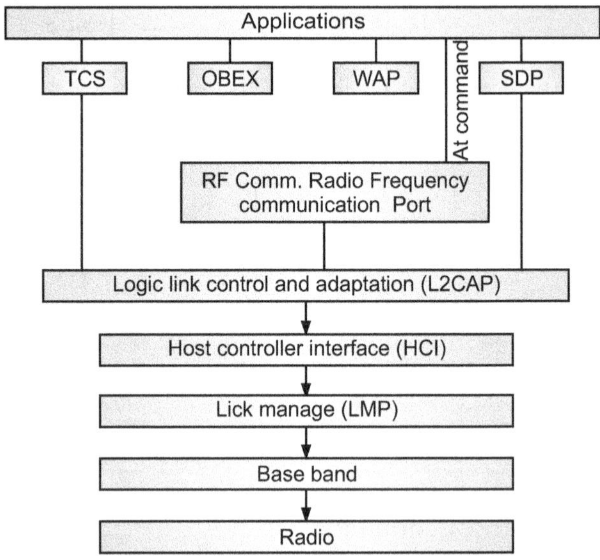

Fig. 6.36

The radio layer is responsible for:

- Modulation/Demodulation of data for transmitting (OR) receiving over air.

The base band layer is responsible for:

- Controlling the physical links via radio
- Assembling the packets
- Controlling frequency hopping.
- The link manager protocol controls and configures links to other devices.
- The host controller interface(HCI) handles communication between the host and the module. For this purpose, it uses several HCI command packets such as the even packets and data packets. The L2CAP layer converts the data obtained from higher layers into packets of different sizes.

- The RF COMM provides a serial interface with wireless application protocol (WAP) and object exchange (OBEX).
- WAP and OBEX provide interface to other communications protocols.
- The TCS(Telephone control protocol specification) provide telephony service.
- The SDP(Service discovery protocol) allows the devices to discover the services available on another Bluetooth enabled device.
- The applications present in the application layer can extract the services of the lower layers by using one of the many profiles available.

6.12 FIREWALLS

- In computing, a firewall is a network security system that monitors and controls incoming and outgoing network traffic based on predetermined security rules. A firewall typically establishes a barrier between a trusted internal network and untrusted external network, such as the Internet.
- Firewalls are often categorized as either network firewalls or host-based firewalls. Network firewalls filter traffic between two or more networks and run on network hardware. Host-based firewalls run on host computers and control network traffic in and out of those machines.
- The term firewall originally referred to a wall intended to confine a fire within a building. Later uses refer to similar structures, such as the metal sheet separating the engine compartment of a vehicle or aircraft from the passenger compartment.
- The term was applied in the late 1980s to network technology that emerged when the Internet was fairly new in terms of its global use and connectivity. The predecessors to firewalls for network security were the routers used in the late 1980s, because they separated networks from one another, thus halting the spread of problems from one network to another.

First Generation: Packet Filters

- The firewall shows its settings for incoming and outgoing traffic. The first reported type of network firewall is called a packet filter. Packet filters act by inspecting packets transferred between computers. When a packet does not match the packet filter's set of filtering rules, the packet filter either drops (silently discards) the packet, or rejects the packet (discards it and generates an Internet Control Message Protocol

notification for the sender) else it is allowed to pass. Packets may be filtered by source and destination network addresses, protocol, source and destination port numbers.
- The bulk of Internet communication in 20th and early 21st century used either Transmission Control Protocol (TCP) or User Datagram Protocol (UDP) in conjunction with well-known ports, enabling firewalls of that era to distinguish between, and thus control, specific types of traffic (such as web browsing, remote printing, email transmission, file transfer), unless the machines on each side of the packet filter used the same non-standard ports.

Fig. 6.37

- The first paper published on firewall technology was in 1988, when engineers from Digital Equipment Corporation (DEC) developed filter systems known as packet filter firewalls. At AT&T Bell Labs, Bill Cheswick and Steve Bellovin continued their research in packet filtering and developed a working model for their own company based on their original first generation architecture.

Second Generation: Stateful Filters

- From 1989–1990, three colleagues from AT&T Bell Laboratories, Dave Presotto, Janardan Sharma, and Kshitij Nigam, developed the second generation of firewalls, calling them circuit-level gateways.
- Second-generation firewalls perform the work of their first-generation predecessors but also maintain knowledge of specific conversations between endpoints by remembering which port number the two IP addresses are using at layer 4 (transport layer) of the OSI model for their conversation, allowing examination of the overall exchange between the nodes.

- This type of firewall is potentially vulnerable to denial-of-service attacks that bombard the firewall with fake connections in an attempt to overwhelm the firewall by filling its connection state memory.

Third Generation: Application Layer

- Marcus Ranum, Wei Xu, and Peter Churchyard released an application firewall known as Firewall Toolkit (FWTK) in October 1993. This became the basis for Gauntlet firewall at Trusted Information Systems.

- The key benefit of application layer filtering is that it can understand certain applications and protocols (such as File Transfer Protocol (FTP), Domain Name System (DNS), or Hypertext Transfer Protocol (HTTP)). This is useful as it is able to detect if an unwanted application or service is attempting to bypass the firewall using a disallowed protocol on an allowed port, or detect if a protocol is being abused in any harmful way.

- As of 2012, the so-called next-generation firewall (NGFW) is a wider or deeper inspection at the application layer. For example, the existing deep packet inspection functionality of modern firewalls can be extended to include:

Intrusion Prevention Systems (IPS)

- User identity management integration (by binding user IDs to IP or MAC addresses for "reputation")

- Web application firewall (WAF). WAF attacks may be implemented in the tool "WAF Fingerprinting utilizing timing side channels" (WAFFle)

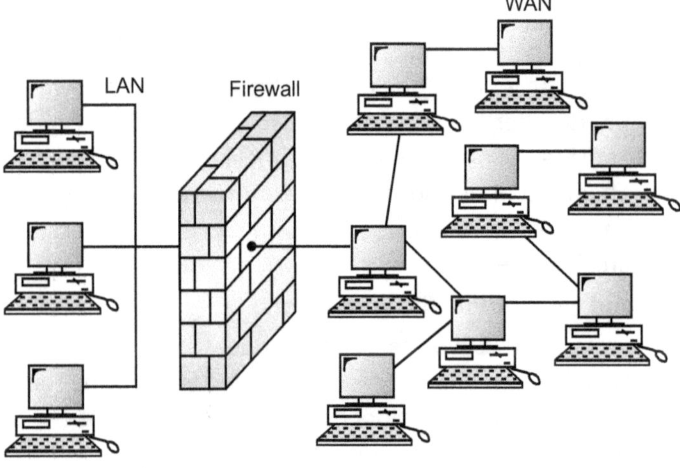

Fig. 6.38

Types

- Firewalls are generally categorized as network-based or host-based. Network-based firewalls are positioned on the gateway computers of LANs, WANs and intranets. They are either software appliances running on general-purpose hardware, or hardware-based firewall computer appliances.

- Firewall appliances may also offer other functionality to the internal network they protect, such as acting as a DHCP or VPN server for that network. Host-based firewalls are positioned on the host itself and control network traffic in and out of those machines. The host-based firewall may be a daemon or service as a part of the operating system or an agent application such as endpoint security or protection. Each has advantages and disadvantages. However, each has a role in layered security.

- Firewalls also vary in type depending on where communication originates, where it is intercepted, and the state of communication being traced.

Network Layer or Packet Filters

- Network layer firewalls, also called packet filters, operate at a relatively low level of the TCP/IP protocol stack, not allowing packets to pass through the firewall unless they match the established rule set. The firewall administrator may define the rules; or default rules may apply. The term "packet filter" originated in the context of BSD operating systems.

- Network layer firewalls generally fall into two sub-categories, stateful and stateless.

Application-Layer

- Application-layer firewalls work on the application level of the TCP/IP stack (i.e., all browser traffic, or all telnet or FTP traffic), and may intercept all packets traveling to or from an application.

- Application firewalls function by determining whether a process should accept any given connection. Application firewalls accomplish their function by hooking into socket calls to filter the connections between the application layer and the lower layers of the OSI model.

- Application firewalls that hook into socket calls are also referred to as socket filters. Application firewalls work much like a packet filter but application filters apply filtering rules (allow/block) on a per process basis instead of filtering connections on a per port basis. Generally, prompts are used to define rules for processes that have not yet received a connection. It is rare to find application firewalls not combined or used in conjunction with a packet filter.

- Also, application firewalls further filter connections by examining the process ID of data packets against a rule set for the local process involved in the data transmission. The extent of the filtering that occurs is defined by the provided rule set. Given the variety of software that exists, application firewalls only have more complex rule sets for the standard services, such as sharing services.

- These per-process rule sets have limited efficacy in filtering every possible association that may occur with other processes. Also, these per-process rule sets cannot defend against modification of the process via exploitation, such as memory corruption exploits. Because of these limitations, application firewalls are beginning to be supplanted by a new generation of application firewalls that rely on mandatory access control (MAC), also referred to as sandboxing, to protect vulnerable services.

Proxies

- A proxy server (running either on dedicated hardware or as software on a general-purpose machine) may act as a firewall by responding to input packets (connection requests, for example) in the manner of an application, while blocking other packets. A proxy server is a gateway from one network to another for a specific network application, in the sense that it functions as a proxy on behalf of the network user.

- Proxies make tampering with an internal system from the external network more difficult, so that misuse of one internal system would not necessarily cause a security breach exploitable from outside the firewall (as long as the application proxy remains intact and properly configured).

- Conversely, intruders may hijack a publicly reachable system and use it as a proxy for their own purposes; the proxy then masquerades as that system to other internal machines. While use of internal address spaces enhances security, crackers may still employ methods such as IP spoofing to attempt to pass packets to a target network.

Network address translation

- Firewalls often have network address translation (NAT) functionality, and the hosts protected behind a firewall commonly have addresses in the "private address range", as defined in RFC 1918. Firewalls often have such functionality to hide the true address of computer which is connected to the network.

- Originally, the NAT function was developed to address the limited number of IPv4 routable addresses that could be used or assigned to companies or individuals as well as reduce both the amount and therefore cost of obtaining enough public addresses for every computer in an organization. Although NAT on its own is not considered a security feature, hiding the addresses of protected devices has become an often used defense against network reconnaissance.

6.13 BASIC CONCEPTS OF CRYPTOGRAPHY

- We need to keep information about every aspect of our lives. Information is extremely important part of our lives. This information needs to be secured from attacks. To be secured, information needs to be hidden from unauthorized access (confidentiality), protected from unauthorized change (integrity), and available to an authorized entity when it is needed (availability).

- Computer networks have created a revolution in the use of information. Authorized people can send and retrieve information from a distance using computer networks. Also information should be confidential when it is stored; there should also be a way to maintain its confidentiality when it is transmitted from one computer to another.

- Network security has to be implemented to achieve following three goals.

1. **Confidentiality :**

- Confidentiality is probably the most common aspect of information security. We need to protect our confidential information.

- An organization needs to protect against those malicious actions that hamper the confidentiality of its information. Confidentiality not only applies to the storage of the information, it also applies to the transmission of information. When we send a piece of information to be stored in a remote computer or when we retrieve apiece of information from a remote computer, we need to conceal it during transmission.

2. **Integrity**

- Information needs to be changed constantly. In a bank, when a customer deposits or withdraws money, the balance of her account needs to be changed. Integrity means that changes need to be done only by authorized entities and through authorized mechanisms.

- Integrity violation is not necessarily the result of a malicious act; an interruption in the system, such as a power surge, may also create unwanted changes in some information.

3. Availability

- The information created and stored by an organization needs to be available to authorized entities. Information is useless if it is not available. Information needs to be constantly changed, which means it must be accessible to authorized entities.

- The unavailability of information is just as harmful for an organization as the lack of confidentiality or integrity. Imagine what would happen to a bank if the customers could not access their accounts for transactions.

6.14 CRYPTOGRAPHY

- Cryptography is a Greek word which means "secret writing." However, we use the term to refer to the art of transforming messages to make them secure and immune to attacks.

- Although in the past cryptography referred only to the encryption and decryption of messages using secret keys, today it is defined in the context of three different mechanisms symmetric-key encipherment, asymmetric-key encipherment, and hashing.

6.14.1 Traditional Ciphers

- Confidentiality can be achieved using ciphers. Traditional ciphers are called symmetric-key ciphers because the same key is used for encryption and decryption and the key can be used for bidirectional communication.

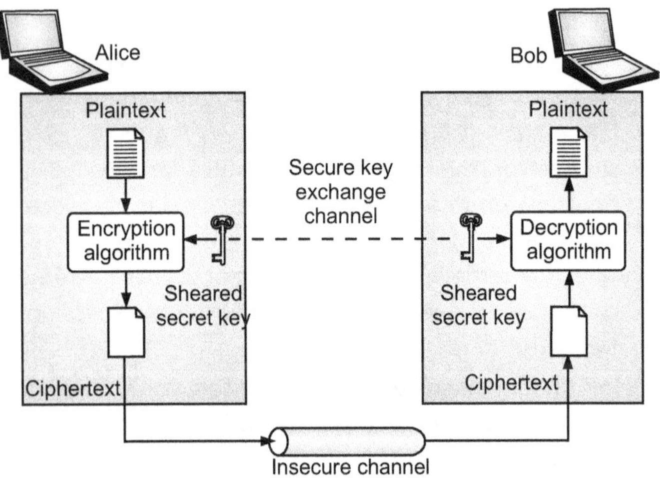

Fig. 6.39 : Traditional ciphers

- Consider an example in which Alice, can send a message to another entity, Bob, over an insecure channel with the assumption that an adversary, Eve, cannot understand the contents of the message by simply eavesdropping over the channel.

- The original message from Alice to Bob is called **plaintext;** the message that is sent through the channel is called the **ciphertext.** To create the cipher text from the plaintext, Alice uses an **encryption algorithm** and a **shared secret key.** To create the plaintext from cipher text, Bob uses a **decryption algorithm** and the same secret key and encryption and decryption algorithms as **ciphers.** A **key** is a set of values(numbers) that the cipher, as an algorithm, operates on.

- Note that the symmetric-key encipherment uses a single key (the key itself may be a set of values) for both encryption and decryption. In addition, the encryption and decryption algorithms are inverses of each other. If P is the plaintext, C is the ciphertext, and K is the key, the encryption algorithm Ek(x) creates the ciphertext from the plaintext; the decryption algorithm Dk(x) creates the plaintext from the ciphertext. We assume that Ek(x) and Dk(x) are inverses of each other: they cancel the effect of each other if they are applied one after the other on the same input. We have in which, Dk(Ek(x)) = Ek(Dk(x)) = x. We need to emphasize that it is better to make the encryption and decryption public but keep the shared key secret. This means that Alice and Bob need another channel, a secured one, to exchange the secret key. Alice and

- Bob can meet once and exchange the key personally. The secured channel here is the face-to-face exchange of the key. They can also trust a third party to give them the samekey.

Key

Encryption can be thought of as locking the message in a box; decryption can bethought of as unlocking the box. In symmetric-key encipherment, the same key locks and unlocks and asymmetric-key encipherment needs two keys, one for locking and one for unlocking.

We can divide traditional symmetric-key ciphers into two broad categories: substitution ciphers and transposition ciphers.

6.14.2 Substitution Ciphers

A **substitution cipher** replaces one symbol with another. If the symbols in the plaintext are alphabetic characters, we replace one character with another. For example, we can replace letter A with letter D, and letter T with letter Z. If the symbols are digits (0 to 9), we can replace 3 with 7, and 2 with 6.

Substitution ciphers can be categorized as either monoalphabetic ciphers or polyalphabeticciphers.

1. Monoalphabetic Ciphers

- In a monoalphabetic cipher, a character (or a symbol) in the plaintext is always changed to the same character (or symbol) in the ciphertext regardless of its position in the text. For example, if the algorithm says that letter A in the plaintext is changed to letter D, every letter A is changed to letter D. In other words, the relationship between letters in the plaintext and the ciphertext is one-to-one.

- The simplest monoalphabetic cipher is the additive cipher (or shift cipher).

- Assume that the plaintext consists of lowercase letters (a to z), and that the ciphertext consists of uppercase letters (A to Z). To be able to apply mathematical operations on the plaintext and ciphertext, we assign numerical values to each letter (lower- or uppercase).

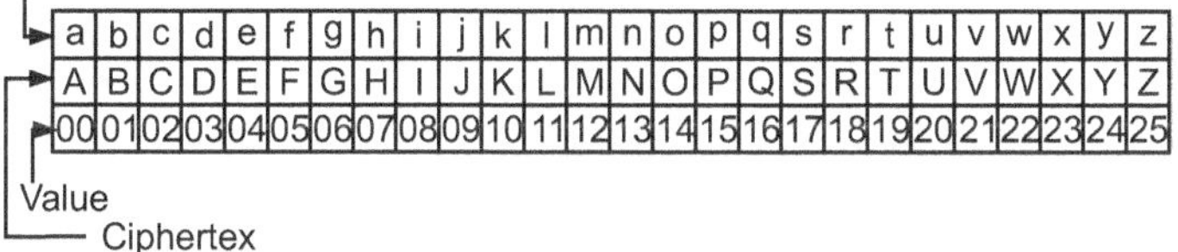

**Fig. 6.40 : Representing plaintext
and ciphertext in modulo 26**

- Each character (lowercase or uppercase) is assigned an integer in modulo 26. The secret key between Alice and Bob is also an integer in modulo 26. The encryption algorithm adds the key to the plaintext character; the decryption algorithm subtracts the key from the ciphertext character. All operations are done in modulo 26.

2. Polyalphabetic Ciphers

- In **polyalphabetic substitution,** each occurrence of a character may have a different substitute. The relationship between a character in the plaintext to a character in the ciphertext is one-to-many. For example, "a" could be enciphered as "D" in the beginning of the text, but as "N" at the middle. Polyalphabetic ciphers have the advantage of hiding the letter frequency of the underlying language. Eve cannot use single-letter frequent cystatistic to break the ciphertext.

- To create a **polyalphabetic cipher,** we need to make each ciphertext character dependent on both the corresponding plaintext character and the position of the plaintext character in the message. This implies that our key should be a stream of subkeys,in which each subkey depends somehow on the position of the plaintext character that uses that subkey for

encipherment. In other words, we need to have a key stream k = (k1, k2, k3, ...) in which key is used to encipher the ith character in the plaintext to create the ith character in the ciphertext.

- To see the position dependency of the key, let us discuss a simple polyalphabeticcipher called the **autokey cipher.** In this cipher, the key is a stream of subkeys, in which each subkey is used to encrypt the corresponding character in the plaintext. The first subkey is a predetermined value secretly agreed upon by Alice and Bob. The second subkey is the value of the first plaintext character (between 0 and 25). The third subkey is the value of the second plaintext.

6.14.3 Transposition Ciphers

- A **transposition cipher** does not substitute one symbol for another, instead it changes the location of the symbols. A symbol in the first position of the plaintext may appear in the tenth position of the ciphertext. A symbol in the eighth position in the plaintext may appear in the first position of the ciphertext. In other words, a transposition cipherre orders (transposes) the symbols.

- Suppose Alice wants to secretly send the message "Enemy attacks tonight" to Bob.

- The encryption and decryption is shown in Fig. 6.41.

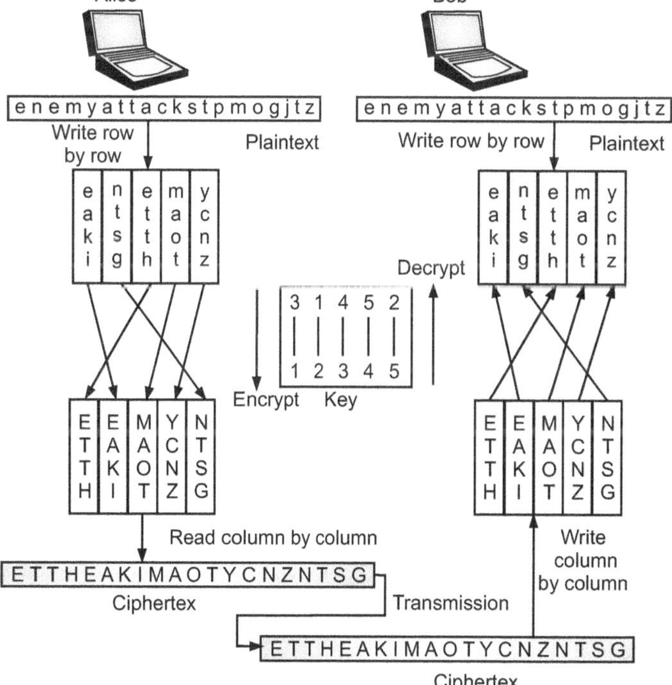

Fig. 6.41 : Transposition cipher

- The first table is created by Alice writing the plaintext row by row. The columns are permuted using a key. The ciphertext is created by reading the second table column by column. Bob does the same three steps in the reverse order. He writes the ciphertext column by column into the first table, permutes the columns, and then reads the second table row by row. Note that the same key is used for encryption and decryption, but the algorithm uses the key in reverse order.

- A modern block cipher is made of a combination of transposition units (sometimes called P-boxes), substitution units (sometimes called S-boxes), and exclusive-or operations, shifting elements, swapping elements, splitting elements, and combining elements.

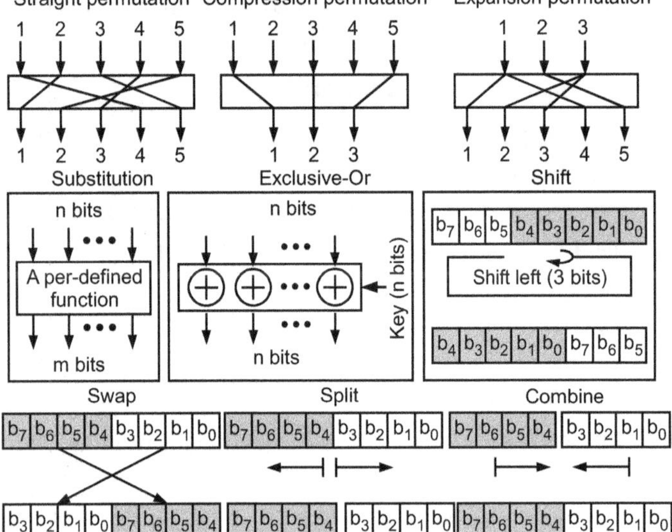

Fig. 6.42 : Modern cipher

- A **P-box** (permutation box) parallels the traditional transposition cipher for characters, but it transposes bits. We can find three types of P-boxes in modern block ciphers:straight P-boxes, expansion P-boxes, and compression P-boxes. An **S-box** (substitution box) can be thought of as a miniature substitution cipher, but it substitutes bits. Unlike the traditional substitution cipher, an S-box can have a different number of inputs and outputs. An important component in most block ciphers is the exclusive-or operation, in which the output is 0 if the two inputs are the same, and the output is 1 if the two inputs are different. In modern block ciphers, we use n exclusive-or operations to combine ann-bit data piece with an n-bit key. An exclusive-or operation is normally the only unit where the key is applied.

- Another component found in some modern block ciphers is the circular shiftoperation. Shifting can be to the left or to the right. The circular left-shift operation shifts each bit in an n-bit word k positions to the left; the leftmost k bits are removed from the left and become the rightmost bits. The swap operation is a special case of the circular shift operation where the number of shifted bits k = n/2.

- Two other operations found in some block ciphers are split and combine. The splitoperation splits an n-bit word in the middle, creating two equal-length words. The combine operation normally concatenates two equal-length words to create an n-bitword.

6.14.4 Asymmetric-Key Ciphers

- The conceptual differences between symmetric key cipher and asymmetric key cipher are based on how these systems keep a secret. In symmetric-key cryptography, the secret must be shared between two persons. In asymmetric-key cryptography, the secret is personal (unshared); each person creates and keeps his or her own secret.

- In a community of n people, n(n −1)/2 shared secrets are needed for symmetric key cryptography; only n personal secrets are needed in asymmetric-key cryptography. symmetric-key cryptography is based on substitution and permutation of symbols (characters or bits), asymmetric-key cryptography is based on applying mathematical functions to numbers. In symmetric-key cryptography, the plaintext and ciphertext are thought of as a combination of symbols. Encryption and decryption permute these symbols or substitute a symbol for another. In asymmetric-key cryptography, the plaintext and ciphertext are numbers; encryption and decryption are mathematical functions that are applied to numbers to create other numbers.

Keys

- Asymmetric key cryptography uses two separate keys: one private and one public. If encryption and decryption are thought of as locking and unlocking with keys, then the entity that is locked with a public key can be unlocked only with the corresponding private key

6.15 RSA CRYPTOSYSTEM

- There are several asymmetric-key cryptosystems, one of the common public key algorithms is the RSA cryptosystem, named for its inventors (Rivest, Shamir, and Adleman). RSA uses two exponents, e and d, where e is public and d is private.

- Suppose P is the plaintext and C is the ciphertext. Alice uses $C = P^e \bmod n$ to create ciphertext C from plaintext P; Bob uses $P = C^d \bmod n$ to retrieve the plaintext sent by Alice. The modulus n, a very large number, is created during the key generation process.

Procedure

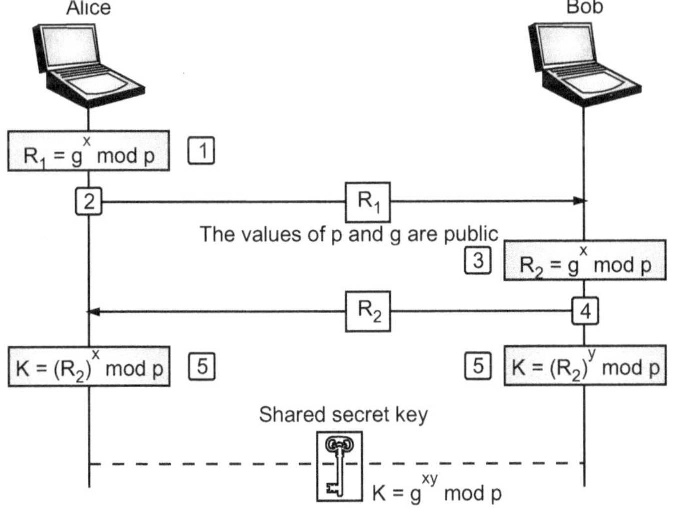

Fig. 6.43 : RSA cryptosystem

- Bob choose two large numbers, p and q and calculates n = p * q and Φ=(p −1) * (q −1). Bob then selects e and d such as (e * d) modΦ= 1. Bob advertises e and n to the community as the public key; Bob keeps d as the secret key. Anyone, including

- Alice, can encrypt a message and send the ciphertext to Bob using $C = P^e$ mod n; Only Bob can decrypt the message using $P = C^d$ mod n. An intruder such as Eve cannot decrypt the message if p and q are very large numbers.

6.16 DIFFIE-HELLMAN ALGORITHM

- In the Diffie-Hellman protocol two parties create a symmetric session key without the need of a KDC. Before establishing a symmetric key, the two parties need to choose two numbers p and g. These two numbers have some properties discussed in number theory, but beyond the scope of this book. These two numbers do not need to be confidential. They can be sent through the Internet; they can be public.

Fig. 6.44 : Diffie-Hellman algorithm

- The steps are as follows:
1. Alice chooses a large random number x such that 0<=x<=p-1 and calculates
   $R_1 = g^x$ mod p.
2. Alice sends R_1 to Bob.
3. Bob chooses another large random number y such that 0<=y <=p-1 and calculates
   $R_2 = g^y$ mod p.
4. Bob sends R_2 to Alice.
5. Alice calculates K = $(R_2)^x$ mod p. Bob also calculates K = $(R_1)^y$ mod p.

K is the symmetric key for the session.

Bob has calculated K = $(R_1)^y$ mod p.= $(R_2)^x$ mod p and Alice has calculated K = $(R_2)^x$ mod p = K = $(R_1)^y$ mod p.. Both have reached the same value without Bob knowing the value of x and without Alice knowing the value of y.

EXERCISE

1. What is the use of HTTP ?
2. Write a short note on HTTP messages.
3. Explain methods in URL field.
4. Why HTTP is a stateless protocol ?
5. Explain HTTP connection types.
6. Write a short note on WWW.
7. Explain browser architecture in detail.
8. What are the types of web documents ? Explain each one of them.
9. Write a short note on HTML.
10. Explain the structure of web page.
11. Write down some of the HTML tags that you know.
12. What is common gateway interface ?
13. Write a short note on Active Documents.
14. Explain File Transfer Protocol in detail.
15. Write down the various FTP commands that you are aware of.
16. What is Trivial File Transfer Protocol ? How it is different from normal File Transfer Protocol ?
17. Compare FTP and TFTP.
18. What is Simple Mail Transfer Protocol ? Why to use it ?
19. Write a short note on Post Office Protocol (POP)
20. Explain IMAP4 in detail.
21. What is Multipurpose Internet Mail Extensions (MIME) ?
22. Write a short note on the working of Domain Name System.
23. What is domain name space ?
24. How domain name-address resolution take place ?
25. Explain following terms :
    (a) Recursive resolution
    (b) Iterative resolution.

26. Explain the types of DNS messages.
27. What is Resource Records (RRs) ?
28. Write a short note on Simple Network Management Protocol (SNMP) ?
29. Explain various packet types of SNMPv3.
30. What is caching ?
31. What do you mean by statelessness and cookies ? Explain.
32. What are dynamic and active pages ?
33. Explain how name resolution happens in DNS. Enlist all the resource records and its function.
34. Differentiate between FTP and TFTP.
35. Discuss various management categories used in network management.
36. Explain SNMP model with its major components.
37. In SMTP, if we send a one line message between two users, how many lines of commands and responses are exchanged ? Give the example.
38. Why Common Gateway Interface (CGI) is required in dynamic web pages ?
39. Discuss the role of SMI in SNMP. Give the data types supported by SMI.
40. Explain the MIB along with its structure
41. List and describe seven message types in SNMP.
42. How to receive the information about device interface and routing information of remote host by using SNMP protocol ?
43. Can a computer have two DNS names that fall in two different top level domains ? If so, give an example. If not, explain why not.
44. Compare between FTP and TFTP.
45. Why cookies are important ? What will happen if cookies are omitted?
46. POP3 allows users to fetch and download e-mail from a remote mailbox. Does this mean that the internal format of mailboxes has to be standardized so any POP3 program on the client side can read the mailbox on any mail server ? Discuss your answer.
47. What is the need of SMI ? Describe the structure of SMI.
48. How SNMP messages are used to monitor and to control the network elements ?
49. What is FTP ? Where and when it is used ? Why does not require 2 ports ? Explain at least 5 user commands used in FTP ?

50. Differentiate between persistent and non-persistent HTTP connection.
51. What is the difference between IMAP and POP 3 protocols ? Explain when and where they are used ?
52. What is the purpose of SMI and MIB in relation to SNMP ?
53. Explain the terms : managing entity, managed device, management agents, MIB in network management context.
54. List the five areas of network management and explain the necessity of each.
55. In SMTP, if we send a one line message between two users, how many lines of commands and responses are exchanged ? Give the example.
56. Explain how DNS service works.
57. Why Common Gateway Interface (CGI) is required in dynamic web pages ?
58. Compare between FTP and TFTP.
59. Explain the MIB along with its structure.
60. Discuss the role of SMI in SNMP. Give the data types supported by SMI.
61. Compare and contrast FTP and TFTP.
62. Explain at least 8 commands of FTP in brief.
63. Explain how DNS service works.
64. Where and why do we use MIME ?
65. List the similarities and differences between POP3 and IMAP. Which protocol is better and why ?
66. Explain RSVP. Why this protocol is needed ?
67. Explain what is MIB along with its structure.
68. What is the purpose of SMI and MIB in relation to SNMP ?
69. How SNMP messages are used to monitor and to control the network elements ?
70. List and explain the principle components of network management architecture.
71. List the 5 areas of network management and explain the necessity of each.
72. What is CGI ? Where and how it is used ?
73. Compare between FTP and TFTP.
74. Explain how DNS works.
75. Draw and explain the architecture of Bluetooth
76. What is firewall? Explain the working of firewall in detail. Draw suitable diagram to illustrate the concept
77. Explain any two encryption algorithm for cryptography.

Model Question Paper for
End-Semester Examination

Time : 3 Hours **Max. Marks : 100**

Instructions to the candidates :

(1) Each Question carries 20 Marks.

(2) Attempt any five questions to the following.

(3) Illustrate your answers with neat sketches, diagram etc., wherever necessary.

(4) If some part or parameter is noticed to be missing, you may appropriately assume it and should mention it clearly.

1. **(a)** Explain the function of each layer of OSI-ISO reference model. What are the different benefits of layered design ?

    [10]

    (b) Write short notes on : **[10]**

      (i)   Similarities of VC switching with circuit switching and datagram networks

      (ii) VC switching between source and destination

      (iii) Virtual Circuit Identifier (VCI)

      (iv) Phases in VC switching

      (v)   Delays in VC switching networks.

2. **(a)** Mention the advantages and disadvantages of FDMA, TDMA and CDMA **[10]**

    (b) Give frame format of HDLC and PPP protocol. Explain each field. **[10]**

3. **(a)** Draw and explain the Bluetooth architecture. **[10]**

    (b) Explain the following connecting devices. **[10]**

      (i)   Repeater

      (ii) Hub

      (iii) Bridge

      (iv) Router

      (v)   Switch

4. **(a)** What is routing? Explain any two routing protocols in brief. **[10]**

    (b) How does CIDR work ? How does it differ from Classfull IP Addressing ? **[10]**

5. **(a)** Explain UDP Header? The following is a dump of UDP header in hexadecimal format. 06 32 00 0D 00 IC E2 17 **[10]**

    (i)　What is source port number ?

    (ii)　What is destination port number ?

    (iii)　What is the total length of the user datagram ?

    (iv)　What is the length of the data ?

    (v)　Is the packet directed from a client to a server or vice versa ?

    (vi)　What is the client process ?

  (b) Explain the three-way handshake algorithm for TCP connection establishment.　　　**[10]**

6. **(a)** Compare OSI and TCP/IP model　　　**[10]**

  (b) What are carrier sense multiple access protocols ?　　　**[10]**

IMPORTANT NOTE

For the current semester the End Semester Examination will be conducted for 100 marks. The question paper template is being provided to all the paper setters. After the evaluation, marks obtained out of 100 in semester examination will be converted for 60 marks and then grade will be calculated.

✠ ✠ ✠